# EVOLUTION AND EMERGENCE

# Evolution and Emergence

*Systems, Organisms, Persons*

*Edited by*
NANCEY MURPHY *and*
WILLIAM R. STOEGER, SJ

**OXFORD**
UNIVERSITY PRESS

Great Clarendon Street, Oxford OX2 6DP

Oxford University Press is a department of the University of Oxford.
It furthers the University's objective of excellence in research, scholarship,
and education by publishing worldwide in

Oxford New York

Auckland Cape Town Dar es Salaam Hong Kong Karachi
Kuala Lumpur Madrid Melbourne Mexico City Nairobi
New Delhi Shanghai Taipei Toronto

With offices in

Argentina Austria Brazil Chile Czech Republic France Greece
Guatemala Hungary Italy Japan Poland Portugal Singapore
South Korea Switzerland Thailand Turkey Ukraine Vietnam

Oxford is a registered trade mark of Oxford University Press
in the UK and in certain other countries

Published in the United States
by Oxford University Press Inc., New York



First published 2007



British Library Cataloguing in Publication Data
Data available

Library of Congress Cataloging in Publication Data
Evolution and emergence : systems, organisms, persons
edited by Nancey Murphy and William R. Stoeger.
Includes bibliographical references and index.
ISBN-13: 978-0-19-920471-7 (alk. paper)
ISBN-10: 0-19-920471-3 (alk. paper)
1. Emergence (Philosophy) 2. Science-Philosophy.
I. Murphy, Nancey C. II. Stoeger, William R.
Q175.32.E44E96 2007
500-dc22 2006102305

Typeset by SPI Publisher Services, Pondicherry, India
Printed in Great Britain
on acid-free paper by
Biddles Ltd., King's Lynn, Norfolk

ISBN 978-0-19-920471-7

1 3 5 7 9 10 8 6 4 2

*to Arthur R. Peacocke*

# *Preface*

We, the editors of this volume, had the privilege of joining a small group of faculty members and administrators at the University of San Francisco to plan a conference on the relations between theology and science. We were given free rein to specify the theme of the conference and to suggest speakers. Given that we have quite different academic backgrounds—one a philosopher and the other an astrophysicist—we expected that we would have very different ideas for the theme. We were pleasantly surprised to discover that we had independently chosen the same topic: that of emergentism, on the positive side, or reductionism on the negative side. Our reasons for the choice were as follows: First, this philosophical issue cuts across many of the more specific controversies regarding science and its implications for theology. Second, this is a problem ripe for solution; recent developments in both science and philosophy have in various ways called reductionism into question, and a variety of resources are at hand for constructing a non-reductive view of reality. One can imagine our excitement, then, in receiving approval from the USF team for making this the focus of the conference, and also in having the resources at our disposal to bring together scholars who, in our judgment, have the best ideas to offer on this collection of related issues.

We express our deep gratitude and appreciation to all those who participated in this stimulating conference, and to those who supported it in any way. These include foremost Stanley Nel, then Dean of the College of Arts and Sciences of the University of San Francisco; Paul V. Murphy, director of the St. Ignatius Institute at USF, and two tireless and immensely competent staff members, Nancy Campagna and Barbara St. Marie. The conference was co-sponsored by the Vatican Observatory and the Center for Theology and the Natural Sciences. We thank the directors of these two institutions as well: George V. Coyne, SJ, and Robert J. Russell, respectively. Thanks to Jeff Phillips and David Brewer, Ph.D. students at Fuller Seminary, for help with the editing and formatting of the book. Rachel Woodforde at Oxford University Press was particularly gracious and helpful seeing the typescript through its transformation into a book.

Finally, we wish to thank one particular participant in the conference, Arthur R. Peacocke. While Peacocke was one presenter among many, we realized that he had done more than anyone else to provide the background against which the ideas here were developed. Among his many contributions to the current dialogue between science and theology, two are particularly relevant. First, he has been in the forefront in criticizing positivist views of science. The logical positivists' manifesto for the unification of the sciences was based on recognition that the sciences could be ordered hierarchically, from physics to the social sciences, on the basis of the complexity of the systems they study; they called for the reduction of each science to the one below, finally intending to show that physics determines the whole of reality. Peacocke's contributions include the resuscitation of earlier attempts to counter the positivists' worldview—early writings on emergence and downward causation. The positivists' assumption was that in the hierarchy of increasingly complex systems, all causation is bottom-up, or, in other words, from part to whole. 'Downward causation,' then, means that the whole has a reciprocal causal influence on its own parts. Peacocke has done a great deal to explicate this concept of downward causation, or whole-part constraint, and to justify its employment by means of examples, especially from his own field of biochemistry.

Second, Peacocke has offered a model for understanding the relations between theology and the sciences that provides the background for many of the contributors' work in this area. Beginning with a nonreductive account of the sciences, Peacocke argues that theology belongs at the top for this reason: The hierarchy is ordered in terms of increasing complexity; God in relation to everything else is necessarily the most complex system possible; therefore, theology, as the study of this system, must be the top-most science. This model is valuable for many reasons. First, the relation between theology and any science can be understood by analogy to the relation of any science to the sciences below it. Second, it provides a way of understanding the claims of many scientists who use science to combat religion: they are simply attempting to replace theology with a materialist metaphysic.

In gratitude, then, for these immensely important contributions to the theology-science conversation in general and to the work at the USF conference in particular, the participants happily agreed to dedicate this book to Arthur Peacocke. Sadly, Arthur passed away as we were finishing the editorial process. He will be missed by many friends and colleagues, but his legacy of scholarship, vision, and creative dialogue between the natural sciences and theology continues.

# *Contents*

# *Abbreviations*

| | |
|---|---|
| *CC* | Russell, R. J., Nancey Murphy, and Arthur R. Peacocke, eds. *Chaos and Complexity*. Vatican City State: Vatican Observatory, 1997. |
| *EMB* | Russell, R. J., William R. Stoeger, and Francisco J. Ayala, eds. *Evolutionary and Molecular Biology.* Vatican City State: Vatican Observatory, 1998. |
| *NP* | Russell, R. J., Nancey Murphy, Theo C. Meyering, and Michael A. Arbib, eds. *Neuroscience and the Person*. Vatican City State: Vatican Observatory, 1999. |
| *PPT* | Russell, R. J., William R. Stoeger, and George V. Coyne, eds. *Physics, Philosophy, and Theology: A Common Quest for Understanding.* Vatican City State: Vatican Observatory, 1988. |
| *QCLN* | Russell, R. J., Nancey Murphy, and C. J. Isham, eds. *Quantum Cosmology and the Laws of Nature.* Vatican City State: Vatican Observatory, 1993. |
| *QM* | Russell, R. J., Philip Clayton, Kirk Wegter-McNelly, and John Polkinghorne, eds. *Quantum Mechanics: Scientific Perspectives on Divine Action*. Vatican City State: Vatican Observatory, 2001. |

# *Contributors*

**Warren S. Brown,** Director of the Travis Research Institute and Professor of Psychology, Fuller Theological Seminary

**Philip Clayton,** Ingraham Professor of Philosophy and Religion, Claremont Graduate University

**Terrence W. Deacon,** Professor of Anthropology, University of California, Berkeley

**George F. R. Ellis,** Professor of Applied Mathematics and Distinguished Professor of Complex Systems, University of Cape Town

**Niels Henrik Gregersen,** Professor of Theology, University of Copenhagen

**John F. Haught,** Landegger Distinguished Professor of Theology, Georgetown University

**Martinez J. Hewlett,** Professor Emeritus, Department of Molecular and Cellular Biology, University of Arizona, and adjunct professor at the Dominican School of Philosophy and Theology, Berkeley

**Don Howard,** Director of the History of Philosophy of Science Graduate Program, and Professor of Philosophy, University of Notre Dame

**Nancey Murphy,** Professor of Christian Philosophy, Fuller Theological Seminary, Pasadena

**Arthur Peacocke,** Formerly Director, Ian Ramsey Centre and Faculty of Theology, University of Oxford

**Alwyn Scott,** Professor Emeritus of Mathematics, University of Arizona

**William R. Stoeger,** Staff Astrophysicist, Vatican Observatory, and Adjunct Associate Professor of Astronomy, University of Arizona, Tucson

**Robert Van Gulick,** Professor of Philosophy and Director of the Cognitive Science Program, Syracuse University.

# Introduction

*Nancey Murphy*

Major change is afoot in current understandings of both the natural and human worlds. A central image, throughout the modern era, has been that of the *hierarchy of the sciences*—the picture of physics supporting chemistry, chemistry supporting the various levels of biology, and perhaps biology supporting psychology and then the social sciences. This image represents a conception of the physical world as itself a *hierarchy of levels of complexity*, with the true simples described by basic physics at the bottom, and everything else understood as larger and more complex organizations of such entities.

Added to this assumption has often been a variety of reductionist theses—to the effect that only the simples are *really* real (whatever that might mean), and that it is these basic constituents that do all of the causal work. There have always been movements that rejected reductionism. One was the emergentist movement in philosophy of biology early in the twentieth century. For various reasons, though, these have remained minority positions—until today.

We, the editors and contributors, believe that something akin to a paradigm change is in progress in philosophy and across the sciences. Developments from quantum physics, to biology, to cognitive science are, in various ways, calling into question the predominant reductionism of modern science. In addition, the development of cybernetics, systems theory, information theory, complexity studies, and mathematical study of nonlinear dynamics are all contributing conceptual resources for understanding how complex (higher-level) entities become causal players in their own right, over and above the effects of their components.

The term 'emergence,' once associated with the postulation of something 'spooky' or anti-scientific, is now being used in a variety of contexts. The purpose of this book is to examine this concept and the (potential) worldview change it represents. Part I (Philosophy) discusses, clarifies, and defends the concept of emergence; Part II (Science) offers a sampling of relevant issues across the range of scientific disciplines; Part III (Theology) investigates the theological implications of this changing worldview.

## Part I

All three authors in this section see the various emergence theses on offer to be roughly equivalent to antireductionist theses; all three focus on causal reductionism. Nancey Murphy and Robert Van Gulick concentrate on the defense of downward causation as an alternative to bottom-up determinism. Terrence Deacon takes the intelligibility of downward causation for granted and explores more complex instances of emergence in which bottom-up and top-down causation involving a system, its components, and its environment, interact in dynamic causal loops.

Murphy begins Chapter 1, 'Reductionism: How Did We Fall Into It and Can We Emerge From It?' with the historical development of the modern hierarchical view of the sciences, the atomist metaphysic that attributed 'ontological priority' to the atoms, and some of the emergence theses that have been offered as antidotes. She distinguishes a variety of interrelated reductionist theses, including epistemological and causal, and two senses of ontological reductionism: one (harmless) version that merely denies the necessity of the addition of new kinds of metaphysical entities (souls or life forces) as one goes up the hierarchy; the other, 'atomist reductionism,' that grants 'ontological priority' to the atoms and sees more complex entities as mere aggregates.

In Chapter 2, 'Reduction, Emergence, and the Mind/Body Problem,' Robert Van Gulick provides a much more elaborate mapping of both reductionist and emergentist theses now being used in the philosophy of mind. The notion of reduction is ambiguous along two dimensions: answers to the question of *what it is* that the reduction relation is supposed to relate, and answers to the question of *how* the items must relate in order to count as reduction. Answers to the first question can be divided into two categories: real world items (objects, events, properties, processes) versus representational items (theories, concepts, models). Answers to the second question depend on answers to the first. For example, real world items may be related to one another as components to whole; theories by logical derivation.

If the basic idea of emergence is more or less the converse of reduction, and the core idea of reduction is that *X*s are nothing more than *Y*s, then the core idea of emergence is that *X*s are something over and above *Y*s. Thus, emergence theses can also be divided into two types: metaphysical and epistemological. Epistemic emergence may pertain either to prediction and explanation or to representations. Within the metaphysical category, two main classes of emergents can be distinguished: properties and causal powers (although these are not strictly distinct if having causal powers is a criterion for the existence of properties).

In another dimension Van Gulick identifies three (rough) categories of emergence theses, distinguished according to the increasingly radical degree of difference between the emergent and the base from which it emerges. 'Specific value emergence' is uncontroversial but uninteresting, in that it refers to cases in which the whole has features of the same type as the parts, but only different values; for example, a statue and the particles making it up both have mass, but the statue has more mass than a particle. 'Modest kind emergence' postulates that the whole has features different in kind from those of its parts. Life is emergent in this sense. 'Radical kind emergence' postulates that the whole has features that are both different in kind from those of the parts and of a kind whose nature and existence is not necessitated by the regularities governing its parts. Radical kind emergence is highly controversial, and would require a robust sort of downward causation, yet without violating or overriding lower-level laws.

In Chapter 3, 'Who's In Charge Here? And Who's Doing All the Work?' Van Gulick constructs a robust account of downward causation that indeed maintains the integrity of the lower-level sciences. While the events and objects picked out by the special sciences are composites of physical constituents, their causal powers are determined not only by the physical properties of the constituents and the laws of physics, but also by the organization of the constituents. It is such patterns of organization that are picked out by the higher-level sciences. Downward causation consists in the higher-level pattern of organization exercising a selective effect on the causal powers of the constituents. This provides an explanation of how new causal powers can emerge as we go up the hierarchy of complexity, but without any objectionable interference with the laws of physics.

Murphy's chapter endorses Van Gulick's account of downward causation via selection, but undertakes two further tasks. First, she notes that Arthur Peacocke has developed a competing antireductionist account, beginning with the literature on downward causation, but finally formulating his thesis in terms of 'whole–part constraint.' One goal, then, is to compare and reconcile these two approaches. Second, she believes that both Van Gulick's and Peacocke's accounts are open to the charge of begging the question because the selecting and constraining must be done by means of ordinary physical forces, and is therefore all bottom-up after all. Murphy endorses the response to this objection developed by Alicia Juarrero: top-down causes operate in nonlinear dynamic systems, whose future behavior is affected by their past, and whose components are defined by their *relations* to one another, not by their physical properties. Global changes in the system at one time affect the components by changing their relations to other components and thus the probability space representing future behaviors of the components. For example, changes in the global property of an ant colony—the density of its forager population—change the probability that an individual ant will meet enough foragers to trigger its return to the nest. Thus, Van Gulick's *selection* is not by means of ordinary physical causes but by means of *constraints.*

Terrence Deacon, in 'Three Levels of Emergent Phenomena' (Chapter 4), investigates interactions between top-down and bottom-up causation, and distinguishes three levels of emergence. First-order emergence occurs in systems (as opposed to mere aggregates) in which relational properties of the constituents (cf. Murphy/Juarrero) constitute higher-order properties—that is, properties dependent on configurational and distributional information (Van Gulick's higher-order patterns).

Second-order emergence occurs in nonlinear systems when a lower-level fluctuation is amplified and affects the probability of future development of the system; in these systems, history matters. An example is the formation of a snowflake; the fact that a water molecule bonds to a particular location on the crystal restricts the possible locations for the bonding of the next set of molecules. All life is (at least) second-order emergent.

Third-order emergence involves the interactions among three levels of complexity: components of a system, the system itself, and the system's interactions with its environment. Here a variety of second-order forms emerge and are constrained (selected) by the environment in such a way that a *representation* of the form is introduced into the next generation; thus, third-order emergence involves some sort of memory. Examples are the evolutionary process (once DNA has appeared) and mental/brain processes.

## Part II

George Ellis's chapter, 'Science, Complexity, and the Natures of Existence' (Chapter 5), provides a transition from the philosophical work of Part I to the scientific discussions in Part II. Ellis assumes the cogency of downward causation and also describes systems with tangled, multi-level causal loops, characterized by Deacon as second-order emergence.

Ellis then offers examples of emergence and downward causation from a variety of levels in the hierarchy of the sciences. Downward causation occurs at the most basic levels of physics. For example, in quantum physics, design of a measurement device constrains the set of possible micro-state outcomes. The appearance of an arrow of time at the macro-level, allowing only entropy-increasing changes, acts as a top-down constraint on microprocesses.

At the level of biology Ellis mentions the evolutionary process and the context-sensitivity of gene expression in developing organisms. He emphasizes the causal efficacy of the mental in the physical world, not only the effect of consciousness on the rest of the organism, but also the effects of human decisions on society and even on the biosphere as a whole.

Ellis then shifts to a more speculative tack. If we give up the atomist-reductionist metaphysic (cf. Murphy), then what is needed by way of a fuller ontology? First, of course, is recognition that everyday objects (molecules, organisms, tables) are as real as the entities studied by basic physics, but what else? Ellis's criterion is to recognize as 'real' or 'existent' anything that has a demonstrable causal effect on this everyday reality. This leads him to identify four 'worlds.' The first is the ordinary physical world. The second is the products of human consciousness: information, ideas, goals, social constructions. These are all embodied or realized in brains or books (world 1) but are not to be identified with or reduced to their physical embodiments.

The third world is the world of physical and biological possibilities. One way to account for these constraints would be to reify the laws of physics and biology. Closely related is the fourth world, a realm of abstract (Platonic) entities such as the mathematical forms that are embodied in the basic laws of physics. Ellis recognizes that these metaphysical speculations may call for assessment of the theological implications of these apparently necessary conditions for the basic operations of the universe.

The remaining papers in this part focus more concretely on particular issues in science. Don Howard, in 'Reduction and Emergence in the Physical Sciences: Some Lessons from the Particle Physics and Condensed Matter Debate' (Chapter 6), notes that bold claims are being made to the effect that condensed matter phenomena (such as superconductivity) are emergent relative to particle physics, and that these claims are then extrapolated to higher levels in the hierarchy of the sciences. Howard evaluates these claims first by distinguishing two senses of emergence. One is *failure* of inter-theoretical reductionism; the other is lack or failure of a supervenience relation between two levels. The failure to reduce theories in condensed matter physics to particle physics does not amount to an interesting claim for emergence, he argues, because such reduction is almost never possible, and this has no implications for supervenience claims.

Howard defines supervenience as a relation between structures of different levels in which the higher-level structure is composed of entities of the lower level and all of the properties and relations of the higher-level structure are determined by those of the lower. Interesting cases of emergence ('S-emergence') would be cases in which the supervenience relation *fails* to obtain. However, it has not been shown that superfluidity or superconductivity are emergent in this sense.

The most basic and important instance of S-emergence is the phenomenon of entanglement: the joint state of two quantum systems that have interacted is not equivalent to any product of the separate states. Here is holism of a very deep kind; the quantum correlations of entangled joint states have a better claim to the status of emergent properties than any other properties in nature. To the extent that condensed matter physics appears to be a case of emergence, it is actually 'borrowing' from the emergence already found at the lower level.

Martinez Hewlett, in 'True to Life?: Biological Models of Origin and Evolution' (Chapter 7), describes the multi-level character of contemporary biological science, from biological molecules at the interface with chemistry, to the level of the informational macromolecules, to cells, organisms, and finally ecology. Cutting across these many layers is the paradigm of the neo-Darwinian synthesis.

Biology began with a determinist and reductionist framework, due to the influence of Newtonian physics on both Charles Darwin and Gregor Mendel. Some current biologists, such as Richard Dawkins and E. O. Wilson, still pursue a reductionist approach. However, in the work of others, such as Stephen Jay Gould, an emergentist paradigm is beginning to appear. While Gould's work still incorporates reductionist elements in its over-emphasis on the role of genes, he also recognizes the (downward) selective effect of the environment. Of even more interest is Gould's and Richard Lewontin's concept of a *spandrel* (developed metaphorically from the triangular architectural elements filling in where arches meet in a dome). Spandrels, in their terms, are nonadaptive structures that set constraints on possible evolutionary pathways for the organism as an integrated whole. Hewlett argues that these and other recent biological models point strongly to the conclusion that the tendency toward complexification is an emergent property of life, just as life is emergent from nonlife.

Hewlett's own research (with Christopher Langton) provides a model of an emergent system. He distinguishes between typical networks, in which the number of connections among nodes varies randomly, and what is called a 'scale-free network,' in which a few nodes are much more highly connected than the rest. A scale-free network develops if there is a preference for new nodes to attach to those that are already more highly connected. They exhibit global properties not seen in ordinary networks, and illustrate an important aspect of emergent systems, namely that it is not the parts in themselves (the nodes) but rather the particular *pattern* (cf. Van Gulick) of *relations* among the parts (cf. Murphy) that is crucial. This pattern of development fits Deacon's account of second-order emergence in that history (how interconnected particular nodes have become in the past) constrains future expansion of the network.

Alwyn Scott, in 'Nonlinear Science and the Cognitive Hierarchy' (Chapter 8), pursues a theme that has run through many of the preceding chapters: the characteristics of nonlinear systems, and the role that nonlinearity plays in the emergence of higher levels of complexity and causal processes. A central aim of the chapter is to canvass a wide variety of reasons why cognitive functions cannot be reduced to neurobiology.

A linear system is one in which two causes acting together produce an effect that is the sum of the two causes acting alone. A nonlinear system is one in which this is not the case; rather, multiple causes *interact* among themselves. Nonlinear mathematics applies to a wide array of phenomena: emergent structures (e.g. cell assemblies in the brain), chaotic effects (e.g. turbulence), spontaneous pattern formation (e.g. the Gulf Stream), and many others.

Scott argues for the nonreducibility of cognitive functions by means of an *a fortiori* argument: if reductionism fails even in some aspects of physics (cf. Ellis and Howard), and in a variety of ways in biology (cf. Hewlett), it should all the more be expected to fail in the cognitive neurosciences. A major difference between the physical sciences and the biological sciences in general is that physics and chemistry deal with homogeneous sets of entities (e.g. all hydrogen atoms are alike), while biology and, especially, neuroscience deal with immensely large heterogeneous sets. For example, at the level of molecular biology there are something like $20^{200}$ possible protein structures. This in itself indicates the need to consider downward causation: The actual number of proteins in existence is a very tiny fraction of this number, and the question of why these and not others can only be answered by considering accidents of evolutionary history. As one goes up the biological hierarchy, greater complexity makes for larger numbers of possible forms, and the need to invoke downward causation increases. This is particularly true of the brain: the particularities of individual brains require explanation in terms of learning, and especially the influence of cultural patterns.

Scott argues that downward causation in nonlinear systems can best be understood in terms of something like Aristotle's concept of a formal cause. In mathematical terms, a formal cause might arise from values of dynamic variables at higher levels of description that enter as a boundary condition at the lower level. On this basis he rejects any notion of downward causation ('strong downward causation') that involves the higher-level entity acting as an efficient cause on its own lower-level constituents (cf. Murphy). He recognizes two other possibilities: weak and medium downward causation. In both cases, the process is described in terms of the attractors (i.e. 'maps' in phase space representing a probability distribution of possible behaviors of a nonlinear system). Weak downward causation is understood as a higher-level phenomenon *moving* components from one basin of attraction to another. Medium downward causation he understands as a higher-level structure *modifying* local features of a component's phase space. Medium downward causation opens the possibility of closed causal loops spanning several levels in a hierarchy. Note the similarities to Deacon's second-order emergence and to Van Gulick's moderate kind emergence.

In 'The Emergence of Causally Efficacious Mental Function' (Chapter 9), Warren Brown pursues the issue of downward causation in the cognitive domain, presupposing points made by Scott regarding the role of multi-level hierarchies with feedback control loops. Brown describes these control loops in greater detail, as well as the role of social factors in human cognition.

Philosophy of mind has been plagued by 'the problem of mental causation,' that is, how to account for the causal efficacy of the mental. One source of this problem is lack (heretofore) of an adequate account of downward causation in general, but this is exacerbated by Cartesian tendencies to view the mental as inner and passive. Rather, mind needs to be understood as embodied and embedded in action–feedback–action loops in the environment.

Brown agrees with Deacon in describing much of cognition as third-order emergent. For example, previously occurring higher-order neural patterns (i.e. memories) exert a cumulative influence over the future of the system by being repeatedly elicited by new, but similar, situations, and thereby re-entered into lower-order neural dynamics.

While Brown begins his account of the efficacy of cognition by considering lower organisms, his focus is on the added levels of processing and evaluation available to humans due to a variety of factors: consciousness, the ability to run behavioral scenarios in imagination in order to evaluate their likely effects, and language. Language is a central example of what Brown calls 'external scaffolding'—the resources provided by culture that allow for increasingly sophisticated mental processes, including higher levels of evaluation, such as evaluation in light of moral concepts.

So we see in Part II a sampling of the role of emergence and downward causation from a variety of levels of the hierarchy of the sciences, as well as detailed exemplification of the conceptual resources provided in Part I.

## Part III

William R. Stoeger, SJ, in 'Reductionism and Emergence: Implications for the Interaction of Theology with the Natural Sciences' (Chapter 10), provides a general introduction to issues where current understandings of reduction and emergence intersect the dialogue between theology and the natural sciences. The primary question is whether our provisional conclusions regarding the power and limits of reduction, and an appreciation for genuine modes of emergence, leave room for the realities and relationships that theology claims to discover.

Stoeger describes a variety of issues wherein this question of consonance needs to be pursued. The most general issue is that of divine action. He distinguishes between God's original act of creation and God's special acts in nature, history, and human consciousness. There is no problem in reconciling traditional accounts of God's creation with these new developments in science and philosophy—God is creator of the laws of nature, in fact, the 'fine-tuner' of laws that are suited to bring forth a universe with capacities and dynamisms to evolve rich diversity.

The more difficult issue since the rise of modern science has been to give an account, consonant with science, of God's acts of special providence and revelation in history. Stoeger notes that there are both bottom-up and top-down accounts of God's action, as well as the traditional account based on the distinction between primary and secondary causation. Increased understanding of the levels at which nature is, and is not, deterministic, along with further understanding of the possibilities and limitations of causal reductionism, will contribute insights here.

Another important locus for discussion is the contribution that emergence theories make to a philosophy of nature and, in turn, to a theology of nature. The created order can now be seen as incomplete, and as developing increasing levels of purposiveness.

A third area is anthropology. An appreciation of emergence and the limits of reduction opens the way for a better understanding of 'soul' and 'spirit' in non-dualistic terms, of human religious awareness and openness to the transcendent; it has a bearing on the problem of free will. It also improves our understanding of the relation of humankind to the rest of nature, including the ways in which we, along with other sentient creatures, suffer at the hands of nature.

In each of these cases (divine action, theology of nature, and anthropology) developments are likely to lead to richer conceptions of God, especially of God's immanence in creation, and to a better appreciation of the direction and purpose of nature and history.

John Haught argues in his 'Emergence, Scientific Naturalism, and Theology' (Chapter 11) that the resources emergentist theories bring to theology are inadequate. Haught observes that the basic methodological assumptions of science are such as to proscribe focal attention to the human experiences of *subjectivity* and of *purposive striving*. Furthermore, many scientific readings of emergence are actually attempts to reduce 'emergent' phenomena to simpler components and laws. Science can only *watch* novelty as it appears, noting that lower-level laws are not violated, but it cannot *explain* novelty. This raises the question of whether 'emergent' phenomena are understood by many scientists merely as simplicity masquerading as novelty.

Haught argues that, while science is legitimate in its own sphere, a *broader empiricism* is called for, based on all of human experience, including an appreciation for genuine novelty, as well as the experience of subjectivity and striving. Philosophers and theologians who have already contributed to this richer empiricism are Alfred North Whitehead, in his development of a metaphysic that takes the 'insidedness' of the universe seriously; Bernard Lonergan, with his focus on 'intelligent subjectivity'; Michael Polanyi, with his recognition that a personal component is operative in scientific knowledge; and Pierre Teilhard de Chardin, with his shift from an understanding of reality based entirely on the past to an understanding on the basis of the future. In particular, an adequate worldview requires the concept of a power of attraction up ahead that lovingly and persuasively gathers the multiple cosmic particulars into a convergent unity—rather than seeing cosmic processes driven forward entirely by a dead and inertial past.

In 'Emergent Realities with Causal Efficacy: Some Philosophical and Theological Applications' (Chapter 12), Arthur Peacocke sets forth an emergentist-monist account of reality: There are no extra entities or forces at higher levels of the hierarchy; the key to the emergence of higher-level entities is the organization of components into new systems of relationships. The higher-level whole then exercises constraints on its components (cf. Murphy). Peacocke argues for the reality of higher-level entities on the grounds that concepts describing them are essential for giving a complete causal account (cf. Ellis).

Peacocke's primary philosophical application of his insights regarding emergence and whole–part constraint is to the problem of mental causation. Typically, when philosophers conclude that physicalism entails the reduction of mental properties to brain properties it is because they take a microphysical description of the relevant neural state to be a *sufficient* description. Peacocke emphasizes that the mental can only be understood as a description of events involving the human-brain-in-the-human-body-in-social-relations (cf. Brown). There are multiple levels of structure from the individual neuron to mental states so understood (cf. Scott).

Peacocke approaches doctrinal issues by arguing, first, that theology should be understood in relation to the sciences by analogy to the relation of a higher-level science to the ones below. Theology studies the most comprehensive and complex level of reality: the relations among God, nature, and humankind. Accordingly, a 'panentheist' theology takes the universe to be internal to a system that is God in the world and the world in God. An approach to the problem of special divine action (cf. Stoeger) follows: God's action in the world can be modeled on the concept of whole–part constraint.

Peacocke considers two theological issues, the Incarnation and the Eucharist. The Incarnation can be understood as an expression of early Christians' perceptions of Jesus: through his divinely inspired response to God he becomes a God-informed human being, instantiating both a *continuity*, in this case with human nature, and a *discontinuity*, since he was a unique manifestation of the divine with distinctive causal efficacy. A ritual such as the Eucharist is an interacting complex of God, nature, and humanity, involving specific actions, words, things (bread and wine), past experiences and shared traditions. As such, it could be regarded as a new kind of reality and as having a particular kind of causal efficacy on human beings.

In 'Reduction and Emergence in Artificial Life: A Theological Appropriation' (Chapter 13), Niels Henrik Gregersen first assesses the implications of artificial-life projects (AL) for the topic of reductionism versus emergentism. If it were indeed possible for the 'strong program' to succeed—the actual creation of life in computer programs—this would count strongly in favor of causal reductionism because it would show that deterministic physical processes following simple rules could be the basis of biological life. Gregersen sees the 'weak' AL program—the attempt only to model biological processes—as more important and as supportive of emergence claims, insofar as they involve downward causation.

The relevance of AL for theology is, first, that it provides new metaphors and models for theological language; for example, God the Watchmaker may become God the Networker. Second and more important, AL provides support for a view of nature as essentially undergirded by *patterns of information*; the substance of matter, the energy of matter, and the organization of matter—pattern understood in informational terms—are independent but inseparable aspects of reality. Gregersen uses this insight to develop a new theology of nature (cf. Stoeger). What is one to make of the fact that Christians hold a triune view of God if one also sees creation as essentially constituted by the relations among the triad of matter, energy, and information? Gregersen presents a view of the world in which God the Father makes room for the otherness of material substance, God the Son (Logos) is the source of all creative pattern (information), and God the Spirit energizes creation.

The theology of nature has implications in turn for trinitarian theology. It shows why the Father must have two 'hands' rather than one: without Spirit/energy there would be structure without life; without the Son/information there would be chaotic activity but no structure.

'Toward a Constructive Christian Theology of Emergence' (Chapter 14) is Philip Clayton's outline for a systematic theology that takes emergence to be a fundamental feature of the natural world and reformulates traditional doctrine in its light. Clayton's starting point is the personal character of God; thus, an emergentist-monist account of human nature (cf. Peacocke) should be used analogously to refine the doctrine of God. Clayton rejects the radical emergentist theism that makes God a product of emergent evolutionary processes. Instead, he adopts a di-polar concept: God's antecedent nature is the Ground of the universe's capacities for emergent novelty; God's consequent nature is a result of God's openness and response to creation. This is consistent with a panentheistic account of God's relation to the world: God in the world and the world in God (cf. Peacocke).

The panentheistic account, focusing on God with us, provides a model for Christology: as Jesus submits his will perfectly to God's will, and God accepts and guides him, so Jesus' and God's actions are identified. Incarnation is located in shared action and attitude rather than an a priori ontology.

The emergence of spirit in creation suggests that matter and divine Spirit cannot be dichotomized. It also points to God's purposes for further creativity and novelty in human life, aiming to fulfill the *imago Dei.* Just as spirit and world are not dichotomized, neither are church and world. The church is an emergent entity wherein the structural differences between God and the world (sin) begin to be overcome (salvation, sanctification) and wherein hope is nurtured for the fullness of God's presence and manifestation in creation (eschatology).

So in a variety of ways these authors have begun to address the issues laid out by Stoeger: divine action (Stoeger, Peacocke, Gregersen, Clayton), a theology of nature (Haught, Gregersen), and theological anthropology (Peacocke, Clayton). They have also begun the task of reinterpreting traditional theological loci in light of the changing worldview intimated in earlier chapters (Peacocke, Gregersen, Clayton). We are in an exciting time of intellectual change; much more lies ahead.

# Part I

# Philosophy

# 1

# Reductionism: How Did We Fall Into It and Can We Emerge From It?

*Nancey Murphy*

## 1.1. Introduction

This essay has two purposes. One is to trace some of the history of debates between reductionists and emergentists, aiming to shed light on the grip reductionist ideas have had on modern thinkers. The second is to suggest in a preliminary way how the concept of *emergence* is best understood and to indicate factors that go into explaining how emergent causal capacities can be reconciled with our best understanding of science.

In the following section I provide some history of the hierarchical and reductionist view of reality. In section 1.3 I survey a list of the sorts of reductionist theses that are on offer, along with my conclusion regarding the central role of *causal* reductionism in motivating other reductionist theses. Causal reductionism is the thesis that, in the hierarchy of the sciences, all causal influences are 'bottom-up'—from part to whole. This thesis has been so pervasive for so long that it should be counted as one of the central metaphysical assumptions of the modern era.

In section 1.4 I trace some history of emergentist ideas, noting that they turn out to be largely the inverse of reductionist theses. If I am right about the central role of causal reductionism, then the emergence of *new causal processes* will be central to any significant emergentist thesis. Thus, the main task of this essay (in section 1.5) is to define and defend a concept of downward causation. I begin with an overview of work to date, and argue that resources are available to show that downward causation is essential for understanding complex systems. I defend the concept against some of the most common objections, and end by relating my account to developments in understanding emergence.

## 1.2. Atomism and the Hierarchy of the Sciences

When we think of the transition from medieval to modern science, the Copernican revolution is most likely to come to mind. However, the transition in physics from Aristotelian hylomorphism to atomism has had equally significant cultural repercussions. Yet, probably because the transition happened more gradually—finally completed in biology only in the twentieth century—this change has received less attention.

Galileo deserves as much credit for this change as for the revolution in astronomy. He was one of the first to reject the Aristotelian theory that all things are composed of 'matter' and 'form' in favor of an atomic theory. His version of atomism took all physical processes to be consequences of the properties of the atoms, which were size, shape, and rate of motion.

Atomism was extended to the domain of chemistry by Antoine-Laurent Lavoisier and John Dalton, who made great headway in explaining the phenomena of chemistry on the assumption that all material substances were composed of atoms. This was a striking triumph for the atomic theory, but also for *reductionism,* the strategy not only of analyzing a thing into its parts, but also of explaining the properties or behavior of the thing in terms of the properties and behavior of the parts.

The increasingly successful reduction of chemistry to physics raised the expectation that biological processes could be explained by reducing them to chemistry, and thence to physics. There have been a series of successes, beginning in 1828, when the synthesis of urea refuted the claim that biochemistry was essentially distinct from inorganic chemistry, and continuing in current study of the physics of self-organizing systems and their bearing on the origin of life. The philosophical question of whether biology could be reduced to chemistry and physics, or whether the emergence of life required additional metaphysical explanation in terms of a 'vital force,' was hotly debated. Vitalism had few supporters in the twentieth century and has now entirely disappeared.

Beginning in the twentieth century, a variety of research projects have carried reductionism into the human sphere. We have evidence for biochemical factors in many mental illnesses; neuropsychologists have localized a surprising variety of cognitive faculties in the brain; theories regarding genetic influence on a wide assortment of human traits have proliferated with the progress of the Human Genome Initiative.

Thus, much of modern science can be understood as the development of a variety of research programs that in one way or another embody and spell out the consequences of what was originally a metaphysical theory. It has been the era in which Democritus has triumphed over Aristotle.

In addition, the reductionist assumption has produced a model for the relations among the sciences. Modern science has increasingly seen the natural world as a hierarchy of levels, from the smallest sub-particles (quarks and leptons), through atoms, molecules, cells, tissues, organs, organisms, and societies, to eco-systems and the universe as a whole. Corresponding to these levels of complexity is the hierarchy of the sciences: physics studying the simplest levels; then chemistry; the various levels of biology; psychology, understood as study of the behavior of complete organisms; sociology; ecology; and cosmology. Versions of this hierarchy go back to Thomas Hobbes at the beginning of the modern period. However, the logical positivists in the 1920s and 1930s, with their program for the unification of science, did a great deal to popularize it.

Causation on this view is 'bottom-up'; that is, the parts of an entity are located one rung *downward* in the hierarchy of complexity, and it is the parts that determine the characteristics of the whole, not the other way around. So *ultimate* causal explanations are thought to be based on laws pertaining to the lowest levels of the hierarchy. The crucial *metaphysical* assumption embodied in this view is the ontological priority of the atoms over that which they compose.[1] This is metaphysical atomism-reductionism. Modern thought, not only in the sciences, but in ethics, political theory, epistemology, and philosophy of language, has tended to be atomistic—that is, to assume the value of analysis, of finding the 'atoms,' whether they be human atoms making up social groups, atomic facts, or atomic propositions.

In addition, modern thought has tended to be reductionistic in assuming that the parts take priority over the whole—that they determine, in whatever way is appropriate to the discipline in question, the characteristics of the whole. In ethics the common good is a summation of the goods for individuals; in the social sciences individual variables explain social phenomena; in epistemology atomic facts provide the justifying foundation for more general knowledge claims; in philosophy of language the meaning of a sentence is a function of the meaning of its parts.

The individualism of much political philosophy in the modern period was based on the attempt to extend Newtonian reasoning to the sphere of the social. One factor that has complicated the modern reductionist program has been the recognition of the determinist implications for human behavior when reductionism is coupled with a deterministic account of the laws of physics: if the body is nothing but an arrangement of atoms whose behavior is governed by the laws of physics, then how can free decisions affect it? From the very beginning of Modernity, there have been two responses. One is a materialist account of the human person that simply accepts determinism. The other is to propose some form of dualism. Dualists have held that essential humanness is associated with the mind, and is independent of the workings of mechanistic nature. Then, of course, the problem of mind–body interaction arises. Modern dualist accounts can be traced to Descartes, who argued that material substance is only part of reality. The other basic metaphysical principle is thinking substance. Human bodies are complex machines, but the person—the 'I'—is the mind, which is entirely free.[2]

[1] Cf. Edward Pols, *Mind Regained* (Ithaca: Cornell University Press, 1998), 64.

[2] This section is a revision of material from my *Anglo-American Postmodernity: Philosophical Perspectives on Science, Religion, and Ethics* (Boulder, CO: Westview Press, 1997), chap. 1.

## 1.3. Forms of Reductionism

There are a variety of reductionist theses, all intimately related:

1. Methodological reductionism: the research strategy of analyzing the thing to be studied into its parts.
2. Epistemological reductionism: the view that laws or theories pertaining to higher levels of the hierarchy of the sciences can (and should) be shown to follow from lower-level laws, and ultimately from the laws of physics.
3. Logical or definitional reductionism: the view that words and sentences referring to one type of entity can be translated without residue into language about another type of entity.
4. Causal reductionism: the view that the behavior of the parts of a system (ultimately, the parts studied by subatomic physics) is determinative of the behavior of all higher-level entities; all causation is 'bottom-up.'
5. Ontological reductionism: this is defined as the thesis that higher-level entities are nothing but the sum of their parts. However, this is ambiguous; we need names for two distinct positions:

   5a. One is the view that as one goes up the hierarchy of levels, no new kinds of metaphysical 'ingredients' need to be added to produce higher-level entities from lower. No 'vital force' or 'entelechy' must be added to get living beings from non-living materials; no immaterial mind or soul needed to get consciousness; no *Zeitgeist* to form individuals into a society.

   5b. A much stronger thesis is that only the entities at the lowest level are *really* real; higher-level entities—molecules, cells, organisms—are only composites made of atoms. This is the assumption, mentioned above, that the atoms have ontological priority over the things they constitute. I'll designate this position 'atomist reductionism' to distinguish it from 5*a*, for which I shall retain the designation of 'ontological reductionism.' It is possible to hold a physicalist ontology without subscribing to atomist reductionism. Thus, one

> might say that higher-level entities are real—as real as the entities that compose them—and at the same time reject all sorts of vitalism and dualism.

I believe that Francisco Ayala was the first to distinguish methodological, epistemological, and ontological forms of reductionism.[3] Arthur Peacocke pointed out the ambiguity in the meaning of 'ontological reductionism' in his *God and the New Biology*.[4]

These theses are interrelated: If atomist reductionism is true, then ontological reductionism is also. Causal reductionism seems to follow from atomist reductionism as well. If causal reductionism is true, then methodological reductionism would be the only sensible research strategy, and epistemological reductionism ought to be true as well. In addition, if it is possible to explain all higher-level phenomena in terms of physics, then it must be possible to relate the language of the higher levels to that of physics, so definitional reductionism must be true.

So what is the crucial reductionist issue? My account of the interrelations among the various theses makes it appear that atomist reductionism is the source of all the others. However, atomist reductionism expresses more of an attitude than a philosophical thesis; it is difficult to state it without employing, as I have done, the nonsense phrase 'really real.' The best candidate for spelling out its import is, in fact, to say that the atoms have *causal* priority over the things they compose. So it may be nothing but a confusing statement of the assumption of causal reductionism.

In principle, if causal reductionism is true, then epistemological and definitional reductionism should follow. In practice, though, both of these forms of reduction are now seen to be impossible in many cases. This means in turn that while methodological reductionism is *often* a useful strategy for research, in practice it needs to be complemented by systems approaches. In short, even if causal reductionism is true, there are a variety of reasons why methodological, epistemological, and logical reduction might still fail. Thus, showing that these three forms of reductionism fail tells us about the

[3] Francisco J. Ayala, 'Introduction,' in Ayala and Dobzhansky, eds., *Studies in the Philosophy of Biology*, vii–xvi.

[4] Arthur R. Peacocke, *God and the New Biology* (Glouster, MA: Peter Smith, 1986).

limitations of our knowledge, but not about what really makes things happen.

I take ontological reductionism to be entirely unobjectionable, so long as it is applied to the cosmos itself and no illegitimate inferences are drawn from it regarding the source of the cosmos. So, by process of elimination, I conclude that causal reductionism is the critical issue. If causal reductionism is true then it is indeed physics that is doing all the work, and human thought and behavior are all epiphenomenal.

There are a variety of important critiques of causal reductionism available in the literature. I mention several of these below. First, however, let us consider some of the history of the idea of emergence.

## 1.4. The Meaning of 'Emergence'

Achim Stephan describes three phases in the development of the concept of *emergence.* The first was in the work of J. S. Mill and George Lewes in the nineteenth century. Mill was interested in effects that were the 'resultant' of intersecting causal laws; Lewes used the term 'emergent' for such effects.

The second phase was a discussion in the philosophy of biology beginning in the 1920s; 'emergent evolutionism' was proposed as an alternative to vitalism, preformationism, and mechanist-reductionist accounts of the origin of life. The authors usually cited are C. Lloyd Morgan, Samuel Alexander, and C. D. Broad in Britain; however, there were also significant contributions made by American philosophers Roy Wood Sellars, A. O. Lovejoy, and Stephen Pepper.[5]

Jaegwon Kim writes that by the middle of the twentieth century emergentism was often trivialized if not ridiculed in philosophy because of the anti-metaphysical stance of logical positivism and

[5] Achim Stephan, 'Emergence—A Systematic View on Its Historical Facets,' in Ansgar Beckermann, Hans Flohr, and Jaegwon Kim, eds., *Emergence or Reduction?: Essays on the Prospects of Nonreductive Physicalism* (Berlin and New York: Walter de Gruyter, 1992), 25–48; 25. Stephan actually divides this movement into two phases, before and after 1926, but I do not see his reason for doing so.

analytic philosophy.[6] Stephan attributes the eclipse of the discussion particularly to the neopositivists Carl Hempel, Paul Oppenheim, and Ernst Nagel.

The third stage of the discussion of emergence began in 1977 with Mario Bunge's and Karl Popper's application to the mind–brain problems. Other contributors include psychologist Roger Sperry and philosopher J. J. C. Smart.[7]

Various types of things have been classified in the literature as emergent: laws, effects, events, entities, and properties.[8] Robert Van Gulick makes a helpful distinction between emergentist theses that pertain to objective real-world items and those that appeal to what we as cognitive agents can or cannot know. He further distinguishes, on the objective, metaphysical side, between two classes of emergents: properties and causal powers or forces. Within the category of epistemological theses, he further distinguishes those pertaining to prediction and those that pertain to understanding. All of these subcategories come in stronger or weaker forms.[9]

Both Van Gulick and I see emergentist theses to be roughly equivalent to anti-reductionist theses. If this is the case, and if I am right in identifying the defeat of causal reductionism as the central issue, then the form of emergence to look for will be one that has to do with causation. I believe that, in any case, we can rule out as relatively uninteresting any epistemological thesis. We know of cases where we can neither predict outcomes nor explain known facts (explain in the sense of 'retrodiction' from laws and initial conditions) simply because the level of complexity or the need for fine-scale measurements goes beyond human capacities. If we attempt to evade this problem by invoking an omniscient predictor, we are unable to *apply* the criterion because we have no way to settle disputes about what the omniscient one would or would not know.

An ontological definition, then, is desirable. But between causal factors and properties, it seems best to focus on causal factors because having causal powers seems to be the best *criterion* for the existence of

[6] Jaegwon Kim, 'Being Realistic about Emergence,' in Philip Clayton and Paul Davies, eds., *The Re-emergence of Emergence* (Oxford: Oxford University Press, 2006).

[7] Stephan, 'Emergence,' 26.

[8] Ibid., 27.

[9] Van Gulick, this vol., chap. 2.

a distinct *property*. Between causal forces and causal powers, I vote for powers because the postulation of new causal forces, over and above those known to physics, would violate the causal closure of physics. So I conclude that what the emergentist needs to show is that as we go up the hierarchy of complex systems we find entities that exhibit new causal powers (or, perhaps better, participate in new causal processes or fulfill new causal roles) that cannot be reduced to the combined effects of lower-level causal processes.

## 1.5. Defending Downward Causation

The central thesis of this essay, then, is that emergence is best understood as the negation of causal reductionism. If causal reductionism is the thesis that all causation in the hierarchy of complexity is bottom-up, then the alternative is the thesis that there is also top-down or downward causation. The cogency of the concept of downward causation is hotly disputed. In this section I first report on some of the history of this debate, and then draw together what I take to be the best resources available for defending it.

### 1.5.1. *History*

Philosophical theologian Austin Farrer was clearly invoking the idea of downward causation in his 1957 Gifford Lectures, though he did not use the term. He distinguished between two types of systems. The familiar type is one in which the pattern of the whole is a simple product of the behavior of its parts. The other sort is one in which 'the constituents are caught, and as it were bewitched, by larger patterns of action.'[10] As examples he cites the molecular constituents of cells, and the cells themselves within the animal body. Furthermore, '[n]ew principles of action come into play at successive levels of organization.'[11] Farrer recognizes that he is denying deep-seated reductionist assumptions, but maintains that 'the intransigence of the [reductionistic] physicists ... need not contradict the claims of the

[10] Austin Farrer, *The Freedom of the Will*, The Gifford Lectures, 1957 (London: Adam and Charles Black, 1958), 57; page references are from the 1966 edition.

[11] Ibid., 58.

biologists to be studying a pattern of action which does real work at its own level, and leads the minute parts of Nature a dance they would otherwise not tread.'[12] In sum:

All we are interested to show is the meaningfulness of the suggestion that a high-level pattern of action may do some real work, and not be reducible to the mass-effect of low-level action on the part of minute constituents. And we are happy if we can show at the same time how the claims to exactitude advanced by minute physics need not stand in the way of our entertaining such a suggestion.[13]

It is unfortunate that Farrer used the metaphors of *bewitchment* and *dancing* in his proposal, as these raise more questions than they answer.

In the 1970s, psychologist Roger Sperry and philosopher Donald Campbell both wrote specifically about downward causation. Sperry wrote that the reductionist view, according to which all mental functions are determined by neural activity and ultimately by biophysics and biochemistry, has been replaced by the cognitivist paradigm in psychology. On this account:

The control upward is retained but is claimed not to furnish the whole story. The full explanation requires that one also take into account new, previously nonexistent, emergent properties, including the mental, that interact causally at their own higher level and also exert causal control from above downward. The supervenient control exerted by the higher over the lower level properties of a system ... operates concurrently with the 'micro' control from below upward. Mental states, as emergent properties of brain activity, thus exert downward control over their constituent neuronal events—at the same time that they are being determined by them.[14]

On some occasions, Sperry wrote, in a manner comparable to Farrer, of the properties of the higher-level entity or system *overpowering* the causal forces of the component entities.[15] The notion of overpowering lower-level causal forces rightly raises worries regarding the compatibility of his account with adequate respect for the basic sciences.

[12] Ibid., 60. [13] Ibid.

[14] Roger W. Sperry, 'Psychology's Mentalist Paradigm and the Religion/Science Tension,' *American Psychologist* 43/8 (August 1988): 607–613; 609.

[15] Roger W. Sperry, *Science and Moral Priority: Merging Mind, Brain, and Human Values* (New York: Columbia University Press, 1983), 117.

In addition, Sperry's use of the concept of *emergent properties* was problematic in that there was no agreed understanding of emergence. Some emergence theses appeared to postulate the existence of spooky new entities; others to threaten the integrity of the basic sciences.

Donald Campbell's work has turned out to be more helpful. Here there is no talk of bewitching or overpowering lower-level causal processes, but instead a thoroughly non-mysterious account of a larger system of causal factors having a *selective* effect on lower-level entities and processes. Campbell's example is the role of natural selection in producing the remarkably efficient jaw structures of worker termites.[16] This example illustrates four theses. The first two give due recognition to bottom-up accounts of causation: First, all processes at the higher levels are restrained by and act in conformity to the laws of lower levels, including the levels of subatomic physics. Second, the achievements at higher levels require for their implementation specific lower-level mechanisms and processes. Explanation is not complete until these micromechanisms have been specified.

The third and fourth theses represent the perspective of downward causation: Third, '[b]iological evolution in its meandering exploration of segments of the universe encounters laws, operating as selective systems, which are not described by the laws of physics and inorganic chemistry.' Fourth:

> Where natural selection operates through life and death at a higher level of organisation, the laws of the higher-level selective system determine in part the distribution of lower-level events and substances. Description of an intermediate-level phenomenon is not completed by describing its possibility and implementation in lower-level terms. Its presence, prevalence or distribution (all needed for a complete explanation of biological phenomena) will often require reference to laws at a higher level of organisation as well.[17]

### 1.5.2. *Downward Causation versus Whole–Part Constraint*

While downward causation is often invoked in current literature in psychology and related fields, it has received little attention in philosophy since Campbell's essay in 1974. However, Arthur Peacocke

[16] Donald T. Campbell, 'Downward Causation,' in Ayala and Dobzhansky, eds., *Studies in the Philosophy of Biology*, 179–186.

[17] Ibid., 180.

has made important contributions to the discussion. In a number of writings he refers to Campbell's and Sperry's concepts of top-down or downward causation, but claims that '[t]here are imprecisions and a lack of generalizability in Campbell's example' of the evolutionary development of the termite jaw structure.[18] He prefers the terminology of 'whole–part constraint' to that of downward causation. His preferred examples are the Bénard phenomenon (the formation of rotating hexagonal convection cells in a heated liquid) and the Belousov-Zhabotinsky reaction (an autocatalytic process producing bands of high and low concentrations of ceric ions). Concerning the Bénard phenomenon, Peacocke says:

> In such instances, [reference omitted] the changes at the micro-level, that of the constituent units, are what they are because of their incorporation into the system as a whole, which is exerting specific constraints on its units, making them behave otherwise than they would in isolation. Using 'boundary conditions' language, [reference omitted] one could say that the sort of relations between the constituent units in the complex whole is a *new* set of boundary conditions for those units.[19]

Thus, the burden of explaining what is happening in instances of whole–part constraint, as well as explaining why it does not abrogate lower-level laws, falls on the concept of *boundary conditions.*

Van Gulick's contributions to the topic of downward causation are particularly important. His account, in contrast to Peacocke's, is based on the notion of selection and is made in the context of an argument for the nonreducibility of higher-level sciences. The reductionist, he says, will claim that the causal roles associated with special-science classifications are entirely derivative from the causal roles of the underlying physical constituents. Van Gulick replies that the events and objects picked out by the special sciences *are* composites of physical constituents, yet the causal powers of such objects are not determined solely by the physical properties of their constituents

[18] Arthur R. Peacocke, 'The Sound of Sheer Silence: How Does God Communicate with Humanity?' in *NP*, 215–248; 220.

[19] Arthur R. Peacocke, 'God's Interaction with the World: The Implications of Deterministic "Chaos" and of Interconnected and Interdependent Complexity,' in *CC*, 263–288; 273.

and the laws of physics. They are also determined by the *organization* of those constituents within the composite; it is just such patterns that are picked out by the predicates of the special sciences. These patterns have downward causal efficacy in that they can affect which causal powers of their constituents are activated. 'A given physical constituent may have many causal powers, but only some subsets of them will be active in a given situation. The larger context (i.e., the pattern) of which it is a part may affect which of its causal powers get activated. ... Thus the whole is not any simple function of its parts, since the whole at least partially determines what contributions are made by its parts'.[20]

Such patterns or entities are stable features of the world, often in spite of variations or exchanges in their underlying physical constituents. Many such patterns are self-sustaining or self-reproducing in the face of perturbing physical forces that might degrade or destroy them (e.g. DNA patterns). Finally, the selective activation of the causal powers of such a pattern's parts may in many cases contribute to the maintenance and preservation of the pattern itself. Taken together, these points illustrate that 'higher-order patterns can have a degree of independence from their underlying physical realizations and can exert what might be called downward causal influences without requiring any objectionable form of emergentism by which higher-order properties would alter the underlying laws of physics. Higher-order properties act by the *selective activation* of physical powers and not by their *alteration*.'[21]

So we have here two recent arguments for the insufficiency of bottom-up accounts of natural processes: Van Gulick's account, specifically relying on selection, and Peacocke's whole–part constraint, which he proposes as an improvement over Campbell's earlier account in terms of selection. Can either, or both, be defended? Can they be reconciled?

### 1.5.3. *Conceptual Developments*

The criticism likely to be leveled against the accounts presented above is something like the following: Van Gulick says that the larger context

[20] Van Gulick, this vol., chap. 3, 83. [21] Ibid.

of which a physical constituent is a part may affect which of its causal powers gets activated. The reductionist will ask *how* the larger system affects the behavior of its constituents. To affect it must be to *cause* it to do something different than it would have done otherwise. Either this is causation by the usual physical means or it is something spooky. If it is by the usual physical means, then those interactions must be governed by ordinary physical laws, and thus all causation is bottom-up after all. The parallel charge to Peacocke would be to ask whether the 'constraints' imposed on the units in the Bénard cells must not also be by ordinary physical processes.

Resources to answer such charges and to reconcile Peacocke's and Van Gulick's accounts can be found in the work of Alicia Juarrero, who introduces the concept of *context-sensitive constraints* in complex dynamic systems.[22] Presupposed in both Peacocke's and Van Gulick's accounts is the fact that we are dealing with complex dynamic (non-linear) systems. Nonlinear systems are those in which the current state affects the development of each future state.[23] Some such systems are wildly unstable, some entirely stable, some (chaotic systems) fluctuate in a patterned manner, and, finally, some (far from thermodynamic equilibrium) jump to higher forms of organization.

Juarrero emphasizes the shift in thinking required to understand such systems. One has to give up the traditional Western philosophical bias in favor of *things*, with their intrinsic properties, for an appreciation of *processes* and *relations*; the components of systems are not things, but processes. Systems are different from both mechanisms and aggregates in that the *properties* of the components themselves are dependent on their being parts of the system in question. So, for example, from a systems perspective, a mammal is composed of a circulatory system, a reproductive system, and so forth, *not* of carbon, hydrogen, calcium. The organismic level of description is *decoupled* from the atomic level.

[22] Alicia Juarrero, *Dynamics in Action: Intentional Behavior as a Complex System* (Cambridge, MA: MIT Press, 1999); parenthetical references in what follows are to this book. For a more complete account of Juarrero's work and its relations to developments in emergence theory and downward causation, see Nancey Murphy and Warren S. Brown, *Did My Neurons Make Me Do It?: Philosophical and Neurobiological Perspectives on Moral Responsibility and Free Will* (Oxford: Oxford University Press, 2007).

[23] See Scott, this vol., chap. 8.

All systems have *boundaries*, without which the system could not be recognized or defined. Systems may be defined as *open* or *closed*. An open system needs to be characterized not only in terms of its own organization, but also in terms of its transactions with its environment. The elements of the environment with which it interacts may be called its *external structure* or its *boundary conditions*. *Inputs* and *outputs* may be material, energetic, or informational.

Systems are related hierarchically. The system that is $S$ in its environment is a higher-order system, $S'$. There is an important difference between what Howard Pattee calls *structural* and *control hierarchies*. Juarrero says: 'When levels of *structural hierarchies* do not interact, they can be described by dynamical equations that deal with only one level at a time. In *control hierarchies*, on the other hand, the upper level exerts an "active authority relation" on the components of the lower level . . .' (114). She quotes Pattee:

> In a control hierarchy, the upper level exerts a specific, dynamic constraint on the details of the motion at the lower level, so that the fast dynamics of the lower level cannot simply be averaged out. The collection of subunits that forms the upper level in a structural hierarchy *now also acts as a constraint on the motions of selected individual subunits. This amounts to a feedback path between levels.* [The description of] the physical behavior of a control hierarchy must take into account at least two levels at a time.[24]

More often, though, three levels are involved: the focal level (e.g. an organism), its components (e.g. its nervous system), and its interactions with the environment.

### 1.5.4. *Meeting Objections*

Juarrero reconciles Van Gulick's and Peacocke's antireductionist accounts, and also responds to typical objections. She describes the role of the system as a whole in determining the behavior of its parts in terms similar to Van Gulick's account of the larger pattern or entity selectively activating the causal powers of its components. She says:

[24] Juarrero, *Dynamics in Action*, 114, quoting Harold Pattee, 'The Physical Basis and Origin of Hierarchical Control,' in Pattee, ed., *Hierarchy Theory* (New York: George Braziller, 1973), 77; Juarrero's emphasis.

The dynamical organization functions as an internal selection process established by the system itself, operating top-down to preserve and enhance itself. That is why autocatalytic and other self-organizing processes are primarily informational; their internal dynamics determine which molecules are 'fit' to be imported into the system or survive. (126)

Juarrero addresses the crucial question of *how* such selection can take place without begging the question of downward causation by giving further consideration to the concept of constraints: it is by means of constraints that the selection takes place, not by physical causal interaction. This requires a development in the understanding of constraints.

While the earliest use of the concept of constraint in science was in physics, as in the motion of a pendulum or an object on an inclined plane, Juarrero says that the important point is not that physical constraints act as an external force that pushes, but rather that they are a means by which one thing's behavior becomes *connected* to the behavior of something else or to its environment (132). More generally, then, constraints pertain to an object's connection with the environment or its embeddedness in that environment. They are relational properties rather than primary qualities in the object itself. Objects in aggregates do not have constraints; constraints only exist when an object is part of a unified system. When two objects or systems are correlated by means of constraints they are said to be *entrained.*

From information theory, Juarrero employs a distinction between *context-free* and *context-sensitive constraints.* First, an example of each: In successive throws of a die, the numbers that have come up previously do not constrain the probabilities for the current throw; the constraints on the die's behavior are context-free. In contrast, in a card game, the chances of drawing an ace at any point are sensitive to history; if one ace has been drawn previously, the odds drop from 4 in 52 to 3 in 51. A nonlinear system can be defined as one that imposes contextual constraints on its components. What has gone on before constrains what can happen next; the history of such a system is essential to its characterization. She says: 'The higher level's self-organization is the change in probability of the lower-level events. Top-down causes cause by changing the prior probability of the components' behavior, which they do as second-order contextual constraints' (146).

Peacocke is right in saying that an autocatalytic process is one of the simplest examples of such systems. A simpler example than the Belousov-Zhabotinsky reaction is the case in which molecule A catalyzes the synthesis of B, and B the synthesis of A. The quantities of both A and B will increase, first slowly and then more rapidly, until the components of A, B, or both are used up. The total state of the system depends on its history. Each synthesis of a molecule of A slightly increases the probability, for each component of B that it will find the other component(s) and a molecule of the catalyst at the same time.

So, Juarrero's reply to the sort of objection raised in the previous section to accounts of downward causation is as follows:

> I have analyzed interlevel causality in terms of the workings of context-sensitive constraints and constraint as alterations in degrees of freedom and probability distributions. It might be objected, however, that 'alteration' presupposes causality and so the entire project is guilty of circularity. In reply, consider the following: assume there are four aces in a fifty-two card deck, which is dealt evenly around the table. Before the game starts each player has a 1/13 chance of receiving at least one ace. As the game proceeds, *once* players A, B, and C have already been dealt all four aces, the probability that player D has one automatically drops to 0. The change occurs because within the context of the game, player D's having an ace is not independent of what the other players have. Any prior probability in place before the game starts suddenly changes because, by establishing interrelationships among the players, the rules of the game impose second-order contextual constraints (and thus conditional probabilities).
>
> ... [N]o external force was impressed on D to alter his situation. There was no forceful efficient cause separate and distinct from the effect. Once the individuals become card players, the conditional probabilities imposed by the rules and the course of the game itself alter the prior probability that D has an ace, not because one thing bumps into another but because each player is embedded in a web of interrelationships. (146)

She notes another common objection to the idea of downward causation or whole–part constraint—that it is a confusion to speak of the whole causally affecting its parts because the whole is nothing other than its parts. For example, Mario Bunge writes that wholes cannot act on their parts because a level of organization 'is not a thing but a set and therefore a concept.... All talk of interlevel action is elliptical or metaphorical.'[25] Juarrero responds by noting the peculiar history of changes in concepts of causation from Aristotle to the present. She laments the contraction of Aristotle's account of causation so as to recognize only efficient causes, but also points out that what is usually not recognized is the fact that modern thinkers retained the Aristotelian principle that *nothing causes or moves itself.* This explains the surprising fact that contemporary thinkers tend to recognize both that the environment has downward effects on the organism and that the organism's parts have bottom-up effects, yet there is no appreciation for ways in which organisms themselves exert downward control over their own biological processes. The fact that this happens all the time should be undeniable: the horse runs toward an open gate and all of its parts go with it!

Juarrero's response to challenges such as Bunge's, then, is to note that the objection betrays philosophers' typical refusal to acknowledge self-cause, with the consequent inability to understand self-organizing, goal-directed systems as (to greater or lesser degrees) the causes of their own behavior—autonomous from the environment and selectively determining the functioning of their own components.

## 1.6. Another Look at Emergence

The thesis of this essay is that emergence needs to be understood as the inverse of causal reductionism. This led me to examine the concepts of downward causation and whole–part constraint. I claimed that Juarrero's use of the concept of *context-sensitive constraints* in dynamic systems provides a necessary piece of the puzzle in defending such moves from the reductionist. To further my argument, this section will attempt to correlate Juarrero's understanding of downward causation via context-sensitive constraints with Terrence Deacon's account of emergence. Deacon's distinctions among levels of emergence will provide insights for classifying the broad spectrum of systems in which downward causation is operative.

[25] Mario Bunge, *Ontology II: A World of Systems* (Dordrecht: D. Reidel, 1979), 13–14; quoted by Juarrero (129).

According to Deacon, there is no emergence in mere aggregates, though an aggregate does have one sort of global properties. For example, the weight of a volume of liquid is a simple addition of the weights of its molecules.

The important difference between an aggregate and a system is that in a system it is *relational* properties of the constituents (as opposed to primary or intrinsic properties) that constitute the higher order. In such cases, additional configurational and distributional information is needed to account for the higher-order properties. Deacon includes here the viscosity of liquids, turbulence in large bodies of water, and typical feedback systems such as a thermostatically controlled heating system. This he calls first-order emergence. In Juarrero's terms, the relations among components impose constraints on the system, but these are context-free rather than context-sensitive constraints. Because fluctuations in such systems are dampened out across time, it is possible to give (rough) reductionistic accounts of their behavior.

Second-order emergence occurs when there is symmetry breaking or the *amplification* of a fluctuation rather than dampening. Systems in which this occurs are nonlinear; their history matters. There are simpler and more complex versions of such systems. The simpler sort is self-organizing, in that higher-order patterns selectively constrain the incorporation of lower-order constituents into the system or select among possible states of the lower-level entities (this is Van Gulick's point, as well). More complex second-order emergent systems are also autopoietic: they change the lower-order constituents themselves. Examples of the simpler sort are the Bénard phenomenon, a thermostat that amplifies rather than dampens feedback, and the development of a snowflake. An autocatalytic cycle (such as the Belousov-Zhabotinsky reaction) is of the more complex sort in that the system manufactures some of its own components. All life involves second-order emergence of the more complex sort.

Deacon distinguishes between first- and second-order (as well as third-order) emergence in terms of what he calls 'amplification logic' or 'the topology' of causal processes. In systems without emergence, global properties are all produced bottom-up (or, I would add, by means of *local* interactions with boundaries—e.g. a water molecule constrained by the presence of the surface of the container). In first-order emergent systems there is 'nonrecurrent' causal architecture: a simple bottom-up and top-down relation in which global properties of the system (e.g. density of components) make a difference to the relations among components and thus to the behavior of the whole system.

Second-order systems have more 'tangled' or 'recurrent' causal architecture as a result of the amplification of lower-level fluctuations. This amplification changes the total state of the system in a way that makes a decisive difference for the future development of the system. This can lead to new orders of complexity. Deacon's second-order emergent systems are the simplest of those that Juarrero describes as being driven by context-sensitive constraints: what happens before changes the probabilities for future behavior of the components.

Third-order emergence involves the interaction among three levels and appears (naturally) only in the biological realm. Here a variety of second-order forms emerge, and are selected (constrained) by the environment, but in such a way that a *representation* of its form is introduced into the next generation. The simplest example is the evolutionary process. The micro-level (the genome), in interaction with the organism's environment, directs the construction of the organism (the mid-level), whose reproductive fate is determined top-down by the environment (top level). The preservation of information regarding the organism's success in the environment is the means by which relatively stable populations of successful organisms can be produced, within which future fluctuations appear. Some of these may be amplified (preserved and re-entered into the system) by means of interaction with the environment, thus enabling the appearance of still higher degrees of complexity. Deacon describes such systems as exhibiting recurrent-recurrent causal architecture: over time, a two-stage process of emergence occurs that results in downward causation not just from top to mid-level, but from top to bottom (environment to genome).

Thus, we can see that Campbell's example of downward selection of termite jaws is in fact an example of Deacon's third-order emergence. The development of systems theory, theories of self-organization, nonlinear mathematics, far-from-equilibrium thermodynamics, and, finally, Juarrero's concept of context-sensitive constraints, gives us the conceptual tools to explain why the whole process could never in principle be reduced to the laws of biochemistry, let alone to the laws of physics.

It is interesting to note that Deacon describes the phenomenon of emergence and downward causation in terms so similar to Farrer's and Sperry's: Deacon's goal was to explain how systems come to be dominated by higher-order causal properties such that they appear to 'drag along' component dynamics. Campbell, Van Gulick, Peacocke, and Juarrero have all contributed to an account that evades Farrer's talk of *bewitchment* and Sperry's talk of *overpowering* lower-level causal forces. In addition, Deacon has provided a clear and noncontroversial definition of *emergent properties.* This represents a remarkable conceptual advance in the space of fifty years.

The essays that follow serve to flesh out and exemplify many of the conceptual resources invoked here, and then, finally, to ask what difference this striking development might make to theology.

# 2

# Reduction, Emergence, and the Mind/Body Problem: A Philosophic Overview

*Robert Van Gulick*

## 2.1. Introduction

My aim in this chapter is to give an overview of the recent discussion in the philosophy of mind that will serve as a map in locating issues and options. I will focus in particular on three central and interrelated ideas: those of emergence, reduction, and nonreductive physicalism. The third of these, which has emerged as more or less the majority view among current philosophers of mind, combines a pluralist view about the diversity of what needs to be explained by science with an underlying metaphysical commitment to the physical as the ultimate basis of all that is real.

The terms 'reduction,' 'nonreductive,' and 'emergence' get used in a bewildering variety of ways in the mind–body literature, none of which is uniquely privileged or standard. Thus, clarity about one's intended meaning is crucial to avoid confusion and merely verbal disagreements. Consequently, much of my mapping will be devoted to sorting out the main versions of reduction and emergence before turning to assess their interrelations and plausibility. My intent is to act largely as a guide and not an advocate; my goal is to lay out the logical geography in a more-or-less neutral way.

## 2.2. Varieties of Reduction

The basic idea of reduction is conveyed by the 'nothing more than ...' slogan. If Xs reduce to Ys, then we would seem to be justified in saying things such as Xs are nothing other (or more) than Ys; Xs are just special sorts, combinations, or complexes of Ys, or Xs are nothing over and above Ys. However, once one moves beyond slogans, the notion of reduction is ambiguous along two principal dimensions: the types of items that are reductively linked and the nature of the link involved. Thus, to define a specific notion of reduction, we need to answer two questions:

- Question of the relata: Reduction is a relation, but *what types of things* does it link?
- Question of the link: *In what way(s)* must the items be linked to count as a reduction?

First, consider the question of the relata. Between *what types of things* might the reduction relation hold? The notion gets interpreted in two distinct ways that involve very different sorts of relata. It can be viewed as

- a relation between real-world items—objects, events, or properties—which we might term *Ontological Reduction* (ONT-Reduction).

or as

- a relation between representational items—theories, concepts, or models—which we can call *Representational Reduction* (REP-reduction).

There are obviously important connections between the two families. But they involve distinct types of relata, and one must not conflate them, as too often happens. Speakers in the reduction debate often talk past one another by failing to distinguish ontological from representational notions, especially in interdisciplinary settings that combine scientists and philosophers.

The distinction is crucial as well for locating nonreductive physicalism in the logical space of options. It typically combines a denial of some form(s) of representational reduction with the acceptance of some type(s) of ontological reduction supposedly adequate to secure its physicalist credentials with an ontological link robust enough to meet the demands of physicalism. Its critics deny that it can be done, but the claim at least locates the view in logical space.

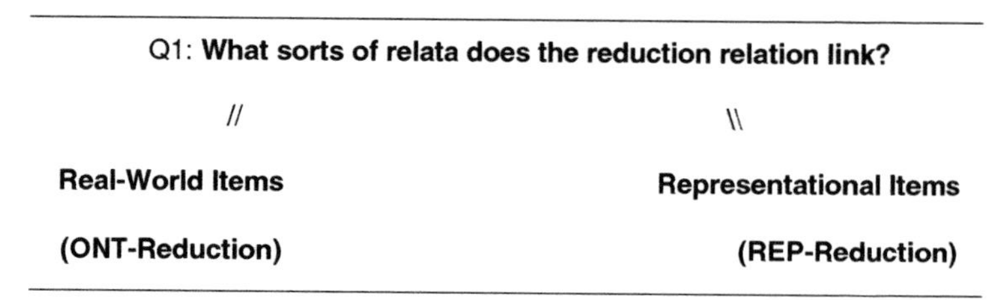


Figure 2.1

The diagram in Figure 2.1 shows the first step in our taxonomy by subdividing the types of reduction into two families based on their answer to our first diagnostic question.

Each family further subdivides based on the specific types of relata involved. Thus, ontological reductions might involve relations between things of various kinds:

- objects in two domains (e.g. minds and brains, or pains and neuron firings),
- properties (e.g. feeling pain and having neural activation of type $N_P$, or having a desire for a cup of coffee and being in a neuro-functional state of type $Nf_d$),
- events (e.g. Bill's having a red visual experience and Bill's brain being globally active in a way that includes neural activities of type $N_{rve}$ in his visual cortex as part of its global focus),
- processes (e.g. my recalling of the cellist's performance and a sequence of reciprocal neural interactions $RNI_c$ between multiple limbic and cortical areas).

REP-reduction similarly divides into more specific subtypes based on the particular relata involved, which might include any of the following:

- concepts (e.g. links between our first-person concept of phenomenal red and concepts from neuroscience),
- theories (e.g. links between theories of conscious experience and theories of global brain function),
- models (e.g. links between models of consciousness and models of reciprocal brain activity),

- representational frameworks (e.g. links between the phenomenal first-person descriptive/explanatory framework and third-person neuroscience frameworks).

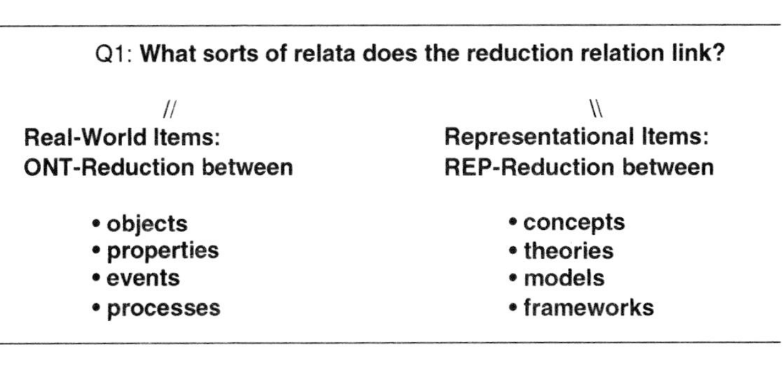


Figure 2.2

Let us move on to our second diagnostic probe, the question of the link: *In what way(s)* must the items be *linked* to count as a case of reduction? Again, there are a variety of answers on both the ontological and the representational side. With regard to ONT-reduction, the question becomes:

Question of the *ontological link:* How must things be related for one to ontologically reduce to the other?

At least five major answers have been championed in the literature:

- elimination
- identity
- composition
- supervenience
- realization

The relative merits and faults of the competing proposals have been intricately debated, but for present purposes suffice it to say a bit about each and give a sense of the range of options.

*Elimination.* One of the three forms of reduction listed by Kemeny and Oppenheim in their classic paper on reduction was replacement, i.e. cases in which we come to recognize that what we thought were Xs are really just Ys.[1] For example, we've come to see that what had been

[1] J. Kemeny and P. Oppenheim, 'On Reduction,' *Philosophical Studies* 7 (1956): 6–17.

thought of as demon possession is just a form of psychosis, perhaps a type of schizophrenia with auditory hallucinations. In such a case we might say that demon possession has been reduced to mental illness; schizophrenia replaces demon possession in our inventory of the world. Eliminativists believe that a similar fate awaits many of the commonly alleged denizens of the mental domain; qualia, beliefs, intentionality, and even consciousness have all been slated for replacement by supposedly more mature scientific alternatives.[2] It is difficult to prove or refute predictions about the future course of inquiry. Nonetheless, the eliminativist shoulders a nontrivial burden to motivate his claim that so much of what we take to be real in the mental domain might go the way of demons. Our sense of what is mentally real seems too intimate and too useful to turn out to be so badly mistaken as to justify an eliminativist judgment. The debate in the literature is extensive, but in this context it is enough to note the controversial status of the issue.

Some contemporary readers may find it odd to describe elimination as a form of reduction since according to the eliminativist there is nothing really there to be reduced. If there are no beliefs, how can they be reduced to anything? Nonetheless, I believe it appropriate to keep it on our list. The eliminativist view is one way to unpack the basic reductive notion of 'nothing but...;' demon possession turns out to be nothing but organic brain disorder.

*Identity.* Identity falls at the opposite extreme from elimination. It involves cases in which we continue to accept the existence of Xs, but come to see that they are identical with Ys (or with special sorts of Ys). Xs reduce to Ys in the strictest sense of being identical with Ys. This most often happens when a later Y-theory reveals the true nature of Xs. Heat is just kinetic molecular energy, lightning is just an atmospheric discharge of static electricity, and genes are just functionally active DNA sequences. However, the 'just are...'

[2] R. Rorty, 'In Defense of Eliminative Materialism,' *Review of Metaphysics* 24 (1970): 112–21; Paul M. Churchland, 'Eliminative Materialism and the Propositional Attitudes,' *Journal of Philosophy* 78 (1981): 67–90; D. Dennett, 'Quining Qualia,' in *Consciousness in Contemporary Science*, eds. A. Marcel and E. Bisiach (Oxford: Clarendon Press, 1988); K. Wilkes, '—, Yishi, Duh, Urn and Consciousness,' in ibid.; and *idem*, 'Losing Consciousness,' in *Conscious Experience*, ed. T. Metzinger (Thorverton: Imprint Academic, 1995).

locution does not lead us to deny the existence of the prior items; we do not deny the reality of lightning, heat, or genes. Rather, we see that two distinct reference routes converge on the same item. In Frege's famous example of the evening star and the morning star, both turned out to be the planet Venus.[3] So, too, identity theorists claim mental states, events, and properties will turn out to be identical with neuroscientifically discovered items. Those mental states are every bit as real as heat and genes, but their true nature is yet to be discovered by a more mature mind/brain theory.

Contemporary physicalism first developed as an identity theory in the 1950s and 1960s,[4] specifically as what came to be called the type–type central state identity theory, which held that types of mental states (e.g. being a stabbing pain) were identical with types of states within the central nervous system (e.g. c-fibre firings or, more realistically, certain patterns of firing in anterior cingulate, somato-sensory cortex and interconnected limbic areas). The theory fell out of favour quickly for a variety of reasons, such as the multiple realizability objection which appeals to the fact that one and the same type of mental state might be realized by neurally quite dissimilar structures in different creatures, in different humans, or even in the same person at different times.[5] Thus, by the late 1960s, most physicalists had moved away from the type–type identity theory in favour of some form of functionalism that treats mental states and properties as higher-order features defined by their higher-level roles but nonetheless realized solely by their neural substrates much as the higher-level program states of a computer are realized by the underlying states of its hardware.

The situation is actually a bit more complicated since some functionalists identify the mental state-type not with the higher-order

[3] G. Frege, 'On Sense and Reference,' in *Translations from the Philosophical Writings of Gottlob Frege*, eds. P. Geach and M. Black, trans. Max Black (Oxford: Oxford University Press, 1952).

[4] U. T. Place, 'Is Consciousness a Brain Process?' *British Journal of Psychology* 47 (1956): 44–50; J. J. C. Smart, 'Sensations and Brain Processes,' *Philosophical Review* 82 (1959): 141–156.

[5] H. Putnam, 'Philosophy and Our Mental Life,' in *Mind, Language and Reality, Philosophical Papers*, Vol. 2, ed. H. Putnam (Cambridge: Cambridge University Press, 1972); J. Kim, 'Multiple Realization and the Metaphysics of Reduction,' in *Supervenience and Mind*, ed. J. Kim (Cambridge: Cambridge University Press, 1993).

or role property but with the specific structural property that plays that role in a given species or population. Though they pick out the property by the role it typically plays, they set the identity conditions in structural terms. In that respect, they are more like classic type–type identity theorists than other functionalists who set the identity conditions in terms of the role itself rather than in terms of the typical occupant of the role. According to the occupant functionalist (e.g. Lewis), if neural state N typically plays the role in humans associated with having a desire for coffee and I am in state N, then I have a desire for coffee even if in my particular case N exhibits none of the causal roles associated with such a desire, for example it does not make me more likely to drink a cup of coffee if offered. Many functionalists find such claims counterintuitive; Lewis concedes that such cases would be odd, but argues they would still be cases of my having coffee desires despite their nonstandard causal profiles.

The type–type identity theory has enjoyed a recent though modest rebirth of interest.[6] Some philosophers have looked to it as a means of solving or dissolving the supposed explanatory gap that baffles those who try to explain how and why any given conscious state correlates with or might be realized by a given neural state. These neo-identity theorists argue that because it is identity that is involved there is no explanatory gap to bridge. There are not two things whose linkage needs to be explained; there is just one thing, and it is necessarily identical with itself. If Brian's pain just is a certain pattern of brain activity in the *identity* sense of 'is,' then there is no gap to be closed any more than there is any case of identity. Some have complained that the explanatory lacuna merely reappears as the unsatisfied demand for some account of how the mental and physical pathways might converge on a common referent, and that complaint seems justified. Nonetheless, it's important to acknowledge that identity versions of reductive physicalism are still alive and being actively defended.

*Composition.* One seemingly plausible alternative to identity is composition. If mental things (e.g. minds) are *composed* entirely of

[6] C. Hill and B. McLaughlin, 'There is Less in Reality than Dreamt of in Chalmers' Philosophy,' *Philosophy and Phenomenological Research* 59 (1999): 445–454.

physical parts, might that not suffice to justify the reductionist claim that they reduce to the physical or are all 'just physical.' If all of a thing's parts are physical, can the reductionist not say that it 'contains nothing over and above the physical?' Moreover, composition is distinct from identity and has a different logical status that easily accommodates the sort of multiple realizability objections raised against identity theory. Higher-level objects can outlive the components of which they are at a given time composed. A marching band which gradually changes its membership over time can continue to exist even after its original members have retired. So, too, a given mental state might persist through changes underlying neural composition. There may be good mentalistic reasons for regarding the memory that I have on Tuesday of the film I saw on Sunday as the same mental state as the memory that I had of it on Monday, even if the two differ nontrivially in their neural components. Brains and the patterned information they encode seem to be quite neurally dynamic; retaining or recalling a given memory need not rely upon the very same ensemble of neuron activations.

Despite its obvious attractions, composition will not suffice to ground reductive physicalism. Indeed, it is compatible with various forms of property dualism, including both fundamental and emergent property dualism. To say that a thing is composed *entirely of physical parts* is not the same as saying that *all its parts are entirely physical*; to assert the former is to say only that all its parts have physical properties, but the second asserts as well that those parts have *only* physical properties. Without the second stronger assertion, nothing excludes the possibility of nonphysical mental properties either at the level of the parts or at the level of the whole (emergent property dualism), and the appeal to composition *per se* will not suffice to support that stronger assertion.

*Supervenience.* Some philosophers have recently appealed to the notion of supervenience as a way of getting beyond mere composition and reductively linking mental and physical properties themselves while still stopping short of strict identity.[7] The issue is complicated

[7] Donald Davidson, 'Mental Events,' in *Experience and Theory*, eds. L. Foster and J. Swanson (Amherst, MA: University of Massachusetts Press, 1970); J. Kim, 'Psychophysical Supervenience,' *Philosophical Studies* 42 (1982): 51–70.

by the lack of any philosophical consensus about the identity and individuation conditions for properties; there is no dominant view in the field about the metaphysical status of properties and the conditions under which one property is the same as another. Nonetheless, various attempts have been made to analyse the relation between mental and physical properties in ways that might legitimate physicalism without invoking a strict type–type identity of properties.

Supervenience, which involves the dependence of one set of properties on another, was first proposed as a way of explaining the relation between normative properties such as moral and aesthetic properties and their non-normative bases.[8] The basic idea is that one set of properties (X-properties) supervenes on another (Y-properties) such that there can be no X-differences without Y-differences, or to put the point the other way round, any two things sharing all their Y-properties must also be alike in all X-properties. For example, the beauty of a painting may not be identical with any of its strictly physical properties such as its distribution of pigments on the surface of the canvas, but any other painting with all the physical properties of the first would also have to share its aesthetic properties. If the first were sublimely beautiful, so too would be the second. In the mind/body domain, the basic view can be conveyed by the slogan, 'No mental difference without a physical difference.'[9] Although supervenience theorists have not generally labelled themselves as reductionists, the dependence relation provides one way to cash out the basic reductionist idea that 'Xs are just Ys.'

Although supervenience enjoyed a brief period of intense interest as a possible way of making sense of ontological physicalism, it has now generally fallen out of favour. Even Jaegwon Kim, who played the largest role in bringing the notion to the centre of discussion,[10] has acknowledged that supervenience is too weak a relation to validate physicalism and is *a fortiori* an inadequate way of analysing the concept of reduction.[11] Kim has conceded that supervenience is compatible with both property dualism and dual aspect theory. Even

[8] G. E. Moore, *Principia Ethica* (Cambridge: Cambridge University Press, 1902).

[9] Davidson, 'Mental Events'; Kim, 'Psychophysical Supervenience.'

[10] J. Kim, *Supervenience and Mind* (Cambridge: Cambridge University Press, 1993).

[11] J. Kim, *Mind in a Physical World* (Cambridge, MA: MIT Press, 1999).

if the mental and physical realms were distinct and separate, supervenience could still hold as long as there were invariant correlations between the two—whether underwritten by natural law (*nomic supervenience*) or some stronger metaphysical link (*metaphysical supervenience*). Indeed, the first sort of correlation would hold in the world of a panpsychic monist like Spinoza. If supervenience is compatible with such explicitly nonphysicalist views, it seems unlikely to provide an adequate account of physicalism and even less of how the mental might reduce to the physical. Some have replied that what is needed is a relation of *logical supervenience*, according to which as a matter of logic it is impossible that mental and physical properties might independently vary.[12] Then the question arises of what sort of link, short of identity, might underwrite such logical necessity. Thus, if an adequate answer is to be found, the real explanatory work will have to be done by the story of that underlying relation rather than by supervenience *per se*.

*Realization*. One way of spelling out that underlying story might be in terms of realization. The multiple realizability of mental states was used as an objection against type–type identity theory,[13] but it has also provided the basis for a positive view of the psycho-physical link that might suffice as a form of ontological reduction. Realization is especially attractive to functionalists. Two systems can manifest the same functional property even if it is realized by different structures in the two cases. Nature can typically build a membrane with a given permeability profile in more than one way, and engineers can design a signal amplifier with the same input/output function using different hardware setups. However, as an ontological matter, the given functional property is fully realized in each case by its underlying physical components and their mode of composition. Once you've fixed all the facts about the structures and processes at the physical level, the facts about the functional properties follow automatically.

Realization appeals to those who favour a mind/computer analogy[14] since the same software or computational processes can be realized by many different types of hardware. However, realization

[12] Chalmers, *Conscious Mind* (Oxford: Oxford University Press, 1996).
[13] Putnam, 'Philosophy.'
[14] Ibid.; J. Fodor, *Representations* (Cambridge, MA: MIT Press, 1981).

is also invoked by some philosophers like John Searle, who are anti-computationalist and explicitly nonfunctionalist about the mental.[15] Searle denies that mental properties can be functionally analysed because of what he regards as their irreducible first-person intrinsic features; he nonetheless classifies his view as physicalist by appeal to his slogan that mental properties are 'caused by and realized in' the physical processes of the brain. The realization claim is essential to avoid property dualism, which would remain an option if he claimed merely that mental properties were caused by brain processes. With the addition of the realization requirement, he regards himself as having shown mental states as metaphysically no more problematic than the liquidity of room-temperature water, which is analogously *caused by and realized in* interactive collections of $H_2O$ molecules.

Realization plays a role as well in many versions of nonreductive physicalism. In her attempt to combine ontological physicalism with a denial of representational reduction (ONT-reduction & Not REP-reduction), the nonreductivist most often appeals to realization to secure her credentials as an ontological physicalist.[16] Can she succeed? Can she give an account of a realization strong enough to vindicate physicalism, yet consistent with robust denial of REP-reduction? We turn to that question below; first we need to shift attention from the ontological side (ONT-Reduction) and consider the equally diverse family of relations that qualify on the representational side as kinds of REP-reduction.

In a case of REP-reduction, one set of representational items is reduced to another. In answer to Question 1, we noted that REP-reductive relations might hold among at least four different kinds of representational items: concepts, theories, models, and frameworks. Those four do not exhaust the relevant options, but for present purposes we can restrict ourselves to them. What then of the representational version of Question 2:

Question of the *representational link:* How must things be related for one to representationally reduce to the other?

[15] J. Searle, *The Rediscovery of the Mind* (Cambridge, MA: MIT Press, 1992).

[16] R. Van Gulick, 'Nonreductive Materialism and Intertheoretical Constraint,' in Beckermann *et al.*, eds., *Emergence or Reduction?*.

There are a diversity of answers given in the literature, some of which make sense only with respect to certain kinds of items in the representational domain; some relations, such as derivability, might make sense as relations between theories but not between models or among concepts. However, certain commonalities run through the family of REP-reductive relations. They all involve some sort of *intentional equivalence*, that is some correspondence in terms of what they can or do represent as opposed to mere correspondence in their form or intrinsic properties. The basic idea is that one representational item (or set of items) reduces to another just if the first is linked to the second in terms of what it can or does say about the world and its features. Thus, REP-reductive relations generally concern either the comparative expressive powers of representational items (what they *can say)* or correlations in their assertoric content (what they *do say*). Representational reductions turn on either relations of meaning or relations of truth. Most of the variants of REP-reduction fall into one of five categories:

- Replacement,
- Theoretical-Derivational (Logical Empiricist),
- *A priori* Conceptual Necessitation,
- Expressive Equivalence (two-term semantic relation),
- Teleo-Pragmatic Equivalence (n-term pragmatic relation).

*Replacement.* The analogue on the representational side of elimination on the ontological side would be replacement. Prior ways of conceptualizing the world might drop out of use and be superseded by newer, more adequate ways of representing reality. For example, many of our mentalistic concepts might turn out not to do a good job of characterizing the aspects of the world at which they were directed, as has happened with demon concepts. If so, future science might develop alternative concepts that would more accurately and effectively represent reality. If adopting those newer systems should lead us to drop our former mentalistic outlook, then in an extreme sense our mentalistic way of speaking and thinking might be regarded as having been reduced to its representationally superior replacement. However, most notions of REP-reduction to be found in the literature are more conservative and involve preserving more of the truth or expressive content of the reduced theory.

*Theoretical-Derivational.* The classic notion of reduction in terms of theoretical derivation, as found in Kemeny and Oppenheim[17] and Ernest Nagel's classic treatment,[18] descends from the logical empiricist view of theories as interpreted formal calculi statable within the resources of symbolic logic. Given the axioms or laws of such a theory together with a formally statable set of actual conditions, one can derive all its consequences, observational and otherwise, by working out its formal implications. Thus, if one such theory T1 could be logically derived from another, T2, then everything T1 says about the world would be captured by T2, and T1 could be said to reduce (REP-reduce) to T2. Because the theory to be reduced, T1, normally contains terms not occurring in the reducing theory T2, the derivation also requires bridge laws or principles to connect the vocabularies of the two theories. These may take the form of strict biconditionals linking terms in the two theories, and when they do, such biconditionals may underwrite an ontological identity claim. If a gas has a given temperature when and only when its molecules have a given average kinetic energy, then we infer that temperature just is average kinetic energy, and that heat is *identical with* molecular motion. However, the bridge principles can take other forms; they need not be strict biconditionals. All that is required is enough of a link between the vocabularies of the two theories to support the derivation.

One other caveat: In most cases what is derived is not strictly speaking the original reduced theory, but an image of that theory within the reducing theory, and that image is typically only a close approximation of the original rather than a precise analogue.[19] In the paradigm case of the reduction of classical thermodynamics to statistical mechanics, the image of the classical laws within the statistical domain allows for possible though extremely improbable deviations from the classical laws. Thus, as a matter of logic, those prior laws are not derived. However, we accept that the classical laws were not strictly speaking true, and the match between the original laws and their statistical analogues is so close that we accept the former as having been reduced to the latter.

[17] Kemeny and Oppenheim, 'On Reduction.'

[18] E. Nagel, *The Structure of Science* (New York: Harcourt, Brace and World, 1961).

[19] P. M. Churchland, 'Reduction, Qualia, and the Direct Introspection of Brain States,' *Journal of Philosophy* 82 (1985): 8–28.

Were we to apply the theoretical-derivational model of REP-reduction to the mind/body case, we would need to find a set of bridge principles that allowed us to derive all the truths of our mentalistic theories of consciousness from the laws and statements of the relevant reducing theory (theories) whatever they might be: neurophysiological, computational, quantum mechanical, or otherwise. As in the thermodynamics case, the derived result might not be an exact analogue, but it would have to be a close enough approximation to our original pre-reductive mentalistic theory to justify the claim of REP-reduction by theoretical derivation.

Such a prospect might have once been viewed as the likely result of eventual scientific progress; logical empiricist views of the unity of science envisioned such an eventual formal integration of our scientific representation of reality.[20] At present, far fewer philosophers expect such derivations even in the long term. Ontological dualists believe that mental and physical theories describe different domains, and thus they expect no reduction of the theories that describe them. But REP-reduction is deemed unlikely also by many physicalists, especially those who accept some form of nonreductive physicalism. Unlike the dualists, they believe there is just one domain of reality which at some level is correctly represented by physics, but they also believe that adequately representing all the complex features that that reality exhibits at its many different levels requires the use of diverse theoretical and representational resources beyond those provided by the formal structures of physical science *per se*.[21] We will inquire further below about why nonreductive physicalists believe that, and about whether such a view can be consistently combined with ontological physicalism. For the present, we need only note that for a diversity of reasons, theoretical REP-reduction by derivation is rejected in the mind/body case by many but not all current philosophers.

[20] P. Oppenheim and H. Putnam, 'Unity of Science as a Working Hypothesis,' in *Minnesota Studies in the Philosophy of Science*, Vol. 1, ed. H. Feigl, M. Scriven, and G. Maxwell (Minneapolis: University of Minnesota Press, 1958).

[21] J. Fodor, 'Special Sciences, or the Disunity of Science as a Working Hypothesis,' *Synthèse* 28 (1974): 77–115; R. Boyd, 'Materialism With Reductionism: What Physicalism Does Not Entail,' in *Readings in Philosophy of Psychology*, Vol. 2, ed. N. Block (Cambridge, MA: Harvard University Press, 1980).

*A Priori Conceptual Necessitation.* Though many reject theoretical reduction as too strong a link, other philosophers regard it as too weak for adequately REP-reducing mind to body. They demand that the bridge principles not merely support the derivation of the reduced theory, but do so by linking the two theories via necessary and *a priori* conceptual links or inter-theoretical definitions.[22] They would not count a derivation based upon mere empirically established links or biconditionals as a successful reduction, since such empirical bridge principles might only describe the correlations between properties in distinct and separate mental and physical domains. This alleged weakness of the derivational account is the representational analogue of the ontological faults charged against supervenience. Law-like covariance, though sufficient to support nomic supervenience on the ontological side and inter-theoretical derivation on the representational, nonetheless seems inadequate to vindicate physicalism or reductionism because it is compatible with both property dualism and dual aspect theory as long as the mental and physical domains are lawfully linked; something which even a Cartesian dualist accepts. Proponents of the a priori view argue that nothing short of logically sufficient conceptual connections will suffice for one theory to reduce another.[23] They often couple that claim with a demand for reductive explanation, that is, for an explanatory account that lets us see in a conceptually necessary way how the conditions described by the reducing theory must as a matter of logic alone guarantee the satisfaction of those described by the reduced theory. They appeal to models such as the liquidity of water. The explanation in such cases supposedly includes two components:

- first, an analysis of the concept to be reduced (e.g. liquidity) in terms of some set of typically functional conditions, and
- second, an account of how those conditions would as a matter of mere logic be satisfied by any underlying system meeting the conditions described by the reducing theory (the micro-interactions of the collection of room temperature $H_2O$ molecules).

[22] J. Levine, 'Materialism and Qualia: The Explanatory Gap,' *Pacific Philosophical Quarterly* 64 (1983): 354–361; *idem*, 'On Leaving Out What It's Like,' in *Consciousness*, ed. M. Davies and G. Humphreys (Oxford: Blackwell, 1993); Chalmers, *Conscious Mind*.

[23] Chalmers, *Conscious Mind*.

The 'apriorists' contend that nothing less would suffice as a theoretical reduction of the mental to the physical.

Again, the issues are numerous and debates in the literature extensive, but we need only note the controversial nature of the apriorists' claim. Though the a priori view has intuitive appeal and its share of supporters, it has a host of critics, some of whom plausibly charge it with setting up—or at least implying—a false dilemma.[24] Most physicalists would agree that mere brute fact mental–physical correlations unsupported by appeal to any underlying explanation of why they were so linked would fall short of providing a reduction even if they sufficed as a bridge to derive our mental theories from our physical ones. Indeed, right at the start of contemporary physicalism, Herbert Feigl disparaged such unsupported links as 'nomological danglers.'[25] However, demanding that the bridge principles provide logically necessary, a priori conceptual links between the two domains seems to swing to the opposite extreme. Surely there must be intermediate cases that involve explanatory rather than merely brute links, but that, nonetheless, fall short of the apriorists' radical requirement for strict logical entailments. There is no consensus about what would count as an adequate explanation of the psycho-physical link and about what sorts of bridge principles would suffice for a satisfactory theoretical derivation, but many physicalists believe that the answer lies somewhere between the two extremes.[26]

*Expressive Equivalence.* The two versions of REP-reduction considered thus far concern what we might call truth preservation, that is, everything that the reduced (mental) theory says about the world is also asserted by the reducing (physical) theory combined with requisite bridge principles. However, REP-reduction might be viewed not as equivalence in what *is* said about the world, but as merely

[24] N. Block and R. Stalnaker, 'Conceptual Analysis, Dualism, and the Explanatory Gap,' *Philosophical Review* 108 (1999): 1–46; R. Van Gulick, 'Conceiving Beyond Our Means: The Limits of Thought Experiments,' in *Toward a Science of Consciousness III*, ed. S. Hameroff, A. Kazniak, and D. Chalmers (Cambridge, MA: MIT Press, 1999); S. Yablo, 'Concepts and Consciousness,' *Philosophy and Phenomenological Research* 59 (1999): 455–463.

[25] H. Feigl, 'The Mental and the Physical,' in *Minnesota Studies in the Philosophy of Science*, Vol. 2, eds. H. Feigl, M. Scriven, and G. Maxwell (Minneapolis: University of Minnesota Press, 1958).

[26] Van Gulick, 'Nonreductive Materialism'; Kim, *Mind*.

a matter of preserving the expressive range of what one *can* say. One representational system, R1, might be regarded as reducible to another, R2, as long as every state of affairs representable by R1 could also be represented by R2.

Though expressive equivalence seems to set a weaker requirement than the derivational or apriorist versions, it is still far from obvious that such a criterion can be met. Many philosophers, dualist and otherwise, have appealed in particular to the alleged special nature of first-person phenomenal concepts and the sorts of experiential facts that we can supposedly know or understand only through their use. Drawing on an old empiricist intuition that goes back at least to John Locke in the seventeenth century, contemporary philosophers such as Thomas Nagel[27] and Frank Jackson[28] have argued that there are facts about consciousness that can be adequately known or understood only from the first-person experiential perspective; for example, one can fully understand what it's like to taste a pineapple only if one has had such an experience. (Although Jackson himself has recently changed his mind on this issue, others continue to invoke his original position and have not followed him in his reversal of opinion.)[29] Thus, some claim that physical theory with its reliance on third-person concepts can never fully achieve the expressive range of our mental modes of representing, especially those that involve experiential concepts. If there are indeed subjective facts that lie beyond the representational power of physical theory, then it may be impossible to REP-reduce the mental to the physical even in the weaker sense of expressive equivalence. There may be things that we can say mentalistically about the world that fall outside the range of what can be said using the resources of physical theory. Unsurprisingly, philosophers disagree about whether or not there are such facts,[30] but for present

[27] T. Nagel, 'What Is It Like to Be a Bat?,' *Philosophical Review*, reprinted in *Mortal Questions*, ed. T. Nagel (Cambridge: Cambridge University Press, 1974).

[28] F. Jackson, 'Epiphenomenal Qualia,' *Philosophical Quarterly* 32 (1982): 127–136; *idem*, 'What Mary Didn't Know,' *Journal of Philosophy* 32 (1986): 291–295.

[29] F. Jackson, 'Postscript on Qualia,' in *Mind, Method and Conditionals*, ed. F. Jackson (London: Routledge, 1998).

[30] D. Lewis, 'Postscript to Mad Pain and Martian Pain,' in *Philosophical Papers Volume I*, ed. D. Lewis (Oxford: Oxford University Press, 1982); P. M. Churchland, 'Reduction, Qualia and the Direct Inspection of Brain States' *Journal of Philosophy* 82 (1985): 8–28; R. Van Gulick, 'Physicalism and the Subjectivity of the Mental,'

purposes we need only note that if there are, then that would seem to preclude REP-reducing the mental to the physical in the expressive equivalence sense.

*Teleo-Pragmatic Equivalence.* The expressive equivalence account of REP-reduction treats representation as primarily a two-term relation between a representation (or set of representations), R, such as words or sentences in a language or theory, and the item (or items), X, represented by R. Moreover, the relation is thought of in terms of familiar semantic notions such as reference and meaning. Thus, if we interpret REP-reduction as a matter of expressive equivalence, the question becomes: For every representing element of T1 that means M or refers to an item X, is there a corresponding element or combination of representing elements in T2, call it R*, that has that same meaning or referent? This is certainly one legitimate way to define REP-reduction, but it may not be the most helpful or revealing. Though representation can be viewed as a two-term relation, doing so involves abstracting away from other significant parameters of representation that are likely to be of importance to understanding the REP-reductive or non-REP-reductive nature of the mind/body link. In particular, what a given representation, R, succeeds in representing is crucially dependent upon the causal structure of the representation-user, U, the social and physical context, C, in which R is applied, and the modes of causal and epistemic access that R affords to U when used in C. Thus, rather than being just a two-term relation between a representation, R, and represented item, X, representation is at least a four-term relation; nor need we stop at four—additional parameters might be added such as the ends toward which the representation is to be applied. Reframing the question of REP-reduction in terms of this more complex relation, the question becomes as follows: If U can use R from theory T1 in context C to represent X, is there an R* from T2 that U can similarly use in C to represent X, or at least some R† from T2 that U can use in some context C† to

*Philosophical Topics* 12 (1985): 51–70; *idem*, 'Nonreductive Materialism'; J. Levin, 'Could Love Be Like a Heat Wave?: Physicalism and the Subjective Character of Experience,' *Philosophical Studies* 49 (1986): 245–261; B. Loar, 'Phenomenal States,' in *Philosophical Perspectives*, Vol. 4: *Action Theory and the Philosophy of Mind*, ed. J. Tomberlin (Atascadero, CA: Ridgeview, 1990); W. Lycan, 'What is the "Subjectivity" of the Mental?' in Tomberlin, ed., *Philosophical Perspectives*.

represent X? To put the matter less abstractly, the problem is to find a way, if possible, to use the contextually embedded resources of the reducing theory to do the equally contextual representational work done by the items in the theory we are trying to reduce. Nor should we ignore the third or pragmatic parameter mentioned above. Success in real-world representation is in large part a practical matter of whether and how fully our attempted representation provides us with practical causal and epistemic access to our intended representational target. A good theory succeeds as a representation if it affords reliable avenues for predicting, manipulating, and causally interacting with the items it aims to represent. This is equally true in the natural realm, when one is judging that a given structure in the rat hippocampus serves as allocentric representation of its spatial environment, as it is when we judge an economist's model successful in representing the effect of interest rate changes on housing markets. In both cases, it is the practical access that the model affords to its user in its context of application that justifies us in viewing it as having the representational content that it does.[31]

Viewed from this more inclusive pragmatic perspective, the question of REP-reduction becomes one about the ability of representation-users to gain the same modes of access with the alleged reducing representations that they do with the representations to be reduced. In the mind/body case, this turns in part on our ability to use the representational resources of physical theory to replicate the functional interactive profile of the access afforded us by our mentalistic resources, including our first-person concepts and theories. Put this way, REP-reduction may seem far less plausible; it seems unlikely that we could use physical theory to gain access to our mental states and processes in ways that afford us the same understanding achieved through first-person and introspective modes of representation. The differences in the contexts are so great that they alone seem sufficient to make such an equivalence for all practical purposes impossible. Many first-person concepts are so directly embedded within our intra-mental processes of self-regulation, self-monitoring, and self-modulation that it is difficult to see how any third-person system of concepts provided by physical theory could

[31] Van Gulick, 'Nonreductive Materialism.'

achieve a pragmatic profile sufficiently similar to support a claim of REP-reduction. Indeed, it is for just this reason that philosophers who adopt a pragmatic view of representation typically also deny the possibility of REP-reduction and are in that sense nonreductionists.[32]

Though much more could be said about the many varieties of ontological and representation reduction and their respective faults and merits, I hope at least to have surveyed the main versions of each, as graphically summarized in Figure 2.3.

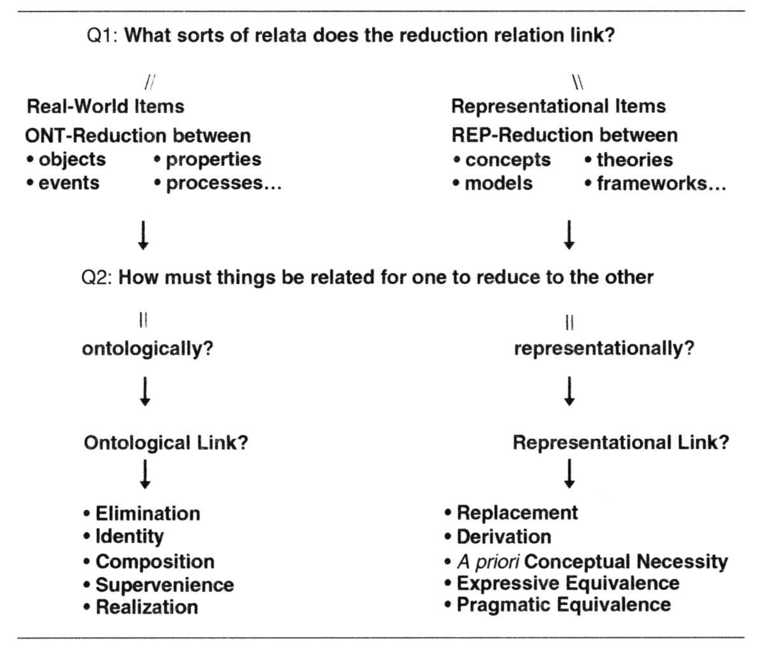


Figure 2.3

## 2.3. Varieties of Emergence

We thus turn our attention to the notion of emergence, which like reduction gets interpreted in diverse ways in the mind/body

[32] H. Putnam, *Meaning and the Moral Sciences* (London: Routledge and Kegan Paul, 1978); Van Gulick, 'Nonreductive Materialism.'

literature.[33] Again, my aim will be to survey the main variants. Then we can investigate relations among the members of these two diverse families of concepts.

The basic idea of emergence is more or less the converse of that associated with reduction. If the core idea of reduction is that Xs are 'nothing more than Ys' or 'just special sorts of Ys,' then the core idea of emergence is that 'Xs are more than just Ys' and that 'Xs are something over and above Ys.' Though the emergent features of a whole are not completely independent of those of its parts since they 'emerge from' those parts, the notion of emergence nonetheless implies that in some significant and novel way they *go beyond* the features of those parts. There are many senses in which a system's features might be said to emerge, some of which are quite modest and unproblematic[34] and others more radical and controversial.[35]

The varieties of emergence can be divided into several groups along lines that are similar in at least some respects to the divisions among the types of reduction. For example, emergence relations might be viewed either as objective metaphysical relations holding among real-world items such as properties, or they might be construed as partly epistemic relations that appeal in part to what we as cognitive agents can explain or understand about such links.

---

Q1: **What sorts of factors figure in emergence relations?**

// \\

| **Metaphysical: relations among real-world items** | **Epistemic: cognitive explanatory relations about real-world items** |
|---|---|

---

Figure 2.4

Relations of the first sort are objective in the sense that they concern the links or lack of links between real items such as properties, independent of considerations about what we, as epistemic subjects, can explain or understand about them. Relations of the second sort are epistemic, and in a sense subjective, because they turn crucially

[33] Searle, *Rediscovery*; W. Hasker, *The Emergent Self* (Ithaca: Cornell University Press, 2001); M. Silberstein and J. McGeever, 'The Search for Ontological Emergence,' *The Philosophical Quarterly* 49 (1999): 182–200.

[34] Searle, *Rediscovery*. [35] Hasker, *Emergent Self*.

on our abilities to comprehend or explicate the nature of the links or dependencies among real-world items rather than just on those links alone. The two sorts of notions often get run together in the discussion of emergence, but it is important to keep them distinct just as with the ontological and representational notions of reduction discussed above.

On the objective side, two main classes of emergents can be distinguished: properties and causal powers or forces. The distinction is not sharp and involves possible overlaps, especially if one individuates properties in terms of their causal profiles. Nonetheless, the issue of emergent causation is critical, so it's worth distinguishing at least initially between emergent properties and emergent powers, even if the line subsequently blurs. Within each of the two classes, there is a continuum of cases running from the extremely modest to the extremely radical. The former involve emergent features, whether properties or powers, that are similar in nature to the features from which they emerge, whereas the emergents in the latter sorts of cases are most unlike their nonemergent bases. Although the cases differ by many variations of degree, we can give some sense of their range by dividing the cases into three rough categories of increasing radicality which we can label: specific-value emergence, modest-kind emergence, and radical-kind emergence (Figure 2.5).

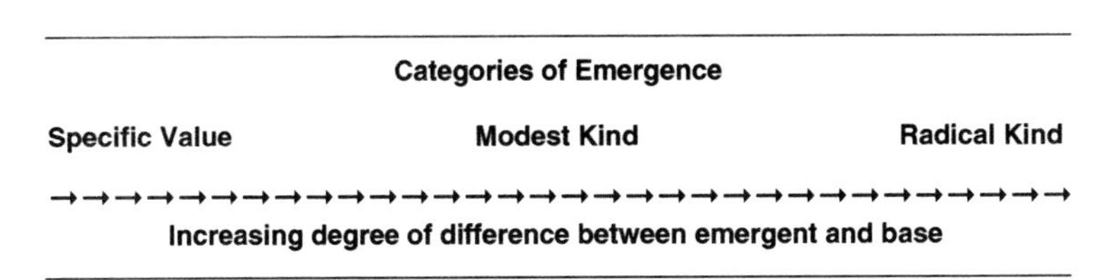


Figure 2.5

We can define the three roughly as follows.

- Specific Value Emergence. The whole and its parts have features of the *same kind*, but have different *specific subtypes or values* of that kind. For example, a bronze statue has a given mass as does each of the molecular parts of which it is composed, but the mass of the whole is different in value from that of any of its proper material parts.

- Modest Kind Emergence. The whole has features *different in kind* from those of its parts (or alternatively that *could* be had by its parts). For example, a piece of cloth might be purple even though none of the molecules that make up its surface could be said to be purple. A mouse might be alive even if none of its parts (or at least none of its subcellular parts) were alive.
- Radical Kind Emergence. The whole has features that are both (1) different in kind from those had by its parts, and (2) of a kind whose nature and existence is not necessitated by the features of its parts, their mode of combination and the law-like regularities governing the features of its parts.

Whether there are cases of radical-kind emergence is controversial. Physicalists who would readily concede to specific-value emergence and modest-kind emergence would most likely deny there are cases of the extreme sort. Accepting radical-kind emergence would be conceding that there are real features of the world that exist at the system or composite level that are not determined by the law-like regularities governing the interactions of the parts of such systems and their features. Doing so would require abandoning the atomistic conception typically embraced by mainstream physicalism. Radical-kind emergence would require giving up at least one of two core principles of atomistic physicalism (AP):

- AP1. The features of macro-items are determined by the features of their micro-parts plus their mode of combination. (In a slogan: Micro-features determine macro-features.)
- AP2. The only law-like regularities needed for the determination of macro-features by micro-features are those governing the interactions of those micro-features in all contexts, systemic or otherwise.

The idea of AP2 is that there are no laws governing micro-features in systemic contexts other than those that govern them outside such contexts. The intent is to exclude special laws that come into play only in restricted systemic contexts. Once the micro-features and their distribution have been fixed, the micro-necessities by themselves suffice to determine the macro-properties. Atomistic physicalism thus includes a commitment to 'bottom-up determination.' An example from Saul Kripke: once God had done the work of fixing the micro-features and laws of the universe, there was no work left to do; in fixing the world's micro nature, He had already determined all its macro properties as well.[36] As we will see below in discussing the dual revolutions position, some physicalists believe we should give up our commitment to atomism and micro-physical determination in favour of a more holistic view of physical reality; such an alternative, though not inconsistent with physicalism *per se*, is certainly far from the mainstream.

If we cross-pair our three rough divisions along the continuum of unlikeness between emergents and bases with our two classes of objective emergents (properties and powers), we then get six versions of metaphysical emergence as shown in Figure 2.6.

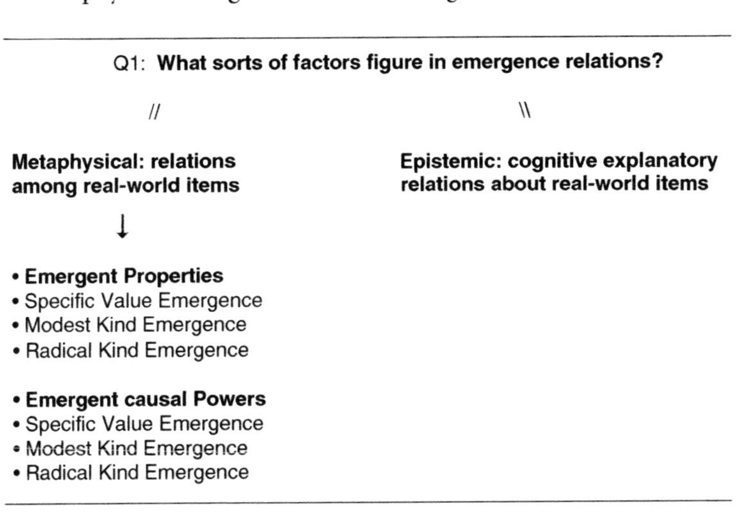


Figure 2.6

The notion that causal powers might exhibit radical-kind emergence merits special attention since it poses the greatest threat to physicalism. If wholes or systems could have causal powers that were radically emergent from the powers of their parts in the sense that those system-level powers were not determined by the laws governing the powers of their parts, then that would seem to imply the existence of powers that could override or violate the laws governing the

[36] S. Kripke, *Naming and Necessity* (Cambridge, MA: Harvard University Press, 1972).

powers of the parts; that is, genuine cases of what is called 'downward causation' in which the macro-powers of the whole 'reach down' and alter the course of events at the micro-level from what they would be if determined entirely by the properties and laws at that lower level.[37] All that is needed to get such a result is the relatively uncontroversial claim that macro-level causal powers have micro-effects. That in itself need cause no trouble for the physicalist as long as the macro-powers are themselves determined by the micro-powers.[38] But, if some macro causal powers were radically emergent, that would free them of determination by the underlying micro-powers, thus allowing them to alter the course of micro-events in ways independent of the micro-level laws. If the physicalist wants to avoid violation of underlying physical laws, she can allow macro-level properties to have micro-effects only if those macro-powers are themselves constrained by the micro-level laws, which is of course just what the radical emergence of causal powers denies.[39] It is in this respect that radically emergent causal powers would pose such a direct challenge to physicalism, since they would threaten the view of the physical world as a closed causal system; that is, the idea that nothing outside the physical causally affects the course of physical events.[40] Unsurprisingly, this feature that makes radical causal emergence so threatening to physicalists[41] is the very one that makes it so attractive to emergentists like Hasker who invoke emergence in support of ontological dualism.[42]

The challenge to those who wish to combine physicalism with a robustly causal version of emergence is to find a way in which higher-order properties can be causally significant without violating the causal laws that operate at lower physical levels. On one hand, if they override the micro-physical laws, they threaten physicalism. On the other hand, if the higher-level laws are merely convenient ways of summarizing complex micro-patterns that arise in special contexts, then whatever practical cognitive value such laws may have, they seem

[37] Sperry, *Science and Moral Priority*; *idem*, 'In Defense of Materialism and Emergent Interaction,' *Journal of Mind and Behavior* 12 (1991): 221–245; J. Kim, '"Downward Causation" and Emergentism and Non-reductive Physicalism,' in Beckermann *et al.*, eds., *Emergence or Reduction?*; *idem*, *Mind*; Hasker, *Emergent Self*.

[38] Kim, *Mind*.

[39] Hasker, *Emergent Self*.

[40] J. Kim, 'Explanatory Exclusion, and the Problem of Mental Causation,' in *Information, Semantics and Epistemology*, ed. E. Villanueva (Oxford: Basil Blackwell, 1990); *idem*, *Mind*.

[41] Kim, 'Downward Causation'; *idem*, *Mind*.

[42] Hasker, *Emergent Self*.

to leave the higher-order properties without any real causal work to do. One possible solution would focus on the respect in which higher-order patterns might involve the selective activation of lower-order causal powers. Micro-properties that were causally irrelevant in most configurations, for example because their random actions cancelled out and had no significant overall effect, might exert a powerful causal influence in a small range of cases involving higher-level patterns that brought those micro-powers into a coherent mode of action making a major difference to the overall operation of the system.[43] For example, the magnetic fields of molecules can be ignored in explaining most materials since they are randomly aligned. However, when they are coherently oriented in a magnet, those micro-properties are crucial to understanding the causal powers and activity of the system that contains them. Such selectional models need pose no problem for atomistic physicalism since they involve no violation of underlying micro-physical causal laws. However, whether they accord a sufficiently potent form of causal efficacy to macro- or system-level properties to justify a claim of emergent causality is open to debate. The selective activation model nonetheless provides an example of how one might try to find a way of reconciling physicalism and causal emergence.

Having given at least six rough options on the metaphysical side, let us turn our attention to epistemic notions of emergence. What makes all such notions epistemic is that they involve some respect in which we are unable to predict, explain, or understand the features of wholes or systems by appeal to the features of their parts. Emergence in this sense is thus, at least in part, subjective, that is, a matter of our cognitive and explanatory capacities and limits rather than just a matter of relations between objective items as in the metaphysical cases. For present purposes, it will suffice to distinguish two versions of epistemic emergence, which focus respectively on different cognitive abilities—the first on prediction and explanation and the second on representation and understanding. Once again, the lines between the two are not sharp, but worth distinguishing.

[43] See my other contribution to this volume in Chapter 3, originally published as R. Van Gulick, 'Who's in Charge Here? and Who's Doing All the Work?' in *Mental Causation*, eds. J. Heil and A. Mele (Oxford: Clarendon Press, 1993).

*Predictive/Explanatory Emergence:* Wholes (systems) have features that cannot be explained or predicted from the features of their parts, their mode of combination, and the laws governing their behaviour.

*Representational/Cognitive Emergence:* Wholes (systems) exhibit features, patterns, or regularities that cannot be represented (understood) using the theoretical and representational resources adequate for describing and understanding the features and regularities of their parts.

Both versions come in weak or restricted forms and in strong or unrestricted forms. On one hand, the relevant cognitive inability to explain or represent might be a restricted fact about our specifically human limitations or, even more restrictively, about our present state of theorizing and scientific progress. Alternatively, the cognitive inability might concern a more general limit that applies universally to all cognitive agents or to all those in some broad category of which we humans are one example among many. Colin McGinn, for example, has argued that humans lack the ability to form the sorts of concepts needed to make the psycho-physical link intelligible.[44] Thus, he believes it will ever remain a mystery to us, even though he accepts consciousness as an aspect of physical reality and allows that cognizers with concept-forming abilities quite different from our own may be able to comprehend the link in intuitively satisfying ways. It's not that we humans are not smart enough; it is that we have the wrong sorts of minds for solving the psycho-physical puzzle. From McGinn's perspective, consciousness is epistemically emergent in the explanatory sense, at least relative to humans and other cognitive agents with our sorts of conceptual capacities.

We can summarize our quick survey of the varieties of emergence in Figure 2.7.

## 2.4. Selected Conflicts, Agreements, and Other Relations

Our taxonomizing survey has thus distinguished among ten varieties of emergence and at least ten versions of reduction. Even so,

[44] C. McGinn, 'Can We Solve the Mind–Body Problem?' *Mind* 98 (1989): 349–366; *idem, The Problem of Consciousness* (Oxford: Blackwell, 1991).

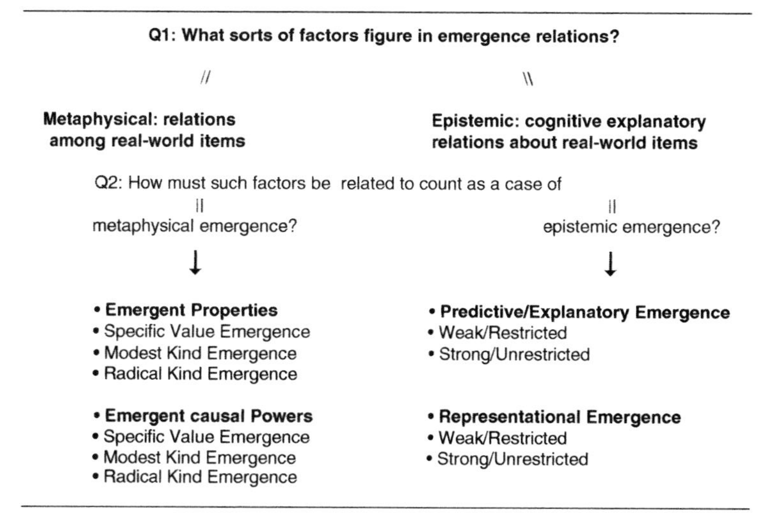


Figure 2.7

it probably does not capture every interesting variant, but I hope it includes all the major ones and gives a fair sense of the range of options.

Since my intent has been to act more as a tour guide than an advocate, I offer just five general observations aimed at further clarifying the lie of the philosophic landscape. It may contain some paths or links between positions other than those that are normally assumed, and being aware of them should help in navigating the terrain.

*1. Pay attention to the 'key'—the need for clarity and the avoidance of conflation.* Labels and terms are used with such a diversity of meanings in the mind/body literature that it is absolutely essential that one be clear about what meaning is intended in a given use. Because central terms such as 'reduction' and 'emergence' have no standard interpretation, one should always make one's own use clear; provide a 'key' for reading the verbal map of your view. And take care in reading others to hear their words as they intended them; otherwise, it's all too easy to mislocate them and for the discussion and to get lost in a fog of misunderstanding. The general need for clarity is platitudinously obvious, but it's so important in this context that it merits restating nonetheless.

*2. Respect the subjective/objective division.* Perhaps nowhere is the need for clarity greater or the threat of conflation more likely than when one is dealing with subjective and objective versions of some notion. As we saw above, the subjective/objective distinction runs through the space of options dividing both the reduction and emergence families into distinct and separate sections, just as a mountain range might separate a chain of plains or valleys through which it runs.

The reduction region divides into objective relations of ontological reduction such as identity, composition, or realization, and subjective regions of representationally reductive relations such as derivability, conceptual necessitation, or pragmatic equivalence. The parallel division within the emergence region is between objective notions of metaphysical emergence such as modest or radical kind emergence, and subjective notions of epistemic emergence concerning the limits on our cognitive abilities to explain or understand the features of wholes in terms of those of their parts.

The most common and controversial moves with respect to this division concern attempts to reach ontological conclusions from subjective premises. Facts about our human incapacity to reductively explain how consciousness might be realized by underlying physical processes cannot by themselves justify us in concluding that consciousness is not a physically realized process. The explanatory gaps may reflect subjective limits in our (current) human conceptual or imaginative capacities rather than any objective divisions in the world. Conceivability arguments for dualism are regarded by physicalists as tripping on this mistake. The apparent possibility to conceive of worlds that contain molecule-for-molecule physical duplicates of humans lacking any conscious mental life does not entitle the dualist to conclude that conscious properties are not physical in any ontologically robust sense (e.g. identity or realization). According to the physicalist critics, making such an inference would require us to pass invalidly across the subjective/objective divide, moving from facts about the limits of our concepts to an objective claim about the distinctness and independence of the real-world features to which we refer by use of those concepts. The dualist making such an inference would need to show that the concepts he employed on both the mental and physical sides of his thought experiment were adequate to support such a metaphysical conclusion;[45] physicalists doubt the dualist can discharge that burden.

Thus, we should not infer that mental properties are ontologically nonphysical just because we cannot representationally reduce our mental concepts or theories to physical ones. Nor should we conclude that mental properties or powers are metaphysically emergent just because they are subjectively emergent relative to our abilities to explain, predict, or understand using resources of physical theory. Additional argument is needed to justify the move from subjective premises to objective conclusion. The dualist champions of conceivability arguments believe additions can be made that validate the move,[46] but physicalist critics argue to the contrary.[47]

3. *Reduction and emergence can overlap (not necessarily disjoint).* Although the notions of reduction and emergence are often paired as polar opposites, there are, in fact, many consistent combinations of views from the two respective families. The slogans associated with the two make them seem mutually exclusive. How could Xs 'just be Ys' or 'merely special sorts of Ys' but also be 'something other than Ys'? Or how could Xs be 'something over and above Ys' but also 'nothing more than Ys'? The contradictions seem immediate and, indeed, they are if one assumes a consistent reading for both conjuncts. But, as we saw above, there are many versions of reduction and perhaps equally as many versions of emergence.

Though some versions of reduction are strictly inconsistent with some versions of emergence, other cross combinations involve no necessary conflict. One could not consistently combine an identity version of ontological reduction with a metaphysical notion of radical-kind emergent properties or powers. But, one might without contradiction accept both a cognitive/explanatory emergence view of mental properties as well as a realization or composition version of ontological reduction.

4. *Nonreductive physicalism lies largely within the intersection of the dual families of emergence and reduction, rather than in a third and separate region.* Nonreductive physicalism is typically regarded

[45] Van Gulick, 'Conceiving Beyond our Means.'

[46] Chalmers, *Conscious Mind.*

[47] S. Yablo, 'Concepts and Consciousness,' *Philosophy and Phenomenological Research* 59 (1999): 455–463; Van Gulick, 'Conceiving Beyond our Means.'

as an option wholly distinct from either reductive physicalism or emergence. However, once one recognizes that reduction and emergence are not mutually exclusive in all their versions, one can see that nonreductive physicalism actually occupies a region within the intersection of the reduction and emergence families. Those two families are to some extent complementaries or duals of each other, especially vis-à-vis the subjective/objective division. Thus, nonreductive physicalism on one hand combines a denial of (subjective) representational reduction with an acceptance of some robust form of (objective) ontological reduction such as physical realization. On the other hand, it pairs the denial of at least the strongest forms of ontological emergence such as radical kind emergence with an acceptance of epistemic emergence in either or both of its forms. Put symbolically:

Nonreductive Physicalism ⇒ (ONT-Reduction & Not REP-Reduction).

Nonreductive Physicalism ⇒ (Epistemic Emergence & Not Radical Metaphysical Emergence).

Despite the 'non' in the name of their position, nonreductive physicalists do accept some forms of objective (i.e. ontological) reduction, while rejecting most forms of subjective (i.e. representational) reduction. In the dual domain of emergence, they take the complementary position. They accept various forms of subjective (i.e. epistemic) emergence but reject the radical versions of objective (i.e. metaphysical) emergence as shown in Figure 2.8.

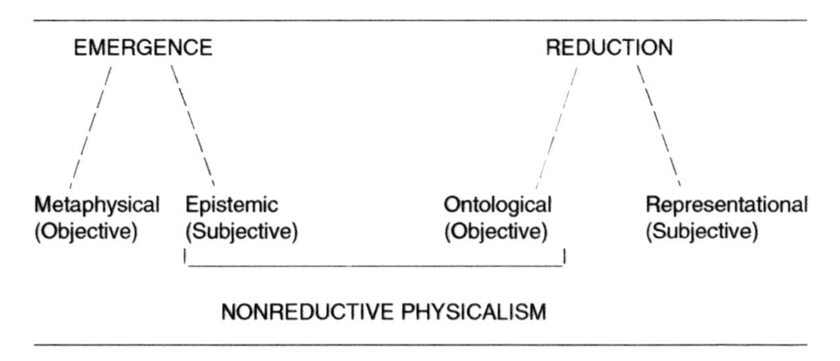


**Figure 2.8**

Thus, contrary to common belief, our map of logical space should not locate nonreductive physicalism in a region disjoint and separate from those occupied by the emergence and reduction families, but rather in a special sub-region of their intersection.

5. *What looks like a gap (an anomaly or a crisis) depends on your viewpoint and location.* Much of the pressure for extraordinary philosophical theorizing about consciousness and the psycho-physical link arises from the sense of what Kuhn would call a persistent anomaly that resists satisfactory resolution by more mainstream physicalist approaches.[48]

There are, no doubt, other factors at work as well. Our current scientific knowledge about the psycho-physical link is primitive, and there are at present no plausible detailed empirical models of how consciousness might arise from a physical substrate. Research has not advanced far enough to generate a consensus paradigm around which a body of 'normal scientific' practice might coalesce. Extraordinary theorizing is the rule at such early stages, as multiple investigators strike off in differing directions in search of some means to gain significant initial progress on the problem. Such an unsettled scientific state might by itself elicit a good deal of extraordinary philosophic speculation, even if the philosophic community remained largely committed to some form of mainstream physicalism as its own normal practice paradigm.

Moreover, extraordinary theorizing need not have any specific trigger at all. Paul Feyerabend, the iconoclastic philosopher of science and champion of revolutionary science, argued that extraordinary science is not and should not be restricted to periods of crisis and anomaly.[49] We need not wait until prevailing research programmes 'break' or fail to make progress before exploring radically alternative approaches to the field. Feyerabend had no sympathy with an 'If it's not broke, don't try to fix it' approach to science. He believed that some level of extraordinary or revolutionary science should be going on at all times. Perhaps a good alternative slogan for the Feyerabendian view might be, 'Even if it's not broke, you might find something better,' or 'If you don't start looking till it is broke, you're not likely to have

[48] T. Kuhn, *The Structure of Scientific Revolutions* (Chicago: University of Chicago Press, 1962).

[49] P. Feyerabend, *Against Method* (London: New Left Books, 1975).

anything to put in its place when it breaks.' Thus, there are at least two possible sources for extraordinary philosophic theorizing about the psycho-physical link other than the supposed perception of a recalcitrant anomaly. Nonetheless, extraordinary activity is far more likely to occur in contexts of anomaly and perceived crisis, and such a sense does seem to exert a lot of pressure on the current state of philosophic play on the mind/body issue.

But, is the perception correct or well founded? Is mainstream physicalism in fact in a crisis caused by its persistent inability to resolve a problem or puzzle that falls clearly within its domain? The answer is not as obvious as it might at first seem. Surely there are puzzles it has not solved and questions to which it has given no satisfying answers, but not every failure to solve a problem counts as an anomaly. As Kuhn made clear from the start, what counts as a problem or its solution is determined on the whole internally by the research community itself through its paradigms of normal practice.[50] Unless a puzzle presents a problem of a sort that the community by its own internal standards counts as one it ought to be able to solve, its failure to do so need not generate any anomaly or sense of crisis. And even if a puzzle meets the standards to count as a valid problem, it need not be open to a quick or easy solution. If so, a string of negative attempts need not indicate anything more than the difficulty of the problem and the need for continuing effort from within the paradigm.

Physicalism's supporters and its critics will obviously differ in their assessments of its status. Its critics will argue that it is no closer to explaining how consciousness might be a physical process than it was three hundred and fifty years ago when Descartes and Leibniz expressed their early scepticism. Nor is it just the explanatory details that we lack. As Thomas Nagel pointed out more then twenty-five years ago,[51] we have no model of how to begin to bring the two sides of the psycho-physical divide together; physicalists still find themselves staring at an explanatory blank wall. Solutions to hard problems rarely come quickly, but if three centuries of failure do not suffice to generate a crisis, what more is needed? Physicalists in reply might compare their inability at the end of the twentieth century to explain the physical nature of consciousness with that of their predecessors

[50] Kuhn, *Structure.* [51] Nagel, 'What Is It Like to be a Bat.'

at the start of the century to do the same for life. Vitalism was still a serious scientific position one hundred years ago, and a sense of mystery and bafflement still attended attempts to explain life, growth, and reproduction as wholly physical. It was only at mid-century that the puzzle was solved when the biochemical revolution provided us with more adequate concepts of both biological processes and their physical substrates. Only when we had a better understanding of both sides of the equation could we see how they fit together.

Just like other normal practice problem solvers, physicalists get a say in what counts as a problem and a solution within their domain, especially when those views are not ad hoc or merely self-protective, but are instead based on independently motivated claims about the nature of explanation. Of course, having an internal say in what counts as a solution does not in itself guarantee that outsiders will or should share your evaluation any more than most of us would be swayed by an astrologer's self-proclaimed success in solving astrological problems in astrologically valid ways. Mainstream physicalists have a responsibility to support and defend their criteria for solving the mind/body problem, but they have plenty of resources for doing so, and their critics have an equal obligation to defend their more reductive standards.

Thus, an accurate map of field should reflect the fact that which problems count as solved, unsolved, or illegitimate depends upon one's own location within the space of possible positions. What links one thinks can or cannot be seen will turn in part upon the lie the land appears to have from where one stands. Nor need there be an outside point of view that shows with neutral truth what links there are; there may be only inside answers to the question.[52]

[52] This chapter is a slightly shortened version of Robert Van Gulick, 'Reduction, Emergence and Other Recent Options on the Mind/Body Problem: A Philosophic Overview,' *Journal of Consciousness Studies* 8/9–10 (September/October 2001): 1–34; reprinted with permission.

# 3

## Who's In Charge Here? And Who's Doing All the Work?

*Robert Van Gulick*

Tom raised the canteen to his lips because he believed it still contained some water and he was very thirsty.

Greta chose her tomatoes carefully, picking only those that looked the reddest; because Henrik was color-blind, he was unable to help her.

Little Sonja pushed away the toy rabbit the day-care worker held out to her; she wanted her own brown teddy bear and nothing else would do.

There is nothing exceptional about such explanations and the situations they describe. In each case, a common sense or folk psychological explanation of someone's behavior is offered that explains why someone acted as he or she did by appeal to one or more of the person's mental states and their mental properties. Tom put the canteen to his lips *because* of his belief and *because* his belief had the propositional content that it did—that there still was water in the canteen.

While the overwhelming majority of philosophers and ordinary folk find such explanations informative and predictive, there is somewhat less unanimity of opinion about how the word 'because' is to be interpreted in such explanations. Due in part to the work of Donald Davidson in the 1960s, the dominant view among Anglo-American philosophers has been that the 'because' in such explanations must be read as involving a causal relation.[1] Beliefs and desires cause the behaviors that they rationalize. There are philosophers who deny this, for example Daniel Dennett[2] and Jonathan Bennett,[3] but it remains the majority view, and since I count myself within that majority, I accept it as a working assumption; my concern will be with the difficulties that supposedly result from adopting the causal interpretation.

It is misleading, however, to speak of *the* 'because' in the relevant explanations; each of them contains at least implicitly a double occurrence of 'because.' Tom raised the canteen to his lips because of his belief *and* because his belief had the intentional content that it did. Such explanations appeal both to mental states or events and to mental properties. If both of these 'becauses' are to be interpreted causally, then what is being claimed about the event of Tom's raising the canteen to his lips is both that it was caused (at least in part) by the event of his having a certain belief and that that belief's having the intentional content that it did was causally relevant to (or causally responsible for) its producing the sorts of effects it did. This may seem an unsurprising claim, one that accords well with common sense or folk intuitions. Surely, if Tom's belief had had a different content, for example that the canteen was empty or that it contained poison, its effects on his behavior would have been quite different.

However, common sense rarely suffices to prevent a good philosophical fight, and there has been of late quite a debate about whether the mental properties invoked in such explanations can be *causally relevant* or *causally potent* properties. To quote Jerry Fodor, there has been an 'outbreak of epiphobia,' a widespread fear of mental property epiphenomenalism—a fear that mental properties are causally irrelevant or inert.[4] The worry is that even if mental events such as believings or desirings have causal effects (including external behavioral effects), they do not do so in virtue of having the mental properties they do. In particular, content-properties and phenomenal properties are alleged to be causally impotent. The view is sometimes

[1] Donald Davidson, 'Actions, Reasons, and Causes,' *Journal of Philosophy* 60 (1963): 685–700.

[2] Daniel Dennett, *The Intentional Stance* (Cambridge, MA: MIT Press, 1987).

[3] Johnathan Bennett, *Linguistic Behavior*, 2nd edn. (Indianapolis, IN: Hackett, 1990).

[4] Jerry Fodor, 'Making Mind Matter More,' *Philosophical Topics* 17 (1989): 59–80.

combined with the token-identity thesis that every mental event-token is identical with some physical event-token. There is just one ontology of events possessing both physical and mental properties. However, with respect to causal potency, the mental properties are distinctly disadvantaged; indeed they are completely excluded from affecting any token event's causal powers or causal relations; those are determined solely by its physical properties.

Mental-property (MP) epiphenomenalism is an unintuitive and not very attractive hybrid. Yet, a variety of arguments may seem to force us to it. The relevant arguments are diverse both in the principles on which they base their claim and in the range of mental properties they allege to exclude from causal relevance. To get a sense of the options, consider the following argument.

1. *Token Physicalism*: Every mental event-token (i.e. every event-token having mental properties) is identical with some physical event-token (i.e. some event-token having physical properties).
2. The causal powers of a physical event-token are completely determined by its physical properties.
3. *The Nonreducibility of the Mental*: Mental properties are neither identical with nor reducible to physical properties. Therefore,
4. A mental event-token's mental properties do not even partially determine its causal powers (from 1, 2, 3). Therefore,
5. Mental properties are not causally potent (from 4). Therefore,
6. Mental properties are not causally relevant; they are epiphenomenal (from 5).

The argument is intended to establish that all mental properties are causally impotent. It relies on two widely held physicalist views expressed by its first and second premises: first, that all events are physical events, and second, that the world of the physical constitutes a closed causal system. Saying just what's wrong with the argument is difficult. It has no obvious point of vulnerability. A direct frontal assault is unlikely to breach its fortifications. To see where it goes wrong, we shall have to dig deeply into the question of causal relevance in the hope of undermining the hidden foundations of the argument.

Let us begin by asking the following question, 'What sorts of things are causal laws, that is what is their ontological status?' I think it's important to ask this question because at times in the mental causation literature there seems to be an almost theological realism about causal laws. The causal relation between two events is said to be 'grounded by the existence of a causal law' or that 'event c caused event e because it is a causal law that F-events cause G-events.' This makes causal laws seem like independent entities over and above the world of events and properties with the power to command events into conformity with their strictures. God said, '*Fiat lux*—let there be light' and there was light. Bodies of matter *must* attract each other gravitationally with a force inversely proportional to the square of their distance *because* were they not to do so they would violate the inverse square law. One might almost conjure up a comic version of Socrates' imagined confrontation in the *Crito* with *hoi nomoi* (the laws), who berate him for contemplating disobedience. I don't suggest that any of the parties to the mental causation debate really hold such a view, but only that a subtle and perhaps unconscious tendency to reify causal laws seems implicit in much of what gets said about the way in which laws make it the case that one event caused another.

How then should one view causal laws and laws of nature in general? Laws are counterfactual sustaining statements (or sentences) in theories, perhaps those which appear in the simplest most comprehensive theories satisfied by the spatio-temporal totality of our world. I will not attempt to define with necessary and sufficient conditions what makes a given statement in such a theory a law; what matters for present purposes is that laws are statements or sentences in our theories of the world, not independent items among the furniture of the world itself. And what in turn are theories? They are organized systems of representation that can be used to structure our cognitive processes and guide our action. And we must not forget this action-guiding function; we are not pure intellects, but agents who must choose, plan, decide, and act to survive. I offer no definition of what makes a theory; I only want to emphasize that theories and the laws they contain are cognitive constructs and that as such they are to be assessed pragmatically in terms of the roles they are designed or expected to fulfill.

What then is the function of causal laws in our cognitive economy? How are they expected to contribute to the organization of thought and the guidance of action? The answer is a familiar one and in a quick and simple way might go like this:

We are biological organisms with needs, goals, and wants. Our ability to succeed in their satisfaction is enhanced by possessing an accurate representation of the environment with which we must interact to achieve those results. In addition to possessing means by which we can pick up information about our present situation and store information picked up in the past, we must be able to make reliable predictions about the future and form plans of action that will enable us to determine or at least influence how the future will develop.

Causal laws provide us with a means of making such projections and of forming such plans of action. Like all laws of nature, they provide us with principles of connection within our representation(s) of the world, establishing connections that mirror stable recurrent patterns in the represented world. As causal laws, they specifically single out the independent variables in such patterns, the levers that can and must be pushed to produce desired changes. One cannot alter the height of a flag-pole by altering its shadow despite the symmetric lawful covariance of the two parameters.[5] All of this I take to be true and obvious. I restate such platitudes only because I believe they are sometimes lost sight of in the debate about mental causation.

Given their cognitive role, what sorts of causal laws would it be useful to include in our representation of the world? Ideally, one would like to have laws that were simple, reliable, and precise, that related properties that were determinate, easily detected, and important to our interests. However, such ideal laws are rarely available, and we must instead accept trade-offs among these various and often competing desiderata.

In constructing causal explanations, precision must often be sacrificed in the face of pragmatic constraints on what we can detect or comprehend. The micro-physical explanation of why my car stopped when I pressed the brake pedal or of why Tom's belief caused him

[5] S. Bromberger, 'Why-Questions,' in *Mind and Cosmos: Essays in Contemporary Science and Philosophy*, ed. R. Colodny (Pittsburgh, PA: University of Pittsburgh Press, 1966).

to bring the canteen to his lips might indeed have a precision not possessed by automotive or intentional explanations of those same sequences of events. And the laws invoked in the micro-physical explanation might be strict laws while those in the alternative explanations are not.

Automotive and intentional explanations may be less precise and reliable, but they have their own advantages and virtues. Most importantly, they are available *in practice*. They relate properties we are readily able to detect and that are relevant to our interests, and their explanation of how those properties interact is one we can survey and comprehend.

The importance of this last point should not be underestimated. As Hilary Putnam illustrated in his peg-in-the-hole example, a *deduction* of a description of an effect from a description of its cause does not necessarily count as an *explanation* of why the effect occurred if the deduction is in practice incomprehensible and hides what is relevant about the causal transaction in a wealth of irrelevant details.[6] Any other object with the same cross-section as the peg and some micro-structure or other that made it rigid under motion would just as easily pass through the hole. Thus, our explanation of the peg's passing through the hole need not, indeed it should not, bring in all the particular micro-physical facts about the specific peg that underlie its being rigid and being of the size and shape it is. Explanations that are not pitched at the right level of abstraction fail to classify events into the similarity classes relevant to our predictive needs.

A proponent of mental property epiphenomenalism might reply as follows: 'I agree with everything you say about causal *explanations*. They have an important pragmatic dimension which will frequently require us to do our explaining in terms of more abstract higher-order properties, including intentional and other mental properties. But the point at issue is whether or not mental *properties* are *causally potent*, that is, whether or not they determine even partially the causal powers of events that have them. And that is an independent question (since causal explanations need not appeal only to causally potent

[6] Putnam, 'Philosophy and Our Mental Life,' in *Mind, Language and Reality, Philosophical Papers*, Vol. 2, ed. Hillary Putnam (Cambridge: Cambridge University Press, 1975).

properties). The fact remains that in each case in which explanatory appeal is made to mental properties, it is still the underlying physical properties that are doing the real causal work; mental properties only appear to determine an event's causal powers.'

This reply may sound plausible, but it brings us back to the central claim of our argument: that is, why should we accept the claim that it is only the underlying physical properties that are really doing any causal work?

Consider an example involving refraction. Imagine three pairs of transparent optically conducting media (*A/B*, *C/D*, and *E/F*) with the following optical densities: $A$: 1.2; $B$: 1.5; $C$: 1.32; $D$: 1.65; $E$: 1.36; and $F$: 1.7. The three pairs of media will have the same refractive index since the optical densities of their members are all in a ratio of five to four. Consider then three cases in which rays of light pass from $A$, $C$, and $E$ respectively into $B$, $D$, and $F$. If the angle of incidence is the same in the three cases, so too will be the angle of refraction since for all three pairs the sine of the angle of incidence will be equal to $1.25 \times$ the sine of the angle of refraction.

Why is the ray of light bent as it is in each case? Is it because the ratio of optical densities is 4:5 in each pair, or must we say that the real reason the *A/B* pair bent the ray x degrees is different from the reason the *C/D* pair also bent it x degrees? Indeed, the defender of the argument may well argue that the real explanation must be in terms of the particular micro-physical transactions between each specific pair of media and the light ray passing through them. It will not suffice to say that the light ray was bent as its leading edge passed from a medium having an optical density of 1.2 into another having an optical density of 1.5 and was thus slowed down before its trailing edge was similarly slowed. The defender of the argument must reject such an explanation as too abstract to provide an account of the causally potent properties. One must at least descend to the physical structures of the two media and how those structures interact with passing photons. For even if we compare the passage of light from $A$ to $B$ with the passage of light from $G$ to $H$ where $G$ like $A$ has an optical density of 1.2 and $H$ like $B$ has one of 1.5, the underlying microstructures of the respective pairs of media may be quite different and their optical densities may result from very different interactions with photons.

If the proponent of the argument takes such a hard line, he will have to concede that none of the properties of the special sciences are causally potent. Not only will mental properties turn out to be epiphenomenal, but so will neurological properties, geological properties, and even biochemical properties. Even many properties of physics proper, such as having a given optical density, will turn out not to be causally potent. As numerous authors have noted (e.g. Fodor[7]), if the sense in which mental properties are epiphenomenal makes them no worse off than biochemical properties, then what's the big deal? Why should psychological properties have to meet a higher standard for causal potency than biochemical, geological, or optical properties? The point is well taken, and if that were the best one could say in response, it might nonetheless suffice to allay worries about the epiphenomenalism of the mental.

However, I believe more can and should be said. For, the epiphenomenalist is still being allowed to claim a special privileged status for fundamental physical properties with respect to causal potency; on his account they remain causally potent in a sense not shared by the properties of the special sciences (except perhaps in those cases in which a special science property can be identified with an underlying basic physical property—that is, in those cases in which the property just is a basic physical property which also has a special science name). At the risk of seeming a demagogic leveler (hardly an attractive option in this post-Marxist era), I propose that the solution to this inequality is not to find a way of giving special science properties (including mental properties) the status claimed for physical properties, but in showing that physical properties have no such special status; it's not a matter of raising the status of special properties but of exposing the pretension of the specialness of the physical. (The physicalist emperor has no clothes.)

Here's how the story goes. Special science explanations work because they classify objects and events so that they share predictable causal roles; they pick out recurrent, stable (if sometimes less than strictly deterministic) patterns of order in the world. Sometimes this is because the classification is explicitly based on causal role (e.g. being a catalyst or a recessive gene), and at other times it is because

[7] Fodor, 'Making Mind Matter More,' 59–80.

the classification is based on features that guarantee a given causal role (e.g. being a certain sequence of nucleic acids or an atmospheric temperature inversion). (Which of the two best fits intentional classification is left as an exercise for the reader.)

Physical explanations work for the same reason; physical classifications also group objects sharing common causal roles (e.g. being a electron with 1/2 negative spin). So far so good—equality. The champion of the physical will reply that the causal roles associated with special science classifications are entirely derivative from the causal roles of the underlying physical constituents of the objects or events picked out by the special sciences. Once again we will be told that it is the physical properties that are doing all the real work.

This is not quite true, however. The events and objects picked out by the special sciences are admittedly composites of physical constituents. But, the causal powers of such an object are not determined solely by the physical properties of its constituents and the laws of physics, but also by the organization of those constituents within the composite. And it is just such patterns of organization that are picked out by the predicates of the special sciences.

In a way, this is just a reminder that physical outcomes are determined by the laws of physics together with *initial boundary conditions.* Special science predicates pick out stable recurring sets of such boundary conditions (or equivalence classes of such boundary conditions). By doing so, they isolate a level of causal order and regularity in the natural world. Thus, we can say that the causal powers of a composite object or event are determined in part by higher-order special science properties and not solely by the physical properties of its constituents and the laws of physics.

'Not so fast,' cries the champion of the physical in reply; 'in any given instance this pattern or organization, these boundary conditions of which you speak, will be nothing more than a strictly physical arrangement of matter in time and space. To treat the pattern as something over and above its strictly physical instantiations would be an exercise in Platonic reification. Surely predicates, like those of the special sciences, that merely pick out such patterns don't pick out anything that's real in a causally relevant sense; what's real and causally relevant are simply the actual instantiations of such patterns, and they are entirely physical. Special science predicates are at best convenient shorthand abbreviations for referring to such physical instantiations; they should not be understood as referring to real and causally potent properties.'

But why should we not regard these patterns as real and causally potent? Consider what might be said on behalf of their reality.

1. Such patterns are recurrent and stable features of the world.
2. Many such patterns are stable despite variations or exchanges in their underlying physical constituents; the pattern is conserved even though its constituents are not (e.g. in a hurricane or a blade of grass).
3. Many such patterns are self-sustaining or self-reproductive in the face of perturbing physical forces that might degrade or destroy them (e.g. DNA patterns).
4. Such patterns can affect which causal powers of their constituents are activated or likely to be activated. A given physical constituent may have many causal powers, but only some subset of them will be active in a given situation. The larger context (i.e. the pattern) of which it is a part may affect which of its causal powers get activated. (For example, the activity of a reagent can be affected by the presence of a catalyzing enzyme that forms a composite with the reagent.) Thus, the whole is not any simple function of its parts, since the whole at least partially determines what contributions are made by its parts.
5. The selective activation of the causal powers of its parts (4) may in many cases contribute to the maintenance and preservation of the pattern itself (2, 3).

Taken together these five points illustrate that higher-order patterns can have a degree of independence from their underlying physical realizations and can exert what might be called downward causal influences without requiring any objectionable form of emergentism by which higher-order properties would alter the underlying laws of physics. Higher-order properties act by the *selective activation* of physical powers not by their *alteration*.

With respect to such patterns and their underlying physical constituents, we can ask the first question posed in my title: 'Who's in charge here?' Given 1–5 above, there is a very real sense in which the constituents of the pattern are organized as they are because of the pattern. It is because of the existence and persistence of the pattern that particular constituents of its instances were recruited and organized as they are. Moreover, many such patterns may be (all but) inevitable features of our world. They are among the stable states of order; because of their persistence and self-sustaining character, if given enough time they naturally emerge from the disorderly flux of nature. Their existence is far from accidental.

Still, the champion of the physical might reply that even if such patterns are inevitably occurrent stable features of our world, they are so only because the world has the physical order that it does. The order of higher-level properties or patterns is entirely dependent on and derivative from the world's underlying physical order. Once again, the primacy of the physical looms in impending triumph.

But perhaps the threat is more apparent than real. Though I can't prove it, I strongly suspect this claim of dependency on the physical is false or at least misleading. Let me explain. The standard thought experiment to demonstrate dependence on the physical is try to imagine a possible world just like the actual world in all physical respects but differing in respect of some higher-order property, for example in some mental property. The standard response is that no such worlds are possible, and I am not now inclined to dispute that intuition.

I am more inclined to deny that this is the right test or the only test relevant to determining whether the order among higher-level patterns is dependent on physical order. One might also try to imagine possible worlds in which many of the laws of physics are different than they are in this world but in which many of the same higher-order patterns are present. Indeed, one might try to imagine worlds containing nothing that one would count as physical but that nonetheless shared patterns of higher-order organization with our world.

To make my abstract speculation just a bit more concrete, consider the patterns associated with acquiring, possessing, and exploiting information, patterns that are pervasive throughout at least the biological portion of our world (though I believe they are in fact more widespread). One might turn to Fred Dretske's work for some account of how to explain such informational relations.[8] One would immediately notice that to bake such an informational cake one needs only very simple ingredients (Dretske's metaphor). One needs lawful covariance and perhaps causal connections, but there is no requirement that such lawful or causal connections be physical connections. In some possible worlds they might well be non-physical connections.

I am now ready to state my conjecture. In most if not all of the neighboring worlds that are like our world in having some lawful or causal order but which do not contain any physical matter, patterns exist that are very much like the patterns associated in our world with acquiring, possessing, and exploiting information. Such patterns are all but inevitable consequences of the tendency of worlds over time to settle into patterns that are self-sustaining and self-preserving. If my conjecture is correct, then there is a sense in which even in our world the order of higher-level patterns is not dependent on the physical order of our world. It is a much more pervasive order that simply manifests itself in our world in physical realizations. (How's that for multiple realization?)

Still the champion of the physical might argue that physical properties and laws still enjoy an ontologically privileged status with respect to causality. Higher-order properties are simply stable recurring patterns of organization, nothing more—no matter how inevitable or widespread they are through the space of worlds. Our physicalist champion might go on to argue that physical properties are causally potent in a way that no such pattern ever is. Special science laws or explanations merely pick out higher-order patterns that are present in projectable, counterfactual-sustaining regularities in the world; that is, they describe a level of organization at which the world exhibits systematic order. But, physical explanations are not mere descriptions of order; they tell us why things actually happen. And physical properties are not mere stable recurrent patterns; they are the basic stuff of which all the patterns are made.

I believe we should view any such claim with skepticism. I proposed earlier that we produce equality between physical and special

[8] Fred Dretske, *Knowledge and the Flow of Information* (Cambridge, MA: MIT Press, 1981); *idem*, *Explaining Behavior: Reasons in a World of Causes* (Cambridge, MA: MIT Press, 1988).

science causal explanations by stripping physical explanations of their allegedly special status. Now is the time to try to do so. What properties of the physical world might one suppose to be more than mere recurrent patterns? Being a proton? Being an electron with 1/2 positive spin? Being a quark with a particular color, charm, and strangeness?

Though it is a question whose answer requires more knowledge of physics than I possess, I believe it could be shown that all physical properties that enter into strict exceptionless laws are themselves nothing more than stable self-sustaining recurrent states of the quantum flux of an irreducibly probabilistic and statistical reality. If cosmologists are to be believed (and I believe them), our universe settled into these patterns very early on, some time within the first three minutes after the Big Bang. But, their antiquity does not alter their status.

They are highly stable patterns whose interrelations approximate deterministic regularities to a very high degree. Or perhaps, turning the question the other way round, one could say the transition functions representing their interaction in our model of that level of reality are highly deterministic. (You have to watch out for that latent theological realism.) I think we are often bewitched by their deterministic rigor, which produces in us an illusion of dealing with more than just a representation of some aspect of the organization of space-time. We feel we are dealing with something more tangible, something more real and objective. Their determinism seduces us into seeing physical properties as determining the sequence of events in a way that no other properties do. But, the fact that the transition functions in our model of the physical are deterministic in the mathematical sense certainly does not entail that the properties they model play a unique (or even a special) role of any metaphysical sort in determining the temporal organization of the world.

I suppose that by now we've been digging long enough. So let's climb up and see where we may have undermined the fortifications of the argument. First, I think we can see that the link from 3 to 4 is weakened by what we have uncovered about the roles played by higher-order properties. The over-determination option now looks more promising. The *complete physical descriptions* which determine the causal powers of all the world's events according to 3 will have to include complete specifications of physical boundary conditions.

Since higher-order properties are sets of recurrent boundary conditions, complete physical descriptions will have to refer to the instantiations of any such properties. But they will lack the conceptual resources to represent them as instantiations of the relevant higher-order properties. What we will get instead is an opaque and disorderly representation that in no way makes perspicuous the higher-order property that is being instantiated or the systematic relations that it bears to other higher-order properties.

Special science explanations refer to the same property instantiations, but do so in a perspicuous representation that makes clear how the temporal sequence of events is structured by those higher-order patterns. We have seen that such patterns and their interrelationships can enjoy a substantial degree of independence from their particular physical instantiations and perhaps even from physical instantiation altogether. Thus, we can accept the global supervenience reading that makes 3 plausible, without having to accept 4 on any reading that treats mental or higher-order properties as not playing their own role in structuring the sequence of events.

We have two models of the world which cannot be reduced in the sense that there are no well-ordered complete translation functions from one to the other—a gap which results in part because of the ways their respective concepts are anchored in our specific discriminative and cognitive capacities. But both models can be satisfied by the same world even in those respects in which they model the causal structure of the world. And each model can have its own pragmatic strengths and weaknesses.

Given the causal roles played by higher-order patterns (which as we have seen are not really any different than those played by physical properties, such as being a bottom quark), there is no reason to make the inferential step to 5 and its claim that mental or higher-order properties fail to be causally potent in any sense that physical properties are potent. Thus, there would be no reason to reach the epiphenomenal conclusion of 6 or to deny that mental properties are causally relevant in the most robust sense.

Thus fell the walls of Jericho, or at least I hope they cracked.[9]

[9] This chapter is a condensed version of Robert Van Gulick, 'Who's In Charge Here? and Who's Doing All the Work?' in *Mental Causation*, eds. John Heil and Alfred Mele (Oxford: Clarendon Press, 1995); reprinted with permission.

# 4

# Three Levels of Emergent Phenomena

*Terrence W. Deacon*

## 4.1. Real Emergence?

The term most often used by scientists to describe the spontaneous appearance of novel fitted orderliness is 'emergence.' This usage has been around for over a century (see G. H. Lewes's *Problems of Life and Mind*[1]), and pretty much characterizes the same troublesome explanatory 'gaps' in the same scientific discussions currently as it did over a century ago. How does life arise from non-living molecules, how does a mind arise from electro-chemical processes in material brains? The term connotes something coming out of hiding, coming into view for the first time, something without precedent and perhaps a bit surprising. But it is not applicable to the common everyday novelty of contingent history. Though each new event stands out as idiosyncratic in various ways with respect to all preceding events, what most people mean when they talk of something being an emergent phenomenon isn't merely something different than before. In nontechnical discussions the phrase 'the whole is more than the sum of the parts' is often quoted to convey this sort of novelty. It captures two aspects of the concept: the distinction between a mere quantitative difference and a qualitative one, and an ascent in scale, involving combination of elements in which their interactions become relevant.

[1] G. H. Lewes, *Problems of Life and Mind*, Vol. 2 (London: Kegan Paul, Trench, Tubner and Co., 1875).

An analogy often used to exemplify this 'not the sum of its parts' logic is that between a mere jumble of words and those same words used to make a sentence. Although the meaning of a sentence is dependent on the meanings of its words, obviously the arrangement of the words is also critically important (as is context). For this reason the meaning of a sentence can't typically be deduced from just knowing the meanings of the words being used. It is also usually quite easy to construct different sentences and phrases from the same words with strikingly different meanings. It is the syntax of a sentence—that combinatorial logic that binds words together in different meaningful ways—which is critical to how the words are individually and collectively interpreted. Can we say, then, that the meaning of a sentence is a combination of the meanings of the words and the relationship information supplied by its syntax? Only in a vague sense, which leaves a lot to be explained by the concept 'syntax.' The critical relationships between words are not merely their proximities or positions. In addition, syntax encodes what might be called the meaningful effects that each word has on the others. Placement of the words (or word-modifiers, such as prefixes, suffixes, inflections) can play a unifying, separating, pointing, or prioritizing role, with respect to word meaning and reference. In short, combination is not always merely combination, not merely 'summation.' Exactly what else it is in each kind of phenomenon is the critical question. Assuming comparability of phenomena with radically different kinds of combinatorial logics is like assuming one can compare languages merely on physical combinatorial relationships of letter strings. The underlying logics need to be compared.

As we shift emphasis from explaining how new kinds of atoms were formed, to explaining how aggregate material properties become expressed, to investigating how new biological functions evolve, to demonstrating how sentences come to mean, we need to treat these different aspects of combinatorial creative processes differently. Current controversies about emergence indicate that we have so far failed to make these distinctions as carefully as we should. Emergence in one context may be quite different than in another, yet there are also reasons to think that there may be some general unifying logic to all these intuitions. If the concept is to remain useful and to be more than a mere promissory note to fill gaps in our knowledge, then we need a systematic way of categorizing and analyzing emergent phenomena.

What criteria can we use to differentiate the different senses of 'emergence'? One important lesson from science is that discovering the logic of how a phenomenon is produced tends to provide the most coherent basis for classifying it. Thus, the periodic table of elements, based on the constituent subatomic architecture of atoms, superseded and explained the older phenomenological taxonomies of metals, earths, gases, etc. Analogously, phylogenetic tree depictions of biological lineages are progressively replacing older biological taxonomies, based on lumping species according to 'diagnostic' body characters. Can there be a scheme for distinguishing among types of emergence that is itself based on how these spontaneous constitutive processes themselves are constituted?

We first need to be clear whether we are describing something about the process of science or about certain processes in the world. There will always be gaps in our understanding. Phenomena that we cannot predict often reflect limitations of theory or of modeling and computing power. Something that is only novel with respect to current tools and theories is an artifact of science, not a feature of the world. Difficulty producing commensurate descriptions that are adequate at very different levels of scale is not evidence that the phenomena at these different levels are causally dissociated in some way. For this reason many scientists suspect that many things described as emergent are actually matters of descriptive inadequacy. But most would also accept evidence that new kinds of matter have arisen in cosmic history and novel kinds of living processes and organisms have arisen in the history of life. Though these *historically emergent* phenomena emerge in a different way than higher-order properties of molecules constituting a liquid emerge from molecular interactions, I think it can be shown that these two senses of emergence, as well as others, reflect certain underlying commonalities. This difference in sense of the term when applied to different phenomena, and the problems arising from failure to make such distinctions, is well exemplified in discussions of the nature of mind.

The concept of emergence seems to be more critical for understanding the origins of mind than of any other natural phenomenon. Indeed, many of the ways this term has been applied to various phenomena have actually been generated in the process of using these as lesser examples of the sort of relationship theorized to exist between mind and its physical–chemical–biological constituents. Francis Crick has called the realization that thoughts, feelings, passions are the results of chemical processes in the brain an 'astonishing hypothesis.' One is suspicious, however, that the astonishing part is to be found in what is not stated in this reductionistic claim. Mind can't be 'nothing but,' and yet isn't exactly 'something in addition.' It is common to read statements like the following: 'Consciousness is an emergent property of the cellular and molecular processes within a brain in the same sense as surface tension is an emergent property of the interactions of water molecules.' Yet there is something not quite right about the comparison. By invoking this analogy, an implicit claim about a presumably well-understood relationship (though there are still mysteries in the physics of water) is meant to inform our understanding of the general logic applicable to a much less well-understood (indeed deeply mysterious) relationship. It is precisely the huge number of unstated details on both sides of the analogy that makes this an exercise in question-begging.

What is common in this comparison is that both involve a contrast between levels of description requiring quite different descriptive tools. Whether these are analogous in any other way is the question. By assumption, water molecule collisions (obeying lower-order component properties) stand in relation to surface tension (exhibit a higher-order aggregate property) as neurochemical reactions stand in relation to conscious experience. Without question, activities supported by whole living brains must at the same time be supported by the component processes of these brains, since the brain is also a composite or aggregate dynamic entity. Beyond that of mere aggregate form, however, we are left with little confidence that the kind of causal relationships implicitly being compared actually reflect a similar compositional logic. Notice, for example, that we would be on solid ground comparing the viscosity or opacity of brains to the relative viscosity or opacity of a liquid, and would feel confident treating both cases as similarly emergent from molecular interactional properties. This is because these properties have an emergence of a similar kind. We implicitly recognize that there may be different 'species' of this relationship. But this also suggests that cross-category comparisons of emergent relationships, such as this, may not be as informative as we might hope.

No doubt mental phenomena are emergent from the subordinate neurochemical interactions occurring in a brain in a more complex way than liquid phenomena are emergent from water molecule interactions. The fact that we can identify similar kinds of properties in brains (e.g. viscosity) to those of other kinds of aggregate matter (e.g. liquidity) also suggests that this brain–mind kind of emergence is of a higher order. The same can be said of life. Living metabolism is not emergent from its constituent chemical processes in the same way that a flame's dynamical features are emergent from its component chemical processes, even though a metabolic process shares some features with burning. The critical features that make metabolisms emergent set it apart a level above those features that make flames emergent, because in some sense flame-level emergence is a component in life-like emergence. Similarly, mind-like emergence is to life-like emergence as life-like emergence is to flame-like emergence.

At the very least, then, we need to articulate a more elaborate and hierarchically organized taxonomy of emergence relationships to aid in avoiding too-simplistic comparisons. More than this, we need a principled way to distinguish between mere newness, descriptive complexity, predictive difficulty, and the true emergence of superordinate forms of orderliness.

## 4.2. The Terminological Problem

So one persistent problem with accounts of emergent phenomena is that the concept of emergence is both ambiguous and used in different ways in different contexts. Another is that it is often used in a merely negative sense, to point to something missing in reductionistic explanations. In explicitly anti-reductionistic criticisms of standard accounts of such phenomena as life and mind, it has become a code word identified with a complex-systems theoretic perspective. Here the concept is a place marker to indicate points where standard reductionistic accounts fail or seem to incompletely explain apparent discontinuities in properties exhibited at different levels of physical scale. This negative usage has unfortunately led many more orthodox thinkers to suspect that there is no underlying phenomenon to be described. This suspicion is also fostered by vagueness of the concept itself. In many examples where it has been more precisely described (e.g. the emergence of liquidity or surface tensions from the interactions of water molecules), it is seen as adding nothing of empirical significance to standard reductionistic accounts. In contrast, where it is used to describe more complex phenomena (e.g. emergence of life or mind), the details and logic are sufficiently obscured by incomplete scientific investigation that it has little explicit content. These incautious uses allow critics to claim rightfully that it mostly serves as a philosophically motivated promissory note.

Though this is not the place to review philosophical debates about the reality of irreducible emergent phenomena, it is worth trying to provide explicit positive definitions and categories of emergence to counter this vagueness. Though by describing as precisely as possible a causal logic distinguishing these examples we risk sacrificing the strong sense of emergence implied by antireductionism, I think we do not necessarily lose the constructive causal novelty that is central to the idea. At the very least an explicit analytic categorization of types of emergence will help to distinguish among theoretical claims commonly confused in discussions of evolutionary and mental phenomena.

Emergent phenomena are often described as having novel properties not contained in their constituent components and exhibiting regularites that cannot be deduced from laws affecting their constituents. However, although explanations of emergent phenomena cannot be completely rendered in terms of the physical properties of their constituents, the physical laws governing these constituents are not superseded or violated in emergent phenomena. What is not provided by these physical properties and laws and is critical to emergent properties is an additional account of the *configurational regularities affecting constituent interactions.*

A related point is made by researchers simulating complex systemic dynamics in computers. In many of these models the large-scale configurations that are produced more often reflect configurational properties of the whole ensemble (e.g. number of elements and their interactional connectivity), rather than properties derived only from interactional dynamics of constituents. The influence of what are sometimes called 'holistic' properties on constituent interactions in complex systems is both the critical differentiating feature of emergent phenomena and a persistent source of misunderstandings. The existence and relative autonomy of higher-order ensemble properties of systems and of a kind of top-down influence over the properties and dynamics of system constituents remains both the key defining character and the most criticized claim of arguments for emergence.

Though I offer only a qualitative descriptive account of what I mean by 'levels of emergence,' I believe it can be made sufficiently precise and unambiguous to be easily rendered in mathematical terms using current theoretical tools.

## 4.3. Definitions

What is it that prompts us to consider some physical phenomenon novel and unprecedented? Not the energy that courses through them or the material they are composed of, but rather their form or configuration; that is, how the matter and energy are organized, and how they behave. What most need explaining are cases where the behavior of some system seems almost entirely unrelated to the ways its constituents typically behave, and where 'top-down' organizing effects appear to predominate. This is most evident in mental phenomena. In efforts to explain cognition, theorists either tend to appeal to special or strange kinds of physical causality or else imagine that science may one day reduce thought to 'nothing but' the interactions of molecules comprising neural computations. There must be a middle ground between these unprofitable extremes, but efforts to chart its boundaries have tended to lapse into compromise positions or unspecified possibilities. The concept of emergence falls into this no man's land. I contend that we do not require a 'new physics' to deal with emergent phenomena, nor to appeal to strange features of quantum physics, nor abandon standard notions of physical reduction. *What we need is to trace ways that nature can tangle causal chains into complex knots.* Emergence is about the topology of causality. By focusing on this topological logic we can find a middle ground between strongly reductionistic and holistic approaches to emergence. What results might be described as a sort of 'weak emergence' or 'soft reductionism.' A more careful analysis of emergence forces an abandonment of both caricatures of explanation as simplistic abstractions.

What needs explaining is how some systems come to be dominated by higher-order causal properties such that they appear to 'drag along' component constituent dynamics, even though these higher-order regularities are constituted by lower-order interactions. The secret to explaining the apparently contrary causal relationships is to recognize the central role played by *amplification processes* in the pattern formation occurring in these kinds of phenomena. Wherever it occurs, amplification is accomplished by a kind of repetitive superimposition of similar forms. It can be achieved by mathematical recursion in a computation, by recycling of a signal that reinforces itself and cancels the background in signal processing circuits, or by repetitively sampling the same biased set of phenomena in statistical analyses. In each case, it is the formal or configurational regularities that serve as the basis for amplification, not merely the 'stuff' that is the medium in which it is exhibited. Amplification can be a merely physical process or an informational process (the latter usually depends on the former). Its role in the analysis of emergence is in explaining how certain minor or even incidental aspects of a complex phenomenon can become the source of its dominant features.

Using this amplification logic we can distinguish three hierarchically related classes of emergent phenomena on the basis of their causal architectures. Specifically, I will suggest that the most useful architectonic feature is whether this causal architecture is recurrent or circular across levels of scale.[2] The history of theoretical discussions of complexity and emergence have regularly cited examples with this sort of causal architecture—whether in terms of physical nonlinearity or computational recursion—but to date I know of no effort to formalize this intuition or to use it as a general analytic tool.

A simple and well-known example of 'circular causality' is embodied in a thermostatic control system. By connecting a heating and/or cooling device to a temperature-sensitive switch the coupled devices are configured to respond to temperature in a manner that will

[2] Though it might be more accurate to use the metaphor of 'spiral' causality, I use the terms 'circular' and 'recursive' because they make it easier to visualize more complex convoluted architectures.

eventually invert the state of the switch and with it the state of the device that heats or cools the environment. In other words, there is a self-undermining pattern of cause and effect—so-called negative feedback—which tends to produce behavioral oscillation around some set-point. If this causal linkage is reversed, of course, so that deviation away from the set-point activates mechanisms to cause the environmental temperature to deviate yet further, a very different and unstable behavior results—so-called positive feedback. This runaway effect is checked only by outside constraints. Even simple deterministic engineering devices where a number of such feedback control devices are coupled can produce highly complex quasiperiodic behaviors or even deterministic chaos.

Though feedback effects of this sort can be quite complicated and even deeply unpredictable in complex dynamical systems, I am not prepared to call some of the higher-order quasi-regularities of these behaviors emergent, except in a minimal sense (which I describe below), even though this characterization is often applied because of the seeming unprecedented nature of these regularities when compared to component dynamics. Though the effects I am interested in derive from this same logic of complex nonlinear interactions, it is valuable to be more explicit about the role played by scale in these phenomena as well. Specifically, I want to analyze relationships of recursive causality in which the feedback is from features of a whole system to the very architecture of its components and how these levels interact. Besides complex behaviors, such systems can exhibit a trend toward progressive amplification of certain distributional or configurational features of components and component interactions. I will focus on these comparatively rare dynamical amplification relationships, not the complex or chaotic effects per se, in identifying what I intend to capture by my use of the emergence concept. The feature I focus on might be described as trans-scale causal recursion, that is, circles of causality that operate across levels of scale. My question is: What happens when the global configurational regularities of a locally bounded open physical system are in some way fed back into that system via effects propagated through its 'environment.'

In more colloquial terms, one might describe the taxonomy of emergence as a categorization based on a sort of 'compound interest' of scale: global attributes altering component attributes altering global attributes and so on. Three general categories of emergence can be derived in this way, distinguished from one another by the way their recurrent causal architectures are topologically intertwined across levels of scale. This yields nonrecurrent-, simple recurrent-, and recurrent-recurrent-trans-scale architectures (in the latter, simple recurrent causal architectures are embedded in a yet higher-order recurrent architecture). These produce phenomena that can correspondingly be called first-, second-, and third-order emergence, reflecting this embedding of recurrent deterministic effects. Though I have little doubt that there can be explicit 'reduced' models of these phenomena, rendered only in terms of interacting components, I suspect that the amplification effects make this modeling only very heuristic and not fully computable.

## 4.4. Supervenience

The most basic sense of emergence can be called *first-order* (or *supervenient-*) *emergence.* It corresponds to the way the term 'emergence' is often applied to descriptively 'simple' higher-order properties of an aggregate, such as statistically or stochastically determined ensemble behaviors. One common example is the liquid properties exhibited by large aggregates of water molecules (e.g. laminar versus turbulent flow, surface tension). Statistical thermodynamics and quantum theory have provided a remarkably complete theory of how the properties of water molecules can produce liquid properties in aggregate. Thus, in one sense they are considered to be fully reducible to relational molecular properties. But such relational properties, as opposed to intrinsic molecular properties (e.g. mass, charge, configuration of electron shells), are not symmetric across levels of description. Precisely because they are relational, these higher-order properties are not applicable to descriptions of water molecules in isolation. It is also due to certain regularities of relationships *between* molecules that these aggregate behaviors emerge with ascent in scale.

Philosophers often refer to this relationship as 'supervenience.' Though reducibility is not identity, it does imply a strict correspondence relationship, even if some completeness in description may be lost. The higher-order property 'supervenes' on specific lower-order interactions to the extent that the former always entails the latter. The difference in kind of property may be understood as a descriptive rather than a constituent difference. The difference is apparent when descriptive properties are applied across levels of scale to the same system.

We should distinguish between simple descriptive/componential supervenience and the more complex supervenient emergence. As an example of the former consider weight or density. Weight has a descriptive adequacy almost irrespective of level of analysis.

The force of gravity exerts an effect irrespective of scale, but its aggregate effect may be scale-dependent. So, for instance, it has a relatively minor contribution to make to the trajectory of any individual water molecule in solution and a major effect on the behavior of large bodies of water. Nevertheless, it is the same force at both levels of scale. Nothing more needs to be added to the account of the aggregate behavior. In contrast, other liquid properties 'supervene' on the properties of water molecules in more complex ways. This occurs in cases where the lower-order *relational* properties are the constitutive factor determining some higher-order property. In these cases the account is fundamentally incomplete without including specific 'configurational' and/or distributional information concerning the relationships between the components (e.g. quantity of constituents, their relative positions and momenta, their molecular geometry, their hydrogen bond strengths in different orientations). This typically produces a description with very high dimensionality (many degrees of freedom, roughly corresponding to the relevant properties of each molecule with respect to the others) at the lower level. At the higher level this dimensional complexity can often be ignored. The astronomically many details cancel out due to the relative linearity of the stochastic processes. This results in descriptive simplification with ascent in scale that tends to be non-uniform, i.e. exhibits threshold effects, or steps, with increasing size. Adding this vast amount of information about components and their interaction histories to the aggregate description produces little gain in descriptive adequacy about the whole. In this way, these supervenient properties can be considered to be emergent physical particulars, despite the fact that they can be given a potentially exhaustive reductive description. But there is another reason to consider them emergent, though in a minimal sense: they are the basis for relationships that can be far more than supervenient.

## 4.5. Diachronic Symmetry-Breaking

Philosophical and scientific discussions of the mind–brain mystery often invoke some implicit version of supervenient emergence to model the presumed relationship between higher-order mental phenomena and the lower-order cellular-molecular processes. But the analogy isn't nearly adequate. At the very least it fails to capture an essential distinguishing feature: temporal development or symmetry-breaking. There is a simple self-similarity to liquid properties across time and position that is further 'smoothed' by entropic processes. In contrast, there is a self-differentiating feature to living and mental processes, which both retains and undermines aspects of self-similarity. This characteristic breakdown of self-similarity or symmetry-breaking is now recognized in numerous kinds of complex phenomena, including systems far simpler than living systems. These complex emergent phenomena share this characteristic change of ensemble properties *across time*, and are often computationally unpredictable. So it would be useful to distinguish first-order emergence from these more complex forms of emergent phenomena in which the cumulative stochastic canceling of configurational interactions exhibited by simple entropic systems is undermined, and where this contributes to development and change of both micro- and macro-properties across time.

At first glance, it may appear as though distinguishing between synchronic and diachronic classes of emergent phenomena can capture this difference. Supervenient emergence is typically described in synchronic terms, and although it is recognized that this is a short-hand descriptive trick, there is very little descriptive loss in this heuristic. But in many discussions the synchronic and diachronic uses of the concept must be made explicit. For example, biologists now feel comfortable with the notion that life is an emergent property of organic matter that supervenes on the properties of its constituent molecular interactions. Because life is a continuous property of an organism, one can describe it at any one moment as synchronically supervenient on the properties of the organism's constituent molecules and their relationships. But this fails to capture the vast and convoluted temporal depth of the causal account that would be necessary to understand the configurational details that constitute a given individual's structures and behaviors that we intend to capture by the notion of 'life.' Specifically, biological aggregate properties that we consider 'functions' implicitly require reference to a detailed history of at least the micro- and macro-configurational relationships that link them with interactions with critical substrate variables, and implicitly some reference to an evolutionary ensemble of related individuals and their interactions with these substrates and the relationship of this to reproduction. In short, the synchronic short-hand account assumes far more than a simple diachronic restatement of supervenient emergence.

This can be demonstrated by considering evolutionary examples. Consider the process by which life must have arisen from non-life in the early history of the earth. Both a synchronic and a diachronic sense of emergence must be considered. The higher-order properties of the molecular systems that constituted the first life-like self-organizing proto-metabolisms can be said to have been supervenient on the molecular interactions concurrently involved, since the present state of these interactions is the basis for describing the whole ensemble as having the property of being alive. But prior to the formation of the first molecular configuration of this kind no living phenomena existed. So, in addition to being superveniently emergent from the interactions of lower-order constituents, there is also higher-order emergence involved. The configurational properties of pre-biotic systems are a part of the causal complex upon which the first biotic systems supervene in time. Evolutionary processes must be described as the successive emergence of new supervenient emergent phenomena from old: emergent phenomena constituted of other emergent phenomena. By this I don't mean to suggest that 'life' itself is more than a redescription of this supervenience, but rather to note that there is a component to aggregate emergence over time as well as scale. There is an asymmetry in configuration across time (which asymmetry is often described as opposed to entropy, though in reality it is merely one rare expression of it), a development.

This difference is not merely adding levels upon levels as in the way solid-state physics reduces to atomic interactions that reduce to elementary particle interactions and so on. It rather involves nonlinear systems that exhibit an internal causal recursivity and ultimately a self-undermining dynamic that causes prior states of the system to be irreversibly replaced and superseded. This recursiveness is the basis of their irreducibly diachronic asymmetry. For this reason the distinction is more complicated than distinguishing between synchronic systems with self-similar simple entropic development (e.g. behavior of a gas) and diachronic ones with difference amplifying development.

## 4.6. Chaos and Autopoiesis

Although surface tension is an emergent property, it is characterized by predictability so long as lower-order properties are within certain parameters; hence it is supervenient. But this is not the case under conditions where more or less chaotic or self-organized behaviors are produced. Under chaotic conditions, for example, certain higher-order regularities become unstable, and an unpredictability of higher-order dynamics results. In such chaotic systems, this unpredictability derives from the fact that regularities at lower levels have become strongly affected by regularities emerging at higher levels of organization. This can happen when configurational features at the ensemble level drastically change the probabilities of certain whole classes of component interactions.

This occurs, for example, in shock wave boundaries where relative movements of whole ensembles of gas molecules exceed the rate at which energy can be exchanged by typical elastic collisions between individual molecules. Breaks in symmetry appear that affect component molecular interactions, such that those on one side or the other of this interactive threshold can have an inordinate influence over ensemble behaviors. The specific nonsystematic locations of these micro-symmetry-breaks become important since they become loci of major energy transfer for whole subsystems distinguished by sharing similar global features. The result is that these specific micro-configurations have macro-configurational consequences (e.g. local energetic 'cascades' and irregularities) which in turn affect future micro-configurations, and so on. This hierarchical nonlinearity consequently produces a kind of self-undermining dynamic across levels that is expressed in a unique time series of configurations.

The signature feature of this complication of supervenient emergence is that the configuration of individual components and the unique interactions can exert an organizing effect on an entire ensemble. While in principle these examples still exhibit supervenient emergence, we now must contend with *second-order emergence* of behaviors and ensemble properties. Whereas micro-configuration can be ignored in supervenient systems with minimal loss of descriptive adequacy, this is not so for systems exhibiting second-order emergence. Chaotic and self-organized systems are generally of this type.

Chaotic phenomena have become a major focus of 'complexity' theory in its many forms. Such phenomena cannot be adequately described without incorporating a detailed history of the system and its components, whereas merely supervenient emergent phenomena can. Moreover, as has come to be the hallmark of complex chaotic systems, initial conditions in this system history can play critical roles in the ongoing global dynamics of the system. This is often expressed as the apocryphal 'butterfly effect' in which a butterfly flapping its wings in a particular pattern in one part of the globe is supposedly responsible for monsoons in another part of the globe. Though in reality this is an implausible and misleading just-so story (since the highly entropic dynamics of the atmosphere tend to quickly smooth out such perturbations rather than amplify them), it does capture the possibility that large-scale configurational properties of a system may be substantially affected by extremely miniscule differences in prior micro-configurations of the system under special circumstances. Whereas atmospheric dynamics can be chaotic and unpredictable, they are not systematic and do not tend to consistently amplify specific perturbations. Certain other kinds of highly organized physical systems can, however, exhibit this character. This is reflected in their critical dependency on historical contingency and complex unique individual structure.

Take snow crystal formation (we will ignore many poorly understood aspects, such as quasiliquid physics, which do not substantially alter this account). The structure of an individual snow crystal reflects the interaction of three factors: the hexagonal micro-structural biases of ice crystal lattice growth, inherited from water molecule symmetry, the radial symmetry of heat dissipation, and the unique history of changing temperature and humidity regimes as a crystal falls. Different temperatures and humidities interact with water-molecule binding regularities to generate a handful of distinct patterns of ice lattice formation. Since a snow crystal grows as it falls through many regions in a variable atmosphere, the history of temperature and humidity differences it encounters is captured in variants of crystal structure at successive diameters.

So the crystal is a sort of record of the conditions of its development, but more than merely a historical record. There is a 'compound interest' effect as well. Prior crystal growth constrains subsequent crystal growth. So even identical conditions of temperature and humidity which should otherwise determine identical lattice growth can produce different patterns depending on the current configuration of the crystal. The global configuration of this tiny developing system plays a critical causal role in its microscopic dynamics by excluding the vast majority of possible molecular accretions and growth points and strongly predisposing accretion and growth at certain other sites.

Snow crystals are in this way minimally self-organizing (though not strictly speaking autopoietic, see below). The unique growth history of the crystal cannot be ignored. Indeed, growth history is a dominant factor determining its final configuration. More importantly, these configurational properties self-amplify. Each past configurational state influences all subsequent configurational states. This occurs by virtue of the progressive contraction of the potential growth options that result from past growth.

Both self-undermining (e.g. divergent) chaotic systems, as in turbulent flow, and self-organizing (e.g. partially convergent) chaotic systems, as in snow crystal formation, exhibit this causal circularity linking higher-order and lower-order configurational properties. *These feed-forward circles of cause and effect linking events at different levels of scale are the defining features distinguishing second-order emergence from first.* This is 'second order' because supervenient emergent properties have become self-modifying, resulting in supervenient emergence of new supervenient emergent phenomena.

More complex second-order emergence is exhibited by 'autopoietic' systems. In autopoiesis the interaction dynamics of sets of different components is constrained both by configurational properties of the whole collection and by configuration symmetries and asymmetries that exist between the micro-configurations of the different classes of its components. In snow crystal dynamics the micro-configuration of each component is essentially the same, producing symmetric interactions and strongly constrained structural consequences. When a system is composed of different kinds of components, however, it can also exhibit a more distributed interactional reflexivity, and because of the combinatorial possibilities the resultant properties can be quite complex.

For example, molecules that interact in highly allosteric fashion, that is, they bond selectively with some but not other types of molecules, can constitute interaction sets with more elaborate self-organizing features. Both the configurations of the different classes of individual interactions and the configuration of the whole set of possible interactions become critical organizing influences. This can occur in a chemical 'soup' containing enough different kinds of molecules so that there is a subset of types that can each catalyze the formation of some other member in the set, constituting a closed loop of syntheses. So long as sufficient energy and other raw materials are available to keep reactions going, this 'autocatalytic' character of the set will play an inordinate role in determining both what chemical reactions can take place and how the whole soup will be constituted. It is this higher-order distributed circularity of the interactions of the different classes of constituents that matters. Such a system can generate far more complex micro-dynamics and macro-dynamics than if the interactions were symmetric. Chemical reactions with these features were well-described by Ilya Prigogine a generation ago, and have become the basis for extensive research with both real and simulated chemical systems. Ultimately, the metabolic molecular dynamics that constitute living cells depends on autopoietic system dynamics constituted by numerous fully and partially autocatalytic sets of molecules. What do these examples of second-order emergent phenomena have in common? A kind of tangled hierarchy of causality where micro-configurational particularities can be amplified to determine macro-configurational regularities, and where these in turn further constrain and/or amplify subsequent micro-configurational regularities. In such cases, it is more appropriate to call the aggregate a 'system' rather than a mere collection, since the specific reflexive configurational and recurrent causal architecture is paramount. So although these systems must be open to the flow of energy and components—which is what enables their development—they additionally include a kind of closure as well. These material flows carry configurational constraints inherited from past states to constrain future behaviors of its components. As material and energy flows on through and out again, form also recirculates and becomes amplified. In one sense this form is nothing more than restrictions and biases on possible future material and energetic events; in another sense it is what defines and bounds a kind of higher-order individual entity that we identify as the system.

Not surprisingly, quasi-crystalline and autopoietic second-order emergent systems are often found linked in living systems. Consider the configuration-constraining role of quasi-crystal-like molecular structures such as the polymers and membranes in living cells. These can grow by mechanisms that exhibit some degree of history-dependent constraint and also tend to play a critical role in constraining the interactions critical to living autopoietic chemical reactions. The configuration of such a substrate may both catalyze certain individual reactions and increase the probability of causal circularity of a reflexive set of other catalytic reactions, for example by aligning multiple species of molecules with respect to one another. This kind of interaction between emergent phenomena turns out to be important for yet higher levels of emergent interaction.

## 4.7. Evolution and Semiosis

There is a further difference, however, between merely chaotic/self-organizing emergent phenomena such as snow crystal growth and autopoiesis, and evolving emergent phenomena such as living organisms. The latter additionally involve some form of information or memory (e.g. as represented in nucleic acids) that is not seen in second-order systems. The result is that specific historical moments of higher-order regularity or of unique micro-causal configurations can additionally exert a cumulative influence over the entire causal future of the system. In other words, via memory, constraints derived from specific past higher-order states can get repeatedly re-entered into the lower-order dynamics leading to future states, in addition to their effects mediated by second-order processes. This is what makes the evolution of life both chaotically unpredictable and yet also historically organized, with an unfolding quasi-directionality.

For this kind of phenomena we must introduce a third order, which recognizes an additional loop of recursive causality, enclosing the second-order recursive causality of chaotic or self-organized systems. *Third-order emergence* inevitably exhibits a developmental or evolutionary character. It occurs where there is both amplification of global influences on parts, but also redundant 'sampling' of these influences and reintroduction of them redundantly across time and into different realizations of the system. Under these conditions there can be extensive amplification of lower-order supervenient and also lower-order chaotic/self-organizing relationships due to the nondegradation of their historical traces, which get repeatedly re-entered into the system. Whereas second-order emergent phenomena exhibit locally and temporally restricted whole-to-part influences, third-order evolutionary emergent phenomena can exhibit amplification of these effects as well, doubly co-involving the link between levels and scales of causality.

This can be imagined as a sort of self-referential self-organization; an autopoiesis of autopoieses. Amplification of complexity and of self-organizational dynamics can be enormously complex, forming a maze of causal circularities, because every prior state is a potentially amplifiable initial condition contributing to all later states. Moreover, because there is a remembered trace of each prior 'self' state contributing to the dynamics of future states, such systems can develop with respect to it, rather than just with respect to the immediately prior state of the whole. This contributes to the characteristic differentiation and divergence from and convergence back toward some 'reference' state, as exhibited in organisms.

The representation relationship implicit in third-order emergent phenomena demands a combination of multi-scale, historical, and semiotic analyses. Thus, living and cognitive processes require introducing concepts such as representation, adaptation, information, and function in order to capture the logic of the most salient emergent phenomena. This makes the study of living forms qualitatively different from other physical sciences. It makes no sense to ask about the function of granite. Though the atoms composing a heart muscle fiber or a neurotransmitter molecule have no function in themselves, the particular configurations of the heart and its cell types additionally beg for some sort of teleological assessment. They do something for something. Organisms evolve and regulate the production of multiple second-order emergent phenomena with respect to some third-order phenomenon. Only a third-order emergent process has such an intrinsic identity.

So life, even in its simplest forms, is third-order emergent and its products can't be fully understood apart from either history or functionality. Indeed, it may be that any third-order emergent system must be considered 'alive' in some sense. This may be something like a definition of life. If so, the origins of life on earth must also be the initial emergence of third-order emergent phenomena on the earth. More generally, it constitutes the origination of information, semiosis, and teleology. Its embedded circular architecture of circular architectures definitely marks the boundary to a unit of causal self-reference extended both in space and time. It is the creation of an 'epistemic cut' to use Howard Pattee's felicitous phrase: the point where physical causality acquires (or rather constitutes) significance.

## 4.8. Overview

I have defined three subcategories of emergent phenomena, arranged into a hierarchy of increasing complexity, because higher-order forms are composed of relationships between lower-order forms. Thus third-order (evolutionary) emergence contains second-order (self-organizing) emergence as a limiting case, which in turn contains first-order (supervenient) emergence as a limiting case. In this way, any given example of evolutionary emergence must also involve self-organizing and supervenient emergence as well, but not vice versa. This means that, while it is technically correct to say that life and mind supervene on chemical processes, it is misleading to say that they are 'merely' (nothing but) chemical processes. Moreover, because higher-order emergent phenomena are dependent on and constituted by lower-order emergent phenomena, their probability of formation is substantially lower. Consequently, there are vastly more examples of supervenient emergent phenomena than self-organizing emergent phenomena and vastly more examples of self-organizing emergent phenomena than evolutionary emergent phenomena. Conversely with the replication made possible by third-order emergence the number of emergent phenomena of all types can rapidly increase, though the production of lower-order subordinate emergent phenomena produced to support the third-order will always be far more numerous. So few supervenient phenomena are alive, but every living thing is a supervenient emergent phenomenon.

This hierarchic construction does not exhaust the possibilities of increasing complexity of emergent phenomena. Evolutionary emergent systems can further interact to form multilayer systems of exceeding complexity. This is exemplified in the ascending levels of 'self' that proceed from gene to cell to organism to colony to lineage in the living world. However, all higher orders of emergence can be described as recursive variants of evolutionary emergent processes, and not some new hierarchic class of emergent phenomena. The introduction of referential relationship as the defining feature of third-order emergence creates a spatial and temporal boundedness able to encompass any physical system. There is no upper or outer or past or future bound to what can constitute a third-order emergent phenomenon. For example, we can characterize brain processes as cellular evolutionary emergent processes embedded in an embryological evolutionary emergent process embedded in a phylogenetic evolutionary process.[3] For this reason it is insufficient to describe mental phenomena as merely supervenient on cellular-molecular interactions. Though true it is a trivially incomplete characterization. This also suggests that mental processes cannot be instantiated by determinate computational machine operations (for which there isn't even supervenient causal architecture!), though it does not deny the possibility that third-order emergent processes such as mental experiences might be achievable by non-organic means. The many levels of

[3] This does *not* imply any recapitulative relationship.

embedded evolutionary emergent processes characteristic of brains are what enable them so rapidly and selectively to amplify such a vast range of possible forms of activity. Cognition is a third-order emergent phenomenon constituted by third-order emergent phenomena. It is continually generating novel emergent outputs. Which particular molecular system configurations will become amplified to produce subsequent states is both essentially untraceable and irrelevant. The most salient causal antecedents are themselves idiosyncratic global third-order emergent configurations of neural activity of an exquisitely indirect and astronomically unlikely variety. It is the capacity of evolutionary emergent processes to progressively embed evolutionary processes within one another via representations that amplify their information-handling power in ways that make the mythical butterfly effect trivially simplistic, and what makes mind possible.

The insistent critique from a systems-theoretic perspective of both genetic reductionism in evolutionary theories and computational reductionism in cognitive theories can now be more precisely paraphrased. Life and mind cannot be adequately described in terms that treat them as merely supervenient because this collapses innumerable convoluted levels of supervenient relationships of supervenient relationships. Life is not mere chemical mechanism, and cognition is not mere molecular computation. These analogies miss the most salient and descriptively important dynamics of these phenomena. To collapse descriptions of living and cognitive systems in this way inevitably ignores whole classes of causal relationships that are essential to their comprehensive explanation and which contribute their most robust and important characteristics.

No novel types of physical causes are evoked by this conception of emergence, only novel types of configurations and what might be described as 'configurational causes.' It does, however, offer an amendment to two major oversimplifications about the nature of causality: that causes have simple location and that causes always flow upward and outward in scale. Failure of these simple assumptions is exhibited most clearly in the most complex third-order emergent phenomena.

In living processes, for example, the sources of structure and function can become increasingly distributed in space and time over the course of evolution. Organizing processes may also extend to include progressively higher-level subsystems exhibiting progressively greater top-down control architecture; brains for example. A living organism is the end-product of a very elaborately convolved history of recursive causal processes that cast wider and wider nets to capture and amplify sources of regularity. Consequently, locating any specific antecedent cause of functional organization is essentially impossible. This is one reason that evolutionary 'explanations' of specific traits are inevitably vague and general in form. There is no clear causal trend from more local to more distributed, and from micro- to macro-scales. There are not even clear boundaries in space and time. Indeed, the causes of living organization grade into molecular noise. An organism's functional properties may be currently instantiated by its molecular architecture (first-order, local) and yet this architecture emerges from a vast ensemble of molecular interactions with the world generated by innumerable past members of a lineage (third-order, non-local).

Brain processes produce a further baroque convolution and temporal amplification of this logic. Brains may be characterized as 'emergence machines,' incessantly churning out complex high-level virtual functions, virtual environments, and virtual evolutionary lineages to track and adapt to the complexity of the world. The minds that result are marvels of high-level holistic causal loci. Subjective experience reflects this convergent holism. 'The feeling of what happens' is a feeling of being something that is incessantly and spontaneously emerging. This experience is itself an emerging locus at the center of a vast but only weakly constraining, weakly determinate web of semiotic and physiological influences. Minds that have become deeply entangled in the evolving symbolic communicative processes of culture—as all modern human minds are—may have an effective causal locus that extends across continents and back millennia, and which grows out of the experiences of hundreds of thousands of individuals. This immense convergence of causal determination is coupled with a vast capacity for selective amplification. Human consciousness isn't merely an emergent phenomenon, it epitomizes the logic of emergence in its very form: the locus of an incessant creation of something from nothing.

# Part II

# Science

# 5

## Science, Complexity, and the Natures of Existence

*George F. R. Ellis*

This chapter looks at complexity from a functional viewpoint, considers the nature of causation in the light of this discussion, and comments on the view we need to take as to the nature of existence in order to comprehend the existence of emergent structures (including the human mind). I conclude that physics gives a causally incomplete view of the world, and hence does not by itself provide an adequate ground for metaphysical speculations.

### 5.1. Complexity and Hierarchical Structure

True complexity occurs with the emergence of higher levels of order and meaning; it occurs in modular hierarchical structures because these are the only viable ways of building up real functional complexity. A hierarchy represents a decomposition of a complex problem into constituent parts, and allows for processes to handle constituent parts, each requiring less data and processing, and more restricted operations than the problem as a whole. The success of hierarchical structuring depends on implementing modules to handle lower-level processes, and on integration of these modules into a higher-level structure.

The levels of a hierarchy represent different levels of abstraction, each built upon the other, and each understandable by itself. This is the phenomenon of emergent order, enabling higher-level phenomenological understanding of behavior described in language suitable to that level and theories describing behavior at each level. This is summarized in Figure 5.1.

**Sociology/Economics/Politics**
**Psychology**
**Physiology**
**Cell biology**
**Biochemistry**
**Chemistry**
**Physics**
**Particle physics**

**Figure 5.1.** A hierarchy of structure and causation. Each lower level underlies what happens at each higher level, in terms of physical causation.

As expressed by N. A. Campbell:

> With each upward step in the hierarchy of biological order, novel properties emerge that were not present at the simpler levels of organization. These emergent properties arise from interactions between the components.... Unique properties of organized matter arise from how the parts are arranged and interact... [consequently] we cannot fully explain a higher level of organization by breaking it down to its parts.[1]

One cannot even describe the higher levels in terms of lower-level language.

## 5.2. Bottom-Up and Top-Down Action

From a functional viewpoint, the first issue underlying complex emergent behavior is the occurrence of both bottom-up and top-down action.

*Bottom-up action*: What happens at each higher level is based on causal functioning at the level below; hence what happens at the highest level is based on physical functioning at the bottom level. This is the basis for reductionist worldviews. The successive levels of order entail chemistry being based on physics, material science on physics and chemistry, geology on material science, and so on.

[1] N. A. Campbell, *Biology* (Redwood City, CA: Benjamin Cummings, 1996), 2–3.

*Top-down action*: Boundary effects as well as structural relations at the higher levels control what happens at the lower levels. For example, depressing a light switch causes numerous electrons to flow in wires from a power source to a bulb and to consequent illumination of a room. Top-down action occurs when higher-level structural relations and boundary conditions coordinate action at lower levels, thereby giving the higher levels their causal effectiveness. This affects the nature of causality significantly, particularly because inter-level feedback loops become possible.

Multiple top-down action enables various higher levels to coordinate action at lower levels in a complex system in a coherent way, and so gives them their causal effectiveness. It is prevalent in the physical world and in biology because no physical or biological system is isolated. A series of examples:

1. Nucleosynthesis and the creation of structure in the early universe. The rates of nuclear interactions depend on the density and temperature of the interaction medium. The microscopic reactions that take place in the early universe, and hence the amounts of the elements produced (mostly hydrogen and helium) thus depend on the rate of expansion of the universe. Similarly, the linearized equations for structure formation depend on quantities in the background cosmology (its density and expansion rate, for example), which therefore determine the nature of the perturbation solutions and the resulting formation of structure.

2. Quantum measurement. Top-down action occurs in quantum measurement.[2] The experimenter chooses the details of the measurement apparatus—for example, aligning the axes of polarization measurement equipment—and that decides what set of microstates can result from a measurement process, and so crucially influences the possible micro-state outcomes of the interactions that happen. Thus, the quantum measurement process is partially a top-down action controlled by the observer, determining what set of resultant states are available to the system during the measurement process. Additionally, top-down action occurs in state preparation: choosing

[2] R. Penrose, *The Emperor's New Mind* (Oxford: Oxford University Press, 1989); C. J. Isham, *Lectures on Quantum Theory: Mathematical and Structural Foundations* (London: Imperial College Press, 1997).

and then enforcing the specific state of the system at the start of the experiment.

3. Evolution. Top-down action is central to two main themes of molecular biology. The first is the development of DNA codings (the particular sequence of bases in the DNA) occurring through an evolutionary process which results in adaptation of an organism to its ecological niche. This is a classical case of top-down action from the environment to detailed biological microstructure—through the process of adaptation, the environment (along with other causal factors) fixes the specific DNA coding. There is no way one could ever predict this coding on the basis of biochemistry or microphysics alone.[3] One cannot even ask the appropriate questions in their languages. Top-down action from the environment codes information about the environment into the detailed base sequence in DNA. As a specific example: a polar bear has genes for white fur in order to adapt to the polar environment, whereas a Canadian bear has genes for black fur in order to be adapted to the Canadian forest. The detailed DNA coding differs in the two cases because of the different environments in which the respective animals live.

4. Biological development. A second central theme of molecular biology is the reading of DNA by an organism in the developmental process.[4] This is not a mechanistic process, but is context dependent all the way down, with what happens before having everything to do with what happens next. The central process of developmental biology, whereby positional information determines which genes get switched on and which do not in each cell, so determining their developmental fate, is a top-down process from the developing organism to the cell and is based on the existence of gradients of positional indicators in the body. Without this feature the structured development of an organism would not be possible. Thus, the functioning of the crucial cellular mechanism determining the type of each cell is controlled in an explicitly top-down way.

At a higher level of organization recent research on genes and various hereditary diseases shows that existence of the gene for such

[3] D. T. Campbell, 'Downward Causation,' in Ayala and Dobzhansky, eds., *Studies in the Philosophy of Biology*.

[4] S. F. Gilbert, *Developmental Biology* (Sunderland, MA: Sinauer, 1991); L. Wolpert, *Principles of Development* (Oxford: Oxford University Press, 1998).

diseases in the organism is not a sufficient cause for the disease in fact to occur: outcomes depend on the nature of the gene and on the rest of the genome and on the environment. The micro-features do work mechanistically but in a broader context that largely determines the outcome. The macro-environment includes the result of conscious decisions (the patient will or will not seek medical treatment for a hereditary condition, for example), so these too are a significant causal factor.

5. Mind on body. Top-down action occurs from the mind to the body and thence to the physical world. The brain controls the functioning of the parts of the body through a hierarchically structured feedback control system, which incorporates decentralized control to spread the computational and communication load and increase local response capacity.[5] It is a highly specific system in that dedicated communication links convey information from specific areas of the brain to specific areas of the body, enabling brain impulses to activate specific muscles; this occurs by means of coordinated control of electrons in myosin filaments in the bundles of myofibrils that constitute skeletal muscles. Through this process there is top-down action by the mind on the body, and indeed on the mind itself, both in the short term (immediate causation through the structural relations embodied in the brain and body) and in the long term (structural determination through imposition of repetitive patterns). An example of the latter is the way repeated stimulation of the same muscles or neurons encourages their growth. This is the basis of both athletic training and learning by rote. Additionally, an area of importance that is only now beginning to be investigated by Western medicine is the effect of the mind on health,[6] for example, through interaction with the immune system.[7]

6. Mind on the world. When a human being has a plan in mind (say a proposal for a bridge being built) and this is implemented, then enormous numbers of micro-particles (comprising the protons, neutrons, and electrons in the sand, concrete, bricks, and so forth, that become the bridge) are moved around as a consequence of this

[5] S. Beer, *Decision and Control* (New York: John Wiley, 1966).

[6] B. Moyers, *Healing and the Mind* (New York: Doubleday, 1993).

[7] E. Sternberg, *The Balance Within: The Science Connecting Health and Emotions* (New York: W. H. Freeman, 2000).

plan and in conformity with it. Thus, in the real world the detailed micro-configurations of many objects (which electrons and protons go where) is to a major degree determined by the macro-plans of humans and the way they implement them.

Because of this effectiveness of the human mind, the structural hierarchy, now interpreted as a causal hierarchy, bifurcates (see Figure 5.2). The left side, representing causation in the natural world, does not involve goal choices: everything proceeds mechanically. The right side, representing causation involving humans, involves choice of goals that lead to actions. Ethics is the high-level subject dealing with the choice of appropriate goals. Because this determines the nature of lower-level goals, and thence the nature of resulting actions, ethics is causally effective in the real physical world. This is, of course, obvious as it follows from the causation chain. However, to make the point absolutely clear: a prison may or may not have present in its premises the physical apparatus of an execution chamber; whether this is so or not depends on the ethics of the country in which the prison is situated.

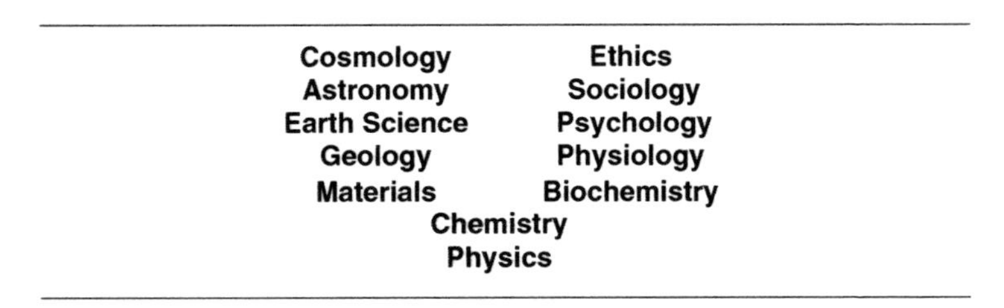


**Figure 5.2.** Branching Hierarchy of Causal Relations: The hierarchy of physical relations (Figure 5.1) extended to a branching hierarchy of causal relations. The left side involves only (unconscious) natural systems; the right side involves conscious choices, which are causally effective. In particular, the highest level of intention (ethics) is causally effective.

## 5.3. Feedback Control Systems and Information

From a functional viewpoint, the second key issue underlying complex emergent behavior is the existence of goals that are causally effective, because they are central to the functioning of feedback control systems.

*Feedback control.* The central feature of organized action is *feedback control*, whereby setting of goals results in specific actions taking place that aim to achieve those goals.[8] A comparator compares the system state with the goals and sends an error message to the system controller if needed to correct the state by making it a better approximation to the goals (Figure 5.3).

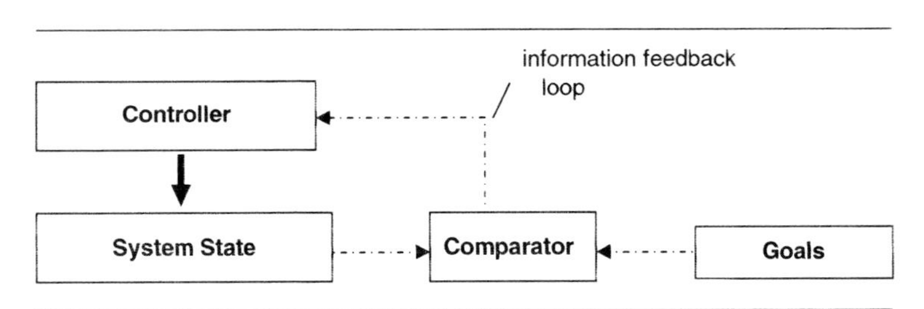


Figure 5.3. The basic feedback control process. The comparator determines the difference between the system state and the goal; an error signal from the comparator activates the controller to correct the error.

The linkages to the comparator and thence to the controller are *information linkages* rather than power or material linkages like that from the activator to the system (the information flow will use a little power, but only that required to get the message to where it is needed). Examples are controlling the heat of a shower, the direction of an automobile, or the speed of an engine.

*The role of goals and information.* The series of goals in a feedback control system are causally effective. They embody information about the system's desired behavior or responses—living systems are goal-seeking ('teleonomic'). The outcome of the system's behavior is determined by these goals, rather than by initial conditions.

These goals are not the same as material states, although they will be represented by material states and become effective through systems that use them. For example, the desired temperature of water may be set on a thermostat and represented to the user on a dial; the thermostat setting is itself a representation of the desired goal. A complete causal description of such systems must necessarily take such goals into account.

[8] R. Ashby, *An Introduction to Cybernetics* (London: Chapman and Hall, 1958); S. Beer, *Brain of the Firm* (New York: John Wiley, 1972); Beer, *Decision.*

Thus, it is here that the *key role of information* is seen: *it is the basis of goal choice in engineering and living systems* that embody feedback control. The crucial issue now is what determines the goals: where do they come from? Two major cases need to be distinguished. The first is homeostasis or inbuilt goals. The second are consciously chosen goals.

*Homeostasis: In-built goals.* There are numerous systems in all living cells, plants, and animals that automatically (i.e. without conscious guidance) maintain homeostasis—they keep the structures in equilibrium through multiple feedback systems that fight intruders (the immune system), and control energy and material flows, breathing and the function of the heart, body temperature and pressure, and so on. They are effected through numerous enzymes, antibodies, regulatory circuits of all kinds, for example, those that maintain body temperature and blood pressure.[9] They have developed in the course of time through evolution, and so are historically determined in particular environmental contexts, and are unaffected by individual history. In manufactured artifacts, the goal may be explicitly stated and controllable (e.g. the temperature setting of a thermostat). Not only are the feedback control systems themselves emergent systems, but also *the implied goals are emergent properties that guide numerous physical, chemical, and biochemical interactions in a teleological way*. They represent distilled information about the behavior of the environment in relation to the needs of life. At the higher levels they include the instinctive behavior of animals.

These feedback control loops are hierarchically structured with maximum decentralization of control from the higher to the lower levels, as is required both in order to handle requisite variety[10] and the associated information loads,[11] and for maximal local efficiency (ability to respond to local conditions). Thus, the human body has literally thousands of control systems. They have been built in through the adaptive process of evolution and so, in essence, embody images of environments in which our ancestors lived.[12]

[9] J. H. Milsum, *Biological Control Systems Analysis* (New York: McGraw Hill, 1966).

[10] Ashby, *Cybernetics*. [11] Beer, *Brain of the Firm*.

[12] Cf. Deacon's account of third-order emergence, in Chapter 4 above.

*Goal seeking: Chosen goals.* At higher levels in animals, important new features come into play: there are now explicit behavioral goals that either are learned or are consciously chosen. *It is in the choice of these goals that explicit information processing comes into play.* Information arrives from the senses and is sorted and either discarded or stored in long-term and short-term memory. Conscious and unconscious processing of this information sets up the goal hierarchy (with ethics the topmost level), which then controls purposeful action.

*Information origin and use.* Responsive behavior depends on purposeful use of information—capture, storage, transmission, recall, and assessment—to control physical functions in accord with higher-level goals. The computations are based on stored variables and structured information flows, so hidden internal variables affect external behavior. Current information is filtered against a relevance pattern, the irrelevant being discarded, the moderately significant being averaged over and stored in compressed form, the important being selectively amplified and used in association with current expectations to assess and revise immediate goals. The relevance pattern is largely determined by basic emotional responses such as those delineated by Panksepp,[13] which provide the evaluation function used in a process of neural Darwinism[14] that determines the specifics of neural connections in the brain.[15] Thus emotional responses underlie the development of rationality. Expectations are based on causal models which in turn are based on past experience. 'Frames,' such as how to behave in a restaurant, are constantly revised on the basis of newer experience and information.

Thus, feedback control systems based on sophisticated interpretation of present and past data enable purposeful (teleological) behavior. *Memory* allows both the long-term past and the immediate environmental context to be taken into account in choosing goals,

[13] J. Panksepp, *Affective Neuroscience: The Foundation of Human and Animal Emotions* (Oxford: Oxford University Press, 1998).

[14] G. Edelmann, *Neural Darwinism* (Oxford: Oxford University Press, 1990).

[15] G. F. R. Ellis and J. A. Toronchuk, 'Neural Development: Affective and Immune System Influences,' in *Consciousness and Emotion: Agency, Conscious Choice, and Selective Perception*, eds. Ralph D. Ellis and Natika Newton (Amsterdam: John Benjamins, 2005), 81–119.

providing historical information used to shape these goals in conjunction with present data. Long-term memory allows a non-local (in time) kind of causation that enables present and future behavior to be based on interpretations of long-past events (e.g. remembering that an individual let one down in important ways years ago). *Learning* allows particular responses to develop into an automatic skill, in particular allowing some responses to become inbuilt and, so, able to be rapidly deployed (e.g. driving a car, many sports moves, and so on).

*Symbolic Representation.* At the highest level, the process of analysis and understanding is driven by the power of symbolic abstraction, codified into language embodying both syntax and semantics.[16] This underpins other social creations such as specialized roles in society and the monetary system on the one hand, and higher-level abstractions such as mathematics, physical models, and philosophy on the other—all encoded in symbolic systems. Information guides all of this and is manifestly real in that it has a commercial value that underlies development of a major part of the international economy (the information technology sector). The meta-question of how context influences behavior is guided and constrained by a system of ethics based on an overall worldview associated with meaning. This will also be encoded in language and symbols.

These are all strongly emergent phenomena that are causally effective. They exist as non-material effective entities, created and maintained through social interaction and teaching, and are codified in books and sometimes in legislation. Thus, while they may be represented and understood in individual brains, their existence is not contained in any individual brain and they certainly are not equivalent to brain states. Rather, the latter serve as just one of many possible forms of embodiment of these features.

Thus, in summary, our actions are governed by hierarchically structured goals at all structural levels in society, which may be explicit or implicit, qualitative or quantitative, but are not physical quantities. They are adaptively formed in response to experience: learning takes place, and the mind responds to the meaning of symbols in the relevant social context.

[16] T. Deacon, *The Symbolic Species: The Co-Evolution of Language and the Human Brain* (London: Penguin, 1997).

## 5.4. The Nature of Causality and Explanation

Multiple layers of causality are always in operation in complex systems. Any attempt to characterize any partial cause as the whole (as characterized by the phrase 'nothing but') is a fundamentally misleading position (indeed, this is the essence of all fundamentalisms). This is important in particular in regard to claims that any of physics, evolutionary biology, sociology, psychology, or whatever are able to give *total* explanations of specific properties of the mind. Rather they each provide partial and incomplete explanations.

### 5.4.1. *The Nature of Explanation*

The key point about explanation is that we take for granted most of the causes in operation in any particular situation and then ignore them, focusing on the particular item of interest that is needed to understand what happens *when all the rest are taken for granted but without which it would not happen*. For example, she died:

- because I sent her to get some cigarettes from the shop on the other side of the road
- because the road was wet so the car could not stop in time
- because she was inattentive to the traffic
- because she saw her dog on the other side of the road
- because of Newton's laws of motion applied to the car
- because her heart stopped beating
- because the ambulance took too long to get here.

All are true aspects of the causal nexus that led to her death.

### 5.4.2. *Causality: Bottom-Up and Top-Down Explanation*

The key point about causality in this context is that simultaneous multiple causality (inter-level, as well as within each level) is always in operation in complex systems. Thus, one can have a top-down system explanation as well as a bottom-up explanation, *both being simultaneously applicable.*

Analysis 'explains' the properties of the machine by analyzing its behavior in terms of the functioning of its component parts (the lower levels of structure). Systems thinking[17] 'explains' the behavior or properties of an entity by determining its role or function within the higher levels of structure.[18] For example, the question: 'Why does an aircraft fly?' can be answered:

- in *bottom-up terms*: it flies because air molecules impinge against the wing and get deflected, leading to a resultant force that counteracts gravity,
- in terms of *same-level explanation*: it flies because the pilot is flying it, and she is doing so because the airline's timetable dictates that there will be a flight today at 4:35 pm from London to Berlin,
- in terms of *top-down explanation*: it flies because it is designed to fly! This was done by a team of engineers working in a historical context of the development of metallurgy, combustion, lubrication, aeronautics, machine tools, computer-aided design, and so forth, all needed to make this possible, and in an economic context of a society with a transportation need and complex industrial organizations able to mobilize the necessary resources for design and manufacture.

These are all simultaneously true explanations. The higher-level explanations rely on the existence of the lower-level explanations, but are clearly of a quite different nature than the lower-level ones, and certainly not reducible to them nor dependent on their specific nature. They are also, in a sense, deeper explanations than the lower-level ones.

*The point is fundamental: The analytic approach ignores the environment and takes the existence of the machine for granted; from that standpoint, it enquires as to how the machine functions. This enables one to understand its reliable replicable behavior, but it completely fails to answer why an entity exists with that specific behavior. Systems analysis in terms of purpose within the higher-level structure, where it is one of many interacting components, provides that answer—giving another equally valid, and in some ways more profound, explanation of why it has the properties it has. This approach determines the rationale, the*

[17] C. W. Churchman, *The Systems Approach* (New York: Delacorte Press, 1968).

[18] R. Ackoff, *Ackoff's Best: His Classic Writings on Management* (New York: John Wiley, 1999).

raison d'être, *of the entity; given that purpose, it can usually be fulfilled in a variety of different ways in terms of structure at the micro level.*

### 5.4.3. *The Effectiveness of Consciousness*

A particularly important aspect of causality is the effectiveness of consciousness. This is also true for thoughts and emotions, through their ability to influence intentions. We can consider there to be three different causally effective aspects of consciousness, namely,

- rationality and understanding
- feelings and intentions
- social constructions, e.g. laws or language.

*The effectiveness of rationality.* Concepts such as the plans for a jumbo jet are not the same as brain states, for they can be represented in many different ways (in words, writing, diagrams, in computer memories associated with computer-aided design programs, etc.). They are causally effective because they affect the nature of physical objects in the world. Indeed, they guide the manufacture of material objects.

*The effectiveness of emotions.* Emotions both influence immediate behavior in obvious ways (e.g. 'She acted in anger') and also underlie brain development and intellect, as explained by the theory of affective neural Darwinism.[19] Higher levels of order and meaning are developed through this process, showing one aspect of how basic emotions can be causally effective: they set up a set of implicit goals in the developing brain. These then structure the further brain development and lead to higher goals.

*The effectiveness of social constructions.* The rules of chess are a clear-cut example. Another one is legislation such as laws on automobile emission limits or speed limits. They both affect what happens on the ground, whether through voluntary compliance or policing actions.

*The effectiveness of ethics.* The highest level of goals is ethics, which is causally effective, as indicated in Figure 5.2 and the related discussion.

[19] G. F. R. Ellis and J. Toronchuk, 'Affective Neural Darwinism,' in *Consciousness and Emotion: Agency, Conscious Choice, and Selective Perception*, eds. Ralph D. Ellis and Natika Newton (Amsterdam: John Benjamins, 2005).

Causal models of the real world will be incomplete unless they include these various effects. Emergent levels of meaning and order hereby control lower levels of developing structure. Physics alone cannot begin to explain this process—it cannot even comprehend the nature of the relevant variables.

## 5.5. The Natures of Existence

In this section, I develop a holistic view of ontology, building on previous proposals by Karl Popper and John Eccles, and Roger Penrose.[20] I clearly distinguish between ontology (existence) and epistemology (what we can know about what exists). They should not be confused: whatever exists may or may not interact with our senses and measuring instruments in such a way as to clearly demonstrate its existence to us.

### 5.5.1. *A Holistic View of Ontology*

I take as given *the reality of the everyday world*—tables and chairs, and the people who perceive them—and then assign a reality additionally to each kind of entity that can have a demonstrable causal effect on that everyday reality. The problem then is to characterize the various kinds of independent reality which may exist in this sense. Taking into account the causal efficacy of all the entities discussed above, I suggest as a possible completion of the proposals by Popper and Eccles, and by Penrose, that the four worlds indicated in Figure 5.4 are ontologically real. These are not different causal levels within the same kind of existence, rather they are quite different kinds of existence, but related to each other through causal links.

The challenge is to show, first, that each is indeed ontologically real and, second, that each is sufficiently and clearly different from

[20] K. Popper and J. Eccles, *The Self and its Brain: An Argument for Interactionism* (Berlin: Springer-Verlag, 1977); R. Penrose, *The Large, the Small and the Human Mind* (Cambridge: Cambridge University Press, 1997). See also G. F. R. Ellis, 'True Complexity and its Associated Ontology,' in *Science and Ultimate Reality: Quantum Theory, Cosmology, and Complexity*, eds. J. D. Barrow, P. C. W. Davies, and C. L. Harper (Cambridge: Cambridge University Press, 2004).

- ***World 1:*** **Matter and Forces**
- ***World 2:*** **Consciousness**
- ***World 3:*** **Physical and biological possibilities**
- ***World 4:*** **Mathematical reality**

Figure 5.4. The different kinds of reality implied by causal relationships can be characterized in terms of four worlds, each representing a different kind of existence.

the others that it should be considered as separate from them. I now discuss them in turn.

## 5.5.2. *World 1: Matter and Forces*

World 1 is the physical world of energy and matter, hierarchically structured to form lower and higher causal levels whose entities are all ontologically real. This is the basic world of matter and material interactions, based at the micro-level on elementary particles and fundamental forces, and providing the ground of physical existence. It comprises three major parts:

- *World 1a: Inanimate objects* (both naturally occurring and manufactured).
- *World 1b: Living things,* apart from humans (amoeba, plants, insects, animals, etc).
- *World 1c: Human beings,* with the unique property of being self-conscious.

All these objects are made of the same physical stuff, but the structure and behavior of inanimate and living things (described respectively by physics and inorganic chemistry, and by biochemistry and biology) are so different that they require separate recognition, particularly when self-consciousness and purposive activity (described by psychology and sociology) occur. The hierarchical structure in matter is a real physical structuration, and is additional to the physical constituents themselves that make up the system. It provides the basis for higher levels of order and phenomenology, and hence of ontology.

There is ontological reality at each level of the hierarchy. Thus, we explicitly recognize as being real: quarks, electrons, neutrinos, rocks, tables, chairs, apples, humans, the Earth, stars, galaxies, and so on. The fact that each is comprised of lower-level entities does not undermine its status as existing in its own right.[21] We can attain and confirm high representational accuracy and predictive ability for quantities and relations at higher levels, independent of our level of knowledge of interactions at lower levels, giving well validated and reliable descriptions at higher levels accurately describing the various levels of emergent non-reducible properties and meanings. An example is digital computers with their hierarchical logical structure expressed in a hierarchy of computer languages that underlie the top-level user programs. The computer has a reality of existence at each level that enables one to deal with it as an entity at that level.[22] The user does not need to know machine code and, indeed, the top-level behavior is independent of which particular hardware and software underlie it at the machine level. Another example is that a motor mechanic does not have to study particle physics in order to ply her trade.

### 5.5.3. *World 2: Consciousness*

World 2 is the world of individual and communal consciousness: ideas, emotions, and social constructions. This again is ontologically real (it is clear that these all exist) and causally effective, as discussed above.

This world of human consciousness can be regarded as comprising three major parts:

- *World 2a: Human Information, Thoughts, Theories, and Ideas,*
- *World 2b: Human Goals, Intentions, Sensations, and Emotions,*
- *World 2c: Explicit Social Constructions.*

[21] R. W. Sellars, *The Philosophy of Physical Realism* (New York: Russell and Russell, 1932).

[22] A. S. Tanenbaum, *Structured Computer Organization* (Englewood Cliffs, NJ: Prentice Hall, 1990).

These worlds are different from the world of material things and are realized through the human mind and society. They are not brain states, although they can be represented as such, for they do not reside exclusively in any particular individual mind. They are not identical to each other: World 2a is the world of rationality; World 2b is the world of intention and emotion, and so comprehends non-propositional knowing; while World 2c is the world of consciously constructed social legislation and convention. Although each is individually and socially constructed in a complex interaction between culture and learning, these are indeed each capable of causally changing what happens in the physical world, and each has an effect on the others. In more detail:

*World 2a: the world of human information, thoughts, theories, and ideas.* This world of rationality is hierarchically structured, with many different components. It includes words, sentences, paragraphs, analogies, metaphors, hypotheses, theories, and indeed the entire bodies of science and literature, and refers both to abstract entities and to specific objects and events. It is necessarily socially constructed on the basis of varying degrees of experimental and observational interaction with World 1, which it then represents with varying degrees of success. World 2a is represented by symbols, particularly language and mathematics, which are arbitrarily assigned and which can themselves be represented in various ways (as sound, on paper, on computer screens, in digital coding, etc.)

Thus, each concept can be expressed in many different ways and is an entity in its own right independent of which particular way it is coded or expressed. These concepts sometimes correspond well to entities in the other worlds, but the claim of ontological reality of entities existing in World 2a makes no claim that the objects or concepts they refer to are real. Thus, this world equally contains concepts of rabbits and fairies, galaxies and UFOs, science and magic, electrons and ether, unicorns and apples; all of these certainly exist *as concepts.* This statement is neutral as to whether these concepts correspond to objects or entities that exist in the real universe (specifically, whether there is or is not some corresponding entity in World 1) or whether the theories in this world are correct (that is, whether they give a good representation of World 1 or not).

All the ideas and theories in this world are ontologically real in that they are able to cause events and patterns of structures in the physical world. Firstly, they may all occur as descriptive entries in an encyclopedia or dictionary. Thus, each idea has causal efficacy as shown by existence of the resulting specific patterns of marks on paper (these constellations of micro-particles would not be there if the idea did not exist, as an idea). Secondly, in many cases they have further causal power as shown by the examples of the construction of the jumbo jet and the destruction of Hiroshima. Each required both an initial idea, and resulting detailed plan and an intention to carry it out. Hence such ideas are indubitably real in the sense that they must be included in any complete causal scheme for the real world. If one denies the reality of this feature one will end up with a causal scheme lacking many features of the real world (one will have to say that the jumbo jet came into existence without a cause, for example!).

*World 2b: the world of human goals, intentions, sensations, and emotions.* This world of motivation and sensations is also ontologically real, for it is clear that they do indeed exist in themselves; for example, they may all be described in novels, magazines, books, and so forth, thus being causally effective in terms of being physically represented in such writings. Additionally, many of them cause events to happen in the physical world—for example, the emotion of hate can cause major destruction both of property and lives, as in Northern Ireland and Israel and many other places. In World 2b, we find the goals and intentions that cause the intellectual ideas of World 2a to have physical effect in the real world.

*World 2c: the world of explicit social constructions.* This is the world of language, customs, roles, laws, and so on, which shapes and enables human social interaction. It is developed by society historically and through conscious legislative and governmental processes. It gives the background for ordinary life, enabling Worlds 2a and 2b to function, particularly by determining the means of social communication (language is explicitly a social construction). It is also directly causally effective, for example speed laws and exhaust emission laws influence the design both of automobiles and of road signs, and so get embodied in the physical shapes of designed structures in World 1; the rules of chess determine the space of possibilities for movements of chess pieces on a chess board. It is socially realized and embodied in legislation, roles, customs, and so on.

### 5.5.4. *World 3: Physical and Biological Possibilities*

World 3 is the world of Aristotelian possibilities. This characterizes the set of all physical possibilities, from which the specific instances of what actually happens in World 1 are drawn. This world of possibilities is ontologically real because of its rigorous prescription of the boundaries of what is possible—it provides the framework within which World 1 exists and operates, and in that sense is causally effective. It can be considered to comprise two major parts:

- *World 3a: The world of physical possibilities*, delineating possible physical behavior.
- *World 3b: The world of biological possibilities*, delineating possible biological organization.

These worlds are different from the world of material things, for they provide the background within which that world exists. In a sense they are more real than that world, because of the rigidity of the structure they impose on World 1. There is no element of chance or contingency in them, and they certainly are not socially constructed (although our understanding of them is so constructed). They rigidly constrain what can happen in the physical world, and are different from each other because of the great difference between what is possible for life and for inanimate objects. In more detail:

*World 3a: the world of physical possibilities.* This delineates possible physical behavior; it is a description of all possible motions and physical histories of objects. Thus, it describes what can actually occur in a way compatible with the nature of matter and its interactions; only some of these configurations are realized through the historical evolutionary process in the expanding universe. We do not know if laws of behavior of matter as understood by physics are prescriptive or descriptive, but we do know that they rigorously describe the constraints on what is possible: one cannot move in a way that violates energy conservation; one cannot create machines that violate causality restrictions; one cannot avoid the second law of thermodynamics; and so on. This world delineates all physically possible actions (different ways particles, planets, footballs, automobiles, aircraft can move, for example); from these possibilities, what actually happens is determined by initial conditions in the universe, in the case of interactions between inanimate objects, and by the conscious choices made when living beings exercise volition.

If one believes that physical laws are prescriptive rather than descriptive, one can view this world of all physical possibilities as being equivalent to a complete description of the set of physical laws (for these determine the set of all possible physical behaviors, through the complete set of their solutions). The formulation given here is preferable, in that it avoids making debatable assumptions about the nature of physical laws, but still incorporates their essential effect on the physical world. Whatever their ontology, what is possible is described by physical laws such as the Second Law of Thermodynamics:

$$dS > 0,$$

Maxwell's laws of Electromagnetism:

$$F_{[ab;c]} = 0,\ F^{ab};b = J^a,\ J^a;_{\ a} = 0,$$

and Einstein's Law of Gravitation:

$$R_{ab} - 1/2R\,g_{ab} = kT_{ab},\ T^{ab};_{\ b} = 0.$$

These formulations emphasize the still mysterious extraordinary power of mathematics in terms of describing the way that matter can behave, and each partially describes the space of physical possibilities.

*World 3b: the world of biological possibilities.* This delineates all possible living organisms. This defines the set of potentialities in biology by giving rigid boundaries to what is achievable in biological processes. Thus, it constrains the set of possibilities from which the actual evolutionary process can choose—it rigorously delineates the set of organisms that can arise from any evolutionary history whatever. This 'possibility landscape' for living beings underlies evolutionary theory, for any mutation that attempts to embody a structure that lies outside its boundaries will necessarily fail. Thus, even though it is an abstract space in the sense of not being embodied in specific physical form, it strictly determines the boundaries of all possible evolutionary histories. In this sense it is highly effective causally.

Only some of the organisms that can potentially exist are realized in World 1 through the historical evolutionary process; thus, only part of this possibility space is explored by evolution on any particular planet. When this happens, the information is coded in the hierarchical structure of matter in World 1 (particularly in the genetic coding embodied in DNA) and so is stored via ordered relationships in matter. This information gets transformed into various other forms until it is realized in the structure of an animal or plant. Doing so encodes both a historical evolutionary sequence, and also structural and functional relationships that emerge in the phenotype and enable its functioning once the genotype is read. This is the way that directed feedback systems and the idea of purpose can enter the biological world, and so distinguishes the animate from the inanimate world. The structures occurring in the non-biological world can be complex, but they do not incorporate 'purpose' or order in the same sense.

Just as World 3a can be thought of as encoded in the laws of physics, World 3b can be thought of as encoded in terms of biological information, a core concept in biology distinguishing the world of biology from the inanimate world.[23]

### 5.5.5. *World 4: Abstract (Platonic) Reality*

World 4 is the Platonic world of (abstract) realities that are discovered by human investigation but are independent of human existence. They are not embodied in physical form but can have causal effects in the physical world.

*World 4a: mathematical forms.* The existence of a Platonic world of mathematical objects is strongly argued by Roger Penrose; major parts of mathematics are discovered rather than invented (rational numbers, zero, irrational numbers, and the Mandelbrot set being classic examples). They are not determined by physical experiment,

[23] B.-O. Küppers, *Information and the Origin of Life* (Cambridge, MA: MIT Press, 1990); C. A. Pickover, *Visualizing Biological Information* (Singapore: World Scientific, 1995).

but are rather arrived at by mathematical investigation. They have an abstract rather than embodied character; the same abstract quantity can be represented and embodied in many symbolic and physical ways. They are independent of the existence and culture of human beings, for the same features will be discovered by intelligent beings in the Andromeda galaxy as here, once their mathematical understanding is advanced enough (which is why they are advocated as the basis for inter-stellar communication). This world is to some degree discovered by humans, and represented by our mathematical theories in World 2; that representation is a cultural construct, but the underlying mathematical features they represent are not. Indeed like physical laws, they are often unwillingly discovered—for example, irrational numbers and the number zero.[24] This world is causally efficacious in the process of discovery and description; one can, for example, print out the values of irrational numbers or graphic versions of the Mandelbrot Set in a book, resulting in a physical embodiment in the ink printed on the page.

A key question is what, if any, part of logic, probability theory, and physics should be included here. In some as yet unexplained sense, the world of mathematics underlies the world of physics. Many physicists at least implicitly assume the actual existence of physical laws.

*World 4b: physical laws*, underlying the nature of physical possibilities (World 3a). Quantum field theory applied to the standard model of particle physics is immensely complex.[25] It conceptually involves, *inter alia*,

- Hilbert spaces, operators, commutators, symmetry groups, higher dimensional spaces;
- particles/waves/wave packets, spinors, quantum states/wave functions;
- parallel transport/connections/metrics;
- the Dirac equation and interaction potentials, Lagrangians and Hamiltonians;
- variational principles that seem to be logically and/or causally prior to all the rest.

[24] J. Seife, *Zero: The Biography of a Dangerous Idea* (London: Penguin, 2000).

[25] M. E. Peskin and D. V. Schroeder, *An Introduction to Quantum Field Theory* (Reading, MA: Perseus Books, 1995).

Derived (effective) theories, including classical (non-quantum) theories of physics, equally have complex abstract structures underlying their use: force laws, interaction potentials, metrics, and so on.

There is an underlying issue of significance: *What is the ontology/nature of existence of all this quantum apparatus, and of higher-level (effective) descriptions?* We seem to have two options.

1. These are simply our own mathematical and physical constructs that happen to characterize reasonably accurately the physical nature of physical quantities.
2. They represent a more fundamental reality as Platonic quantities that have the power to control the behavior of physical quantities (and can be represented accurately by our descriptions of them), i.e. World 4b exists.

On the first supposition, the 'unreasonable power of mathematics' to describe the nature of the particles is a major problem—if matter is endowed with its properties in some way that we are unable to specify, but not determined specifically in mathematical terms, and its behavior happens to be accurately described by equations of the kind encountered in present-day mathematical physics, then that is truly weird! Why should it then be possible that *any* mathematical construct whatever gives an accurate description of this reality, let alone ones of such complexity as in the standard theory of particle physics? Additionally, it is not clear on this basis why all matter has the same properties—why are electrons here identical to those at the other side of the universe? On the second supposition, this is no longer a mystery—the world is indeed constructed on a mathematical basis, and all matter everywhere is identical in its properties. But then the question is how did that come about? How are these mathematical laws imposed on physical matter? And which of the various alternative forms (Schrödinger, Heisenberg, Feynman; Hamiltonian, Lagrangian) is the 'ultimate' one? What is the reason for variational principles of any kind?

### 5.5.6. *Existence and Epistemology*

The major proposal made here is that *all these worlds exist—Worlds 1 through 4 are ontologically real and are distinct from each other*. These claims are justified in terms of the effectiveness of each kind of reality in influencing the physical world. What then of epistemology? Given the existence of the various worlds mentioned above, the proposal here is that, epistemology is the study of the relation between World 2 and Worlds 1, 3, and 4. It attempts to obtain accurate correspondences to quantities in all the worlds by means of entities in World 2a.

This exercise implicitly or explicitly divides World 2a theories and statements into: (1) true/accurate representations, (2) partially true/misleading representations, (3) false/poor/misleading representations, and (4) ones where we do not know the situation. These assessments range from statements such as 'It is true her hair is red' or 'There is no cow in the room' to 'Electrons exist,' 'Newtonian theory is a very good description of medium-scale physical systems at low speeds and in weak gravitational fields' and 'the evidence for UFOs is poor.' This raises interesting issues about the relation between reality and appearance. For example, everyday life gives a quite different appearance to reality than microscopic physics—as Sir Arthur Eddington pointed out, a table is actually mostly empty space between the atoms that make up its material substance, but in our experience is a real hard object.[26]

There is a widespread tendency to equate epistemology and ontology. This is an error, and a variety of examples can be given where it seriously misleads. This is related to a confusion between World 2 and the other Worlds discussed here and seems to underlie much of what has happened in the so-called 'Science Wars' and the Sokal affair.[27] The proposal here strongly asserts the existence of independent domains of reality (Worlds 1, 3, and 4) that are not socially constructed. It implies that we do not know all about them and indeed cannot expect to ever understand them fully. This ignorance does not undermine their claim to exist, quite independently of human understanding. The explicit or implicit claim that they depend on human knowledge means that we are equating epistemology and ontology—just another example of human hubris.

[26] A. S. Eddington, *The Nature of the Physical World* (Cambridge: Cambridge University Press, 1928).

[27] A. Sokal, see http://www.physics.nyu.edu/faculty/sokal/index.html.

## 5.6. Emergence and Fundamental Physics

Fundamental physics underlies this complexity of World 1 by determining the nature and interactions of matter. The basic questions for physicists are: What are the aspects of fundamental physics that allow and enable this extraordinarily complex modular hierarchical structure to exist, in which the higher levels are quite different from the lower levels and have their own ontology? What are the features that allow it to come into being (i.e. that allow its historical development through a process of spontaneous structure formation)? How they do it is not fully clear—what fine-tunings at the lower levels are needed for the entire higher-level hierarchy to exist?

Physics underlying complexity, the physics that allows this independence of higher-level properties from the nature of lower-level constituents has been discussed *inter alia* by P. W. Anderson,[28] Silvan Schweber,[29] and L. Kadanoff,[30] focusing on the renormalization group; however more is needed to understand the appearance of fundamentally different higher order structures. This crucially important property of physics, underlying our everyday lives and their reality, has its source in the nature of quantum field theory applied to the micro-properties of matter (as summarized in the standard model of particle physics). It would be helpful to have more detailed studies of which features of quantum field theory on the one hand, and of the standard particle/field model (with all its particular symmetry groups, families of particles, and interaction potentials) on the other, are the keys to this fundamental feature.[31]

What is clear is that this is a remarkable property, in that *almost all laws of physics of which we can conceive will not allow life of the kind we see around us to exist.* This is the issue of the *fine-tuning of*

[28] P. W. Anderson, 'More is Different,' *Science* 177 (1972): 393–396.

[29] S. Schweber, 'Physics, Community and the Crisis in Physical Theory,' *Physics Today* (November 1993): 34–40.

[30] L. Kadanoff, *From Order to Chaos: Essays Critical, Chaotic and Otherwise* (Singapore: World Scientific, 1993).

[31] See for example C. Hogan, 'Why the Universe is Just So,' *Review of Modern Physics* 72 (2000): 1149–61.

*physics* which lies in that very small region of parameter space that allows complexity to emerge.[32] This leads to all sorts of metaphysical speculation on the possible origin of such fine-tuning, which I will not pursue here except to remark that the key contenders in explanatory terms are either a multiverse (many universes exist) or a designer (this fine-tuning is evidence of purpose). There is not enough evidence attainable to prove either is correct, so the choice between them is not a scientific decision: it is a matter of philosophy and faith.[33]

Some believe that this issue would be solved if a successful physical 'Theory of Everything' were to be developed, showing how all fundamental forces and particles are unified as aspects of some underlying theory, perhaps based on 'M-theory' or superstring theory.[34] Such a theory would have a status logically and causally superior to the rest of physics in the sense of underlying all of physics at a fundamental level. The puzzle regarding complexity, however, would not be solved by existence of such a theory; rather it would be reinforced, for such a theory would in essence have the image of humanity built into it, even though its underlying principles (variational and symmetry principles based on some symmetry group, plus aspects of quantum field theory) have nothing whatever to do with the existence of life. Why that should be so is far from obvious; indeed it would border on the miraculous if a logically unique theory of fundamental physics were also a theory that lay in that tiny corner of parameter space that allows the existence of life. This would be a coincidence of the most extraordinary sort, crying out for some kind of deeper explanation.

*Human thought and physics.* Human thoughts can cause real physical effects; this is a top-down action from the mind to the physical

[32] J. D. Barrow and F. J. Tipler, *The Anthropic Cosmological Principle* (Oxford: Oxford University Press, 1986); M. Tegmark, 'Is the Theory of Everything Merely the Ultimate Ensemble Theory?' *Annals of Physics* 270 (1998): 1–51; M. J. Rees, *Our Cosmic Habitat* (Princeton, NJ: Princeton University Press, 2001).

[33] G. F. R. Ellis, U. Kirchner, and W. Stoeger, 'Multiverses and Physical Cosmology,' *Monthly Notices of the Royal Astronomical Society* 347 (2004): 921–936; 'Multiverses and Cosmology: Philosophical Issues,' submitted for publication.

[34] B. Greene, *The Elegant Universe: Superstrings, Hidden Dimensions, and the Quest for the Ultimate Theory* (New York: Norton, 2003).

world. At present there is no way to express this interaction in the language of physics, even though our causal schemes are manifestly incomplete without this being taken into account. The minimum requirement to do so is to include the relevant variables in the space of variables considered. This, then, allows them and their effects to become a part of physical theories—perhaps even of fundamental physics. The issue of how this all relates to the underlying nature of quantum physics on the one hand and of consciousness on the other remains a challenge.

*The challenge to physics.* The challenge to physics is that the higher levels are demonstrably causally effective. In particular, consciousness is causally effective; but conscious plans, intentions, and emotions are not describable in present-day physical terms. Thus, physics has two choices, either (1) *extending its scope of description to encapsulate such higher-level causal effects*, for example, including new variables representing thoughts and intentions and so enabling it to model the effects of consciousness and its ability to be causally effective in the real physical world, or (2) *deciding that these kinds of issues are outside the province of physics, which properly deals only with inanimate objects and their interactions.*

In the latter case, physics must give up the claim to give a causally complete description of interactions that affect the real physical world: it cannot even account for a pair of spectacles! Hence it does not provide an adequate basis for metaphysical speculations about the nature of existence.[35]

## 5.7. Conclusion

This paper has argued that any scientific attempt to explain the world around us in physical terms necessarily involves acknowledgment of various different kinds of existence as well as that of the materialist world of particles and forces. Without such an extension of our understanding of ontology, any attempt at a theory of causation of

[35] G. F. R. Ellis, 'Physics, Complexity, and Causality,' *Nature* 435 (9 June 2005): 743; 'Physics and the Real World,' *Foundations of Physics* 36/2 (February 2006); http://www.mth.uct.ac.za/~ellis/realworld.pdf.

the world around us will necessarily be incomplete. This loosening up of views of ontology[36] then provides a worldview in which proposals of other aspects of ontology, for example, to do with morality or meaning,[37] may not seem out of place.

[36] Ellis, 'True Complexity.'

[37] Nancey Murphy and George F. R. Ellis, *On the Moral Nature of the Universe: Theology, Cosmology, and Ethics* (Minneapolis, MN: Fortress Press, 1996).

# 6

## Reduction and Emergence in the Physical Sciences: Some Lessons from the Particle Physics and Condensed Matter Debate

*Don Howard*

### 6.1. Introduction: A Note of Caution

The task that I set myself is a mundane philosophical one—getting clear about fundamental concepts, developing a taxonomy of viewpoints, assessing the validity of arguments for those views, and handicapping the odds for one or another of them to emerge triumphant. The arena is the much contested one of questions about reduction and emergence in the physical sciences, more specifically the relationship between particle physics and condensed matter physics. The main point that I wish to make is that we know so little about that relationship, and that what we do know strongly suggests that condensed matter phenomena are *not* emergent with respect to particle physics, that we should be wary of venturing hasty generalizations and of making premature extrapolations from physics to the biosciences, the neurosciences, and beyond.

Caution is the byword. Caution is called for because the academy is yet again seized by an enthusiasm. Seventy years ago, it was complementarity.[1] Thirty years ago, catastrophe theory.[2] Twenty years ago,

[1] Niels Bohr, 'Light and Life,' *Nature* 131 (1933): 421–423, 457–459. Reprinted in *Atomic Physics and Human Knowledge* (New York: John Wiley & Sons, 1961), 3–12.

[2] René Thom, *Structural Stability and Morphogenesis: An Outline of a General Theory of Models*, trans. D. H. Fowler (Reading, MA: W. A. Benjamin, 1975).

fractals.[3] Yesterday it was cellular automata.[4] Today it is complexity theory, cooperative phenomena, and nonlinear dynamics.[5] Enthusiasm is good. It promotes creativity. It stimulates imagination. It gives one strength to carry on in the face of dogmatic opposition. But, Descartes—himself no intellectual wallflower—taught us in the *Meditations* that error is a consequence of the will outrunning the understanding. Like Faust, many of us want to know 'was die Welt am innersten zusammnehält.' Let's just be sure that our desire to solve the riddle of the universe doesn't get too far out in front of what we actually understand.

As mentioned, the specific place where I want today to make the case for caution is at the interface between particle physics and solid state or condensed matter physics. Here is where we find some of the boldest assertions that physics has demonstrated emergence. Various salient physical properties of the *mesorealm*, properties such as superconductivity and superfluidity, are held to be emergent with respect to particle physics. Such properties are said to exemplify coherent states of matter or long-range cooperative phenomena of a kind often associated with systems obeying a nonlinear dynamics. Such coherent states are said not to be explicable in terms of the properties of the molecular, atomic, or still more elementary constituents of superconductors or superfluids. I urge caution here for two reasons: (1) the physics of the mesorealm is not well enough established to license any inferences about the *essential* and *distinguishing* properties of matter at this intermediate scale; (2) that coherent states of matter are not to be explained at the level of particle physics has simply not been demonstrated. On the contrary, it is precisely at the level of particle physics that we *do* find compelling physical arguments and empirical evidence of the holism said—wrongly, I think—to be *distinctive* of mesophysics. We've had a name for such microphysical holism since 1935. That name is *entanglement*. And at least superconductivity and superfluidity, if not also various other phenomena in the realm of condensed matter physics, find their

[3] Benoit Mandelbrot, *The Fractal Geometry of Nature* (New York: W. H. Freeman, 1983).

[4] Stephen Wolfram, *A New Kind of Science* (Champaign, IL: Wolfram Media, 2002).

[5] Alwyn Scott, *Nonlinear Science: Emergence and Dynamics of Coherent Structures* (Oxford: Oxford University Press, 1999); see also Chapter 8 in this volume.

proper explanation as mesoscopic manifestations of microscopic entanglement.

## 6.2. Some Conceptual Preliminaries: Reduction, Supervenience, and Emergence

Contemporary discussions of emergent phenomena often start with a helpful distinction between two different relationships that might obtain between two different levels of description, *intertheoretic reduction* and *supervenience*.[6]

*Intertheoretic reduction* is a logical relationship between theories. In the classic formulation owing to Ernest Nagel, theory $T_B$, assumed correctly to describe or explain phenomena at level $B$, reduces to theory $T_A$, assumed correctly to describe or explain phenomena at level $A$, if and only if the primitive terms in the vocabulary of $T_B$ are definable via the primitive terms of $T_A$ and the postulates of $T_B$ are deductive consequences of the postulates of $T_A$.[7] As normally formulated, this definition of reduction assumes a *syntactic* view of theories as sets of statements or propositions.

*Supervenience* is an ontic relationship between structures. A structure, $S_x$, is a set of entities, $E_x$, together with their properties and relations, $PR_x$. A structure, $S_B$, characteristic of one level, $B$, supervenes on a structure, $S_A$, characteristic of another level, $A$, if and only if the entities of $S_B$ are composed out of the entities of $S_A$ and the properties and relations, $PR_B$, of $S_B$ are wholly determined by the properties and relations, $PR_A$, of $S_A$. One way to understand the relevant sense of 'determination' is as requiring that there be no differences at level $B$, say different values of a parameter such as the temperature of a gas, without there being a corresponding difference at level $A$, say in the mean kinetic energy of the molecules constituting the gas.[8]

[6] Robert Batterman, *The Devil in the Details: Asymptotic Reasoning in Explanation, Reduction and Emergence* (Oxford: Oxford University Press, 2002); and Michael Silberstein, 'Reduction, Emergence, and Explanation,' in *The Blackwell Guide to the Philosophy of Science*, eds. Peter Machamer and Michael Silberstein (Oxford: Blackwell, 2002), 182–223.

[7] Ernst Nagel, *The Structure of Science: Problems in the Logic of Scientific Explanation* (New York: Harcourt, Brace & World, 1961).

[8] Donald Davidson, 'Mental Events,' in *Experience and Theory*, eds. Lawrence Foster and J. W. Swanson (Amherst, MA: University of Massachusetts Press, 1970), 79–101.

There is no straightforward relationship between reduction and supervenience. One might think that reduction implies supervenience, in the sense that, if theory $T_B$ reduces to theory $T_A$, then the structures, $S_B$, assumed correctly to be described or explained by $T_B$, supervene on the structures, $S_A$, assumed correctly to be described or explained by $T_A$. This need not be the case, however, if some of the properties and relations constitutive of $S_B$ depend on boundary conditions. Not all structure is nomic. Think of global metrical structure in big-bang cosmology or 'edge state' excitations in the fractional quantum Hall effect. That supervenience does not imply reduction should be even clearer, for the properties and relations, $PR_B$, constitutive of structure $S_B$ can be wholly determined by the properties and relations, $PR_A$, of $S_A$ without there being laws governing $PR_B$ that are deductive consequences of laws governing $PR_A$, perhaps because there are no exceptionless laws governing $PR_B$.

Emergence can be asserted either as a denial of intertheoretic reduction or as a denial of supervenience. There being no necessary relationship between reduction and supervenience, there will, in consequence, be no necessary relationship between the corresponding varieties of emergence, which must, therefore, be distinguished. What we might term *R*-emergence is a denial of reduction, and what we might term *S*-emergence is a denial of supervenience.

Thinking about the relationship between different levels of description in terms of intertheoretic reduction has the advantage of clarity, for while it might prove difficult actually to determine whether a postulate at level *B* is derivable from the postulates of level *A*—as is the case with the ergodic hypothesis, which is to be discussed shortly—we at least know what we mean by derivability and definability as relationships between syntactic objects like terms and statements, since we know by what rules we are to judge. The chief disadvantage of this way of thinking about interlevel relationships is that one is hard-pressed to find a genuine example of intertheoretic reduction outside of mathematics, so to assert emergence as a denial of reduction is to assert something trivial and uninteresting. Yet, another disadvantage is the restriction to theories represented syntactically as sets of statements or propositions, central among which are statements of laws, for there is reason to think that many important scientific theories—evolution is an often cited example—are not best understood in this way. Later on I will say a word or two about the possible advantages of a *semantic* view of theories, whereupon a theory is conceived as a set of models.

The chief advantage of thinking about interlevel relationships from the point of view of supervenience is that it seems to many to capture well our pre-analytic intuitions, such as those about the relationship between heat and agitated molecular motion. The chief disadvantage of so posing the question of interlevel relationships is that it is not always clear by what general rules we are to assess claims about supervenience and its denial, so in asserting emergence as a denial of supervenience one risks asserting something validated by little more than intuition. There are, however, some reasonably clear paradigm cases of emergence as a failure of supervenience, the most important for our purposes being quantum mechanical entanglement, which is shortly to be addressed.

Distinguishing intertheoretic reduction and supervenience along with the respective notions of emergence is a big step in the direction of clarity and understanding. For example, I will argue that while condensed matter physics does not obviously reduce to particle physics, phenomena characteristic of condensed matter physics such as superfluidity and superconductivity do supervene on physical properties at the particle physics level and hence are not emergent with respect to particle physics. But, we might also find, as I think we do find in the case of condensed matter physics, that neither reduction nor supervenience is the most helpful analytical tool for explicating the truly important and interesting features—both structural and methodological—of interlevel relationships in the physical sciences.

## 6.3. Ergodicity and Entanglement: Two Challenges to Our Presuppositions

That intertheoretic reduction might not be a helpful way to think about interlevel relationships is perhaps best shown by pointing out that everyone's favorite example of a putatively successful reduction—that of macroscopic thermodynamics to classical statistical mechanics—simply does not work. Recall what is required for reduction: the definability of terms and the derivability of laws.

Concede the former in this instance—as with the definition of temperature via mean kinetic energy—and focus on the latter. Foremost among the thermodynamic laws that must be derivable from statistical mechanical postulates is the second law, which asserts the exceptionless evolution of closed non-equilibrium systems from states of lower to states of higher entropy. Providing a statistical mechanical grounding of the second law was Boltzmann's paramount aim in the latter part of the nineteenth century.[9] Did he succeed?

The answer is no. For one thing, what Boltzmann derived was not the deterministic second law of thermodynamics but a statistical simulacrum of that law, according to which closed non-equilibrium systems are at best highly likely to evolve from states of lower to states of higher entropy. More importantly, even this statistical simulacrum of the second law is derived not from mechanical first principles alone but from those conjoined to what was early termed the *ergodic hypothesis*, which asserts that, regardless of its initial state, an isolated system will eventually visit every one of its microstates compatible with relevant macroscopic constraints. The ergodic hypothesis can be given comparably opaque equivalent formulations, such as the assertion of the equality of time and ensemble averages, but the work that it does in the foundations of statistical mechanics is clear: The theory being a statistical one, it must work with averages. The ergodic hypothesis makes the averages come out right. The crucial fact is, however, that for all but a few special cases or for highly idealized circumstances, the ergodic hypothesis and its kin cannot be derived from mechanical first principles. On the contrary, we demonstrate non-ergodic behavior for a large class of more realistic models.[10]

[9] Still one of the most illuminating studies of the conceptual foundations of Boltzmann's program is to be found in the Ehrenfests's splendid monograph of 1911. Paul Ehrenfest and Tatiana Ehrenfest, 'Begriffliche Grundlagen der statistischen Auffassung in der Mechanik,' in *Encyklopädie der mathematischen Wissenschaften, mit Einschluss ihrer Anwendungen*, Vol. 4: *Mechanik*, part 4, eds. Felix Klein and Conrad Müller (Leipzig: Teubner, 1907–1914), 1–90. English translation: *The Conceptual Foundations of the Statistical Approach in Mechanics*, trans. Michael J. Moravcsik (Ithaca: Cornell University Press, 1959).

[10] For a survey of the current state of opinion regarding the ergodic hypothesis, see Lawrence Sklar, *Physics and Chance: Philosophical Issues in the Foundations of Statistical Mechanics* (Cambridge: Cambridge University Press, 1993), 156–195.

Are macroscopic thermodynamic phenomena, therefore, emergent with respect to the mechanical behavior of the individual molecular and atomic constituents of the systems of interest? Yes, if emergence means the failure of intertheoretic reduction. Is that an important fact? Yes, if our aim is to undermine dogmatic reductionist prejudices or to unsettle the presupposition that physics, generally, is a paradigmatically reductionist science. Otherwise, the significance of there not being a reduction of thermodynamics to statistical mechanics is not so clear. Does the lesson of the ergodic hypothesis generalize to other cases of interlevel relationships? I know of no reason to think that it does, though one should also not be surprised to encounter analogous situations elsewhere. Whether the relationship between particle physics and condensed matter physics is thus analogous will be discussed in a moment.

What about emergence in the sense of a failure of supervenience? Does the irreducibility of thermodynamics to statistical mechanics show that thermodynamic phenomena do not supervene on mechanical phenomena? Hard to say. The intuitions of many of us point in the opposite direction, to the conclusion that thermodynamic phenomena do supervene on mechanical ones. On the other hand, if we regard satisfaction of the ergodic hypothesis as a property of systems like a gas, perhaps we should regard that as emergent with respect to more narrowly mechanical properties of the molecular constituents of the gas. But recall my noting that the chief disadvantage of supervenience as a perspective on interlevel relations is precisely that, in the generic case, it is hard to know how to judge whether supervenience obtains.

One case where it is, however, not at all hard to make a judgment about supervenience, or rather its failure, is the case of quantum mechanical entanglement. Start with ordinary (non-relativistic) quantum mechanics. We represent the state of a system by means of a quantum mechanical state function, $\psi$, corresponding, technically, to a ray in some Hilbert space, which is a complex vector space. For many of us, a more comfortable way of representing a state function is as a Schrödinger wave function. The question now is how to represent the joint state, $\psi_{12}$, of a composite system consisting of two (or more) previously interacting systems. Had the two systems not interacted, then quantum mechanics would represent the joint state just as, in effect, all 'classical' theories do (think of Newtonian mechanics and Maxwellian electrodynamics), namely, as the product of two separate states:

$$\psi_{12} = \psi_1 \times \psi_2$$

If, however, systems 1 and 2 have interacted, then, in general, quantum mechanics describes their joint state in such a way as to make it not equivalent to any product of separate states:

$$\psi_{12} \neq \psi_1 \times \psi_2$$

Such joint states are said to be *entangled* joint states, and it is an easy bit of mathematics to show how these entangled states necessarily yield different predictions than do factorizable joint states, especially for certain types of correlations between two entangled systems, such as spin correlations.

The term 'entanglement' has been around since Erwin Schrödinger coined it in 1935 when, in the wake of the famous Einstein, Podolsky, and Rosen argument for the incompleteness of quantum mechanics,[11] he drafted the papers that for the first time presented in a systematic way what is now termed the quantum mechanical interaction formalism.[12] That some such departure from classical assumptions about the mutual independence of interacting systems would be part of the full story of the quantum realm was already clear as early as Einstein's first paper on the photon hypothesis in 1905.[13] That entanglement is, in fact, an essential part of the quantum mechanical formalism and the most important distinguishing feature of the quantum mechanical description of nature was clear by 1927, when Einstein ceased being a contributor to the further development of quantum mechanics after discovering that his own attempt at a

[11] Albert Einstein, Boris Podolsky, and Nathan Rosen, 'Can Quantum-mechanical Description of Physical Reality Be Considered Complete?' *Physical Review* 47 (1935): 777–780.

[12] Erwin Schrödinger, 'Die gegenwärtige Situation in der Quantenmechanik,' *Die Naturwissenschaften* 23 (1935): 807–812, 823–828, 844–849; *idem*, 'Discussion of Probability Relations Between Separated Systems,' *Proceedings of the Cambridge Philosophical Society* 31 (1935): 555–662; *idem*, 'Probability Relations Between Separated Systems,' *Proceedings of the Cambridge Philosophical Society* 32 (1936): 446–452.

[13] Albert Einstein, 'Über einen die Erzeugung und Verwandlung des Lichtes betreffenden heuristischen Gesichtspunkt,' *Annalen der Physik* 17 (1905): 132–148.

hidden variables interpretation of quantum mechanics also required the employment of non-factorizable joint states[14] and when Niels Bohr made entanglement the centerpiece of his complementarity interpretation of quantum mechanics (see Howard 1994, 2004). It was entanglement, which Einstein could not abide, about which Einstein and Bohr (Bohr 1935) were really arguing at the time of the Einstein, Podolsky, and Rosen paper (see Howard 1985).

The fundamental significance of quantum mechanical entanglement has long been understood and appreciated by philosophers working on the foundations of quantum mechanics.[15] Entanglement is a fact not only about non-relativistic quantum mechanics, but about any quantum theory. Entanglement is ineluctably and deeply woven into the fabric of quantum electrodynamics, quantum chromodynamics, and all of the good candidates for a quantum theory of gravity, including string theory and loop quantum gravity.[16] That its fundamental significance has not been so widely appreciated by the mainstream physics community is a historical puzzle that I won't attempt to solve right now, though it is a fact pregnant with implications for the current debate over reduction and emergence in physics. One is cheered by the fact that recent interest in topics such as quantum computing, quantum cryptography, and quantum information theory has finally put entanglement on the mainstream agenda, for now, in effect, we find physicists doing engineering with entanglement.[17]

[14] Don Howard, '"Nicht sein kann was nicht sein darf," or the Prehistory of EPR, 1909–1935: Einstein's Early Worries about the Quantum Mechanics of Composite Systems,' in *Sixty-two Years of Uncertainty: Historical, Philosophical, and Physical Inquiries into the Foundations of Quantum Mechanics*, ed. Arthur Miller (New York and London: Plenum, 1990), 61–111.

[15] See Bernard d'Espagnat, *Conceptual Foundations of Quantum Mechanics*, 2nd edn. (Reading, MA: W. A. Benjamin, 1976) on the role of entanglement in quantum mechanics.

[16] On quantum field theory, see Harvey Brown and Rom Harré, eds., *Philosophical Foundations of Quantum Field Theory* (Oxford: Clarendon Press, 1988); on quantum gravity, see Craig Callender and Nick Huggett, *Physics Meets Philosophy at the Planck Scale: Contemporary Theories in Quantum Gravity* (Cambridge: Cambridge University Press, 2001).

[17] For an accessible recent review, see Barbara Terhal, Michael Wolf, and Andrew Doherty, 'Quantum Entanglement: A Modern Perspective,' *Physics Today* 56/4 (April 2003): 46–52.

For our purposes, entanglement is important because it is the clearest example known to me from any domain of investigation of a failure of supervenience. How and why the properties of a pair of previously interacting and, therefore, entangled quantum systems fail to supervene on the properties of the two individual systems taken separately is perfectly well understood and today routinely demonstrated in the laboratory, as in experimental tests of Bell's theorem. Even with perfect, complete knowledge of the states of the separate systems, one cannot account for the correlations between those systems characteristic of entangled joint states.[18] That it must be so in the quantum domain is shown by a simple and straightforward mathematical demonstration. Here is holism of a very deep kind, and here is emergence in the sense of a failure of supervenience. By my lights, the quantum correlations characteristic of entangled joint states have a better claim to the status of emergent properties than do any of the other properties elsewhere in nature so far nominated for the prize.

Savor the significance of this point. It is at the most fundamental level of description in nature that the clearest instance of emergence is found. Emergence in the guise of entanglement is the most basic fact about the quantum realm. We will speak in a moment about the relationship between particle physics and condensed matter physics. Particle physics is quantum field theory. Entanglement is a fundamental fact about quantum field theory and, therefore, a fundamental fact about particle physics. It is, therefore, simply not true that holism, coherent states of matter, and long-range correlations occur first at the mesoscopic level of condensed matter physics. Nor is complexity the key. It's hard to imagine anything simpler than two charged particles like a proton and an electron interacting electromagnetically, which is to say, the hydrogen atom, or a positron-electron pair resulting from pair creation, or two correlated optical photons. Generating an analytic solution of the Schrödinger equation for the hydrogen atom is so simple that it has long been a homework problem for

[18] Richard Healey, *The Philosophy of Quantum Mechanics: An Interactive Interpretation* (Cambridge: Cambridge University Press, 1989) is a helpful source on many of these issues, as are many of the papers collected in James Cushing and Ernan McMullin, *Philosophical Consequences of Quantum Theory: Reflections on Bell's Theorem* (Notre Dame, IN: University of Notre Dame Press, 1989).

first-semester students of quantum mechanics. Nor is nonlinearity involved in any obvious way, unless one simply defines nonlinearity as a species of holism.[19] For the Schrödinger equation that governs the dynamics of these entangled quantum systems is a linear partial differential equation. It is linear Schrödinger evolution that carries the non-entangled pre-interaction joint state into the entangled post-interaction joint state. Far from nonlinearity engendering the kind of holism evinced as entanglement, some famous attempts to *evade* puzzling consequences of the quantum theory associated with entanglement take the form of proposed *nonlinear* variants of the Schrödinger equation. For example, in some 'solutions' to the measurement problem the addition of a nonlinear term to the Schrödinger equation serves to break the entanglement between instrument and object that is the basis of the measurement problem.[20] What, then, is going on in the relationship between particle physics and condensed matter physics?

## 6.4. Cooper Pairs and $^4$He: Evidence for Emergence in Condensed Matter Physics

With prophetic mien, prominent solid state physicists like Philip Anderson, Robert Laughlin, and David Pines have been heralding the appearance of a new paradigm of emergence in the physics of the mesorealm and arguing that, precisely because condensed matter physics has the conceptual tools for thinking about emergent properties, it is a way of doing physics more likely to hold the key to a future theory of everything than inherently reductionistic particle physics.[21] How sound is the prophecy?

[19] This is how I read Scott in Chapter 8 of this volume.

[20] Helpful surveys of this approach can be found in Gian Carlo Ghirardi, and Alberto Rimini, 'Old and New Ideas in the Theory of Quantum Measurement'; and in Philip Pearle, 'Toward a Relativistic Theory of Statevector Reduction,' in *Sixty-two Years of Uncertainty: Historical, Philosophical, and Physical Inquiries into the Foundations of Quantum Mechanics*, ed. Arthur Miller (New York and London: Plenum, 1990), 167–191, 193–214.

[21] P. W. Anderson, 'More is Different,' *Science* 177 (1972): 393–396; Robert B. Laughlin and David Pines, 'The Theory of Everything,' *Proceedings of the National Academy of Sciences* 97 (2000): 28–31; Robert B. Laughlin *et al.*, 'The Middle Way,' *Proceedings of the National Academy of Sciences* 97 (2000): 28–31.

If thermodynamics does not reduce to classical statistical mechanics, then we should not expect condensed matter physics to reduce to particle physics. If emergence is a failure of reduction, then condensed matter physics would be emergent with respect to particle physics. But I have argued that the question of intertheoretic reduction is not the right question. The right question is the question of supervenience, and what I now want to argue is that there is good reason to think that condensed matter physics supervenes on particle physics, once the latter is understood properly as assuming quantum entanglement as the most fundamental physical property of microphysical systems.

Consider three more or less incontestable facts and consequences thereof.[22]

*Fact 1 (incontestable)*: There is no unified, general theory of condensed matter physics. Some areas are in reasonably good shape, among them superfluidity and low-temperature superconductivity. Elsewhere the picture is spotty. In some important areas, most notably high-temperature superconductivity, few are so bold as to claim any adequate theoretical understanding.

*Consequences*: In the absence of a more unified, general theoretical framework for condensed matter physics and a better understanding of how and when effective Hamiltonian techniques work, it is hard to see how one can draw any general conclusion about emergence as a pervasive, essential, and distinctive feature of the mesorealm. Here the contrast with the microrealm as described by quantum mechanics and quantum field theory is striking, for it is precisely the fact that there we do have a unified, general theoretical framework that makes possible a strong conclusion about the pervasive, essential, and distinctive character of quantum entanglement.

*Fact 2 (incontestable)*: In many nonlinear systems, one encounters striking coherent structures not obviously explicable in microphysical terms. We have all seen long-lived eddies on the surface of a

[22] For a good historical introduction to the development of solid state and condensed matter physics see Lillian Hoddeson *et al.*, eds., *Out of the Crystal Maze: Chapters from the History of Solid-State Physics* (New York: Oxford University Press, 1992).

fast-moving and in other respects seemingly turbulent stream. The generic term for such stable structures is 'solitons.'

*Consequences*: Decidedly unclear. Don't be misled by the suffix '-on,' which suggests a likeness in kind to leptons, baryons, and other elementary particles, for solitons are features mainly of classical, not quantum nonlinear systems, though similar structures can emerge in a nonlinear quantum setting. That such stable structures are emergent (in either sense of the term) with respect to classical particle mechanics is not worthy of dispute. But so what? Classical particle mechanics is not true of the microworld; quantum mechanics is. Whether classical nonlinear phenomena supervene on microstructure as described quantum mechanically is, perhaps, not even a well-posed question, given that we have no good story to tell about the relationship between quantum and classical descriptions. Glib talk of the correspondence principle or of taking a classical limit by letting Planck's constant, $h$, go to zero just obscures the fact that, from a first-principles conceptual point of view quantum mechanics does not go over continuously to classical mechanics in the limit of small $h$. However small we let $h$ become, the difference between quantum and classical descriptions is still the difference between non-commutative and a commutative algebraic structure, which is a big difference.

Don't be misled either by the fact that stability of structure is a hallmark of the quantum realm, as in the existence of stable stationary atomic states. The kind of stability characteristic of the quantum realm, the stability of electron orbits and, therefore, the stability of chemical bonds and molecular structures, is a consequence of the fundamental linearity of quantum dynamics, deeply associated with entanglement. The hydrogen atom is a stable structure because the proton and the electron form an entangled pair.

*Fact 3 (more or less incontestable)*: In those areas of condensed matter physics where we do have a reasonably satisfactory theory—I have in mind mainly superfluidity and low-temperature superconductivity—there is also a reasonably clear connection to microphysical entanglement. This is especially so in the case of superfluidity, where the mechanism long thought to be in play, what is known as Bose-Einstein condensation, is a famous instance of entanglement, the atoms of a $^4$He superfluid, for example, being in an entangled joint state.[23] The connection to entanglement is only a little less straightforward in the case of low-temperature superconductivity, where sets of fermion pairs like the electrons designated Cooper pairs in the BCS (Bardeen-Cooper-Schrieffer) theory are described by coherent macroscopic wave functions, the bosonic fermion pairs in effect forming a condensate.

*Consequences*: The examples of superfluidity and superconductivity suggest that success in explaining phenomena in condensed matter physics will typically depend upon our making clear precisely the connection to quantum mechanical entanglement. That means that, far from such phenomena being emergent with respect to particle physics, they are proven to supervene on particle physics. The properties of entangled composite systems do not supervene on the properties of the individual components, but the molar properties of mesoscopic condensed matter systems, properties like superfluidity and superconductivity, do supervene on the most basic property of the quantum mechanical microrealm, namely, entanglement. The only emergence is, ironically, that found at the particle physics level itself.

The connection of superfluidity and superconductivity to Bose-Einstein condensation and the connection of the latter to entanglement is no secret. I'm not here asserting a radically heterodox point of view. How, then, could the idea that condensed matter physics is emergent with respect to particle physics have become so deeply entrenched in the community of condensed matter physicists? Frankly, I'm puzzled by this phenomenon. My best guess is that folks have been misled by the particle analogy. Intuitively, we regard particles as inherently mutually independent structures of a kind that cannot be entangled with one another. But Einstein recognized that, in general, photons would not behave as mutually independent particles, and de Broglie taught us to associate a similar wave-like

[23] For recent discussions of the physics of superfluidity and superconductivity, see Tony Guénault, *Basic Superfluids* (London: Taylor & Francis, 2003); and Lev Pitaevskii and Sandro Stringari, *Bose-Einstein Condensation* (Oxford: Clarendon Press, 2003). An interesting recent discussion of the place of entanglement in quantum statistics can be found in Michela Massimi, 'Exclusion Principle and the Identity of Indiscernibles: A Response to Margenau's Argument,' *British Journal for the Philosophy of Science* 52 (2001): 303–330.

aspect to massive particles like electrons. Thus, what we, today, call particle physics is, the name notwithstanding, not really a theory of particles. Were we all clear about the fact that particle physics takes entanglement as the most basic attribute of the systems it describes, then we would be unlikely to regard the phenomena of condensed matter physics as emergent with respect to particle physics.

## 6.5. Other Ways to Model Interlevel and Intertheory Relationships

In denying that condensed matter phenomena like superfluidity and superconductivity are emergent with respect to particle physics, I don't mean to deny that there are interesting and important questions about the relationship between the two theoretical realms. On the contrary, that relationship is, and should be even more so, a fertile area of investigation in the foundations of physics. Moreover, I also don't want to disparage the view that condensed matter physics enjoys a measure of explanatory autonomy vis-à-vis particle physics, this for two reasons. First, recall my noting above that supervenience does not imply reduction. Superfluidity can supervene on the entanglement fundamental to particle physics without condensed matter physics reducing to particle physics. Second, and more importantly, the manner in which condensed matter physics explains phenomena like superfluidity and superconductivity is thought by some to differ in crucial respects from the way explanation proceeds in particle physics. Explanatory autonomy of this kind is, to me, far more interesting than dubious claims about emergence.

Philosophers of physics have overcome the logical empiricist prejudice according to which there is one and only one right method for all scientific domains. While the provision of unified explanations of disparate phenomena is still widely prized as a worthy epistemic ideal,[24] the *methodological unity of science thesis* finds rather less support today, the dominant tendency now being to emphasize the

[24] Philip Kitcher, 'Explanatory Unification,' *Philosophy of Science* 48 (1981): 507–531. For a dissenting view see Helen Longino, *Science as Social Knowledge* (Princeton, NJ: Princeton University Press, 1990).

features distinctive of scientific practice in different domains.[25] In that spirit, a small but growing number of philosophers of physics are studying carefully condensed matter physics as well as its relationship to particle physics, trying hard to make clear methodological sense out of the explanatory strategies characteristic of the former.

Illuminating that relationship is one of the principal aims of Ang Wook Yi, who finds the semantic view of theories (wherein they are regarded as sets of models[26]) more helpful in thinking about condensed matter physics. Yi suggests that we distinguish 'global theories' from 'substantial theories.'[27] The former—global theories—model structure common to systems in a wide phenomenal domain; the latter—substantial theories—fill in the structural details for specific phenomenal domains. In sharing the structure encoded in the relevant global theory, different substantial theories would be related to one another by partial isomorphism. But that global theory can and typically will be a theory at a deeper level of description, in which case it is not obvious that emergence is the most felicitous way to characterize the relationship between the substantial theories and the associated global theory. Think of my story about entanglement in particle physics and take entanglement—a fact about the microdomain—to be the global structure incorporated in the substantial theories that condensed matter physics proposes for different mesoscopic phenomena like superfluidity and superconductivity.

## 6.6. In Conclusion: Extrapolations beyond Physics

I have argued that the case for emergence in condensed matter physics has not been made, partly because of confusion over what is being claimed, different meanings of the term 'emergence' not always being clearly distinguished, and partly because of the lack of

[25] Nancy Cartwright, *The Dappled World: A Study of the Boundaries of Science* (Cambridge: Cambridge University Press, 1999).

[26] See Bas van Fraassen for a now classic formulation of the semantic view. Bas C. van Fraassen, *The Scientific Image* (Oxford: Clarendon Press, 1980).

[27] Ang Wook Yi, 'How to Model Macroscopic Worlds: Towards the Philosophy of Condensed Matter Physics' (Ph.D. diss., London School of Economics/University of London, 2000).

any overall theory of condensed matter physics upon which to base general assertions about distinctive features of the mesorealm. But my main argument is that the physical structure that seems actually to do the explaining in condensed matter physics—in those cases where we have good explanations—is the very structure, entanglement, that is the defining trait of the quantum microworld described by particle physics, the microworld upon which condensed matter physics is said not to supervene.

What does any of this have to do with the science/theology dialogue? I think that it has important implications. I warned at the outset that, in the current enthusiasm for viewing emergence as the hallmark of interlevel relations, the will might be outrunning the understanding. Claims for emergence in condensed matter physics constitute one of the most important premises in the argument. But if the case has not been made here, where we have a modicum of theoretical control over the relevant phenomena, then one should be wary of extrapolations to levels of description—to organic life, to the mind, to the soul, perhaps—where our theoretical control of the phenomena is orders of magnitude less secure.

Patience, modesty, and humility are intellectual virtues as well as moral ones. Let us be patient, modest, and humble. Don't let wishful thinking and vague analogies take the place of clear understanding. Let the science lead us where it will.

# 7

## True to Life?: Biological Models of Origin and Evolution

*Martinez J. Hewlett*

### 7.1. Introduction

Erwin Schrödinger, after his participation in the founding of quantum mechanics, turned his attention towards the life sciences. In *What is Life?*, his self-described 'little book,' he wondered about what the physical character of the gene must be.[1] It struck him that the nature of heredity is such that whatever the chemical material of the gene might be, it resists the thermodynamic imperative of the second law towards dissolution. In his book, he focuses on well known examples of strong inheritance, such as the famous Hapsburg lip. His speculation leads him to suggest that there may, indeed, be new physical principles or even laws waiting to be discovered.

His musing had a profound influence on at least one young physicist. Max Delbrück took up this quest and became a guiding light in the new discipline that would bring the quantitative approaches of physics and genetics to bear on discovering the chemical nature of the gene. The birth of molecular biology and Schrödinger's influence on Delbrück are recounted in a delightful and recently updated *Festschrift* volume called *Phage and the Origins of Molecular Biology*.[2]

[1] Erwin Schrödinger, *What is Life?: The Physical Aspects of the Living Cell* (Cambridge: Cambridge University Press, 1944).

[2] John Cairns, Gunther Stent, and James Watson, *Phage and the Origins of Molecular Biology* (Plainview, NY: Cold Spring Harbor Laboratory Press, 1992).

## 7.2. What is Life?

The question that occurred to Schrödinger and that motivated Delbrück is perhaps too simple for our purposes. Rather, like the proverbial newspaper reporter, we need to ask for the complete story—the who, what, when, where, why, and how of life. To do this, we need to acknowledge that not all of these questions are properly within the domain of science; some of these may be meta-questions for biology. I offer Table 7.1 as an explanation of what I mean.

Table 7.1.

| Question | Domain |
|---|---|
| Where is life? | ecology |
| When is life? | evolutionary biology |
| How is life? | biochemistry, molecular biology, cellular biology, developmental biology |
| Who is life? | organismal biology, classification |
| What is life? | ontology/philosophy; a meta-question for biology |
| Why is life? | theology; a meta-question for biology |

The first four questions in this table lead to investigation space that is clearly within the realm of the science of biology. The last two—the what and why questions—lead, instead, to philosophical and perhaps theological investigation space.

The standard hierarchy of biology can be viewed as a road map to the answers for the first four of these questions. The hierarchy begins with biological molecules as chemicals (biochemistry), moves to biological molecules as information (molecular biology), and progresses on through cellular biology, organismal biology, and finally arrives at the study of populations of organisms (ecology). The overarching paradigms of Darwinian evolution and Mendelian genetics are considered to operate at each of these levels, thus producing the neo-Darwinian synthesis.

One way of interpreting this hierarchy is to assume reductionism, both methodological and causal, at all levels. My purpose is to examine this assumption in light of the types of emergence discussed in this volume.

I will ask two general questions:

- Is life an emergent property of non-life?
- Is complexification an emergent property of life?

I will present the current biological models that attempt to address these two questions. I will then discuss future directions in biology, with a focus on models that may challenge the current underlying philosophy of the discipline.

## 7.3. Life from Non-Life

### 7.3.1. *Prebiotic Simulation Models*

In 1953, Stanley Miller, then a student in the laboratory of Harold Urey, reported on an attempt to simulate the pre-biotic conditions on earth and to look for the synthesis of organic molecules under these conditions. He placed methane, ammonia, hydrogen, and water in a flask and repeatedly introduced an electric spark to simulate lightning in the atmosphere of the primitive earth. He was able to recover organic molecules from this experiment, including some forms of amino acids, the subunits of proteins.[3] These results were based upon the assumption that the primitive earth would have a reducing atmosphere in which organic molecules could form in the presence of electric discharge.

In no way, however, can it be said that these molecules were living, yet this experimental approach did much to support the notion that at least the chemicals necessary for life might have been present on the pre-biotic earth. These kinds of experimental approaches continue at present, with an emphasis in exobiology (the study of the possibilities for life in places other than this planet).

The Miller experiment and those deriving from it opened the debate about what the earliest 'living' molecule might be. Although the precursors for proteins are found under these conditions, no one could imagine how proteins would eventually become self-replicating. Proposals were made that would allow living systems to form on the surfaces of silicates, with this surface acting as a catalyst for the reproduction of pre-biotic proteins. However, it was a discovery related to the processing of information in eukaryotic cells that changed the model.

[3] Stanley Miller, 'A Production of Amino Acids Under Possible Primitive Earth Conditions,' *Science* 117 (1953): 528–529.

### 7.3.2. *Self-Replicating Models*

Tom Chech and Sid Altman were among a group of four Nobel Prize recipients (two in medicine and two in chemistry) for work on the way in which informational RNA molecules in the nucleus of eukaryotic cells are processed before being delivered to the cytoplasm for translation into proteins. Part of the sequence of steps includes the removal of certain segments of the RNA through a set of reactions called RNA splicing. Chech observed that some of these reactions are catalyzed by the RNA itself in an event called self-splicing.[4] This discovery led to the notion that RNA molecules can catalyze the kind of reactions that take place within cells. The ultimate function, however, is self-replication. Such self-replicating RNA molecules have indeed been described several times.[5]

The idea that RNA can self-replicate leads to the notion that some kind of RNA might be a good model for a pre-biotic molecule. If such molecules could form under conditions present early in the history of the planet, it could be that replication becomes a selectable property that would allow further propagation of the structures. This has been termed the RNA world and, for a time, dominated the thinking about origin of life issues and exobiology. More recently, however, the popularity of RNA for these models has waned. No one has been able to set up pre-biotic conditions in which RNA can be isolated. Since it is an inherently unstable molecule, the likelihood of its occurrence in such systems is quite slim. At this time, however, proteins have come back to the fore, especially as discussed by Manfred Eigen in his influential book, *Steps Toward Life*.[6] The model currently favored is one in which both protein and nucleic acid formed during the prebiotic stage preceding the formation of cellular life.

### 7.3.3. *Spontaneous Order Models*

In deciding how such molecules formed, the problem seems to be one related to the thermodynamics of the situation. How can it be

[4] Thomas Chech, 'A Model for the RNA-Catalyzed Replication of RNA,' *Proceedings of the National Academy of Sciences, USA* 83 (1986): 4360–4363.

[5] Wendy Johnston *et al.*, 'RNA-Catalyzed RNA Polymerization: Accurate and General RNA-Templated Primer Extension,' *Science* 292 (2001): 1319–1325.

[6] Manfred Eigen, *Steps Towards Life: A Perspective on Evolution* (New York: Oxford University Press, 1992).

that complex structures that can self-replicate would occur in the face of the loss of entropy that this would entail? Stuart Kauffman of the Santa Fe Institute has tackled this problem and has described his conclusions in *Origins of Order*.[7] Kauffman proposes that pre-biotic molecules are such that they will spontaneously increase in complexity, without any violation of thermodynamic principles. Remember that the second law applies to closed systems, and molecules in pre-biotic environments are, in fact, open and exchanging energy with the environment. Kauffman's view is that order emerges from the very properties of the molecules.

Kauffman and other complexity specialists such as Christopher Langton argue that the most interesting transitions occur in systems that are at the edge of chaos. It is under these conditions that life might have developed.

## 7.4. The Complexity of Life

While biologists struggle with modeling how life arose initially on this planet the fact is that life does exist, in all of its glorious complexity. My second question bears on this observation: is complexification an emergent property of life?

For biology, the nineteenth century saw two major achievements: Darwinian evolution and Mendelian genetics. Although initially no one realized the relationship between them, they would be brought together in the twentieth century in the neo-Darwinian synthesis.

### 7.4.1. *The Darwinian Model*

Darwin's great achievement was the recognition that the complexity apparent in the living world could be explained by descent with modification from a common ancestor. He argued that the driving force of this process would be natural selection, favoring those organisms best able to reproduce in a given environment.

The late evolutionary biologist John Maynard Smith defined the Darwinian model as follows:

[7] Stuart Kauffman, *The Origins of Order: Self Organization and Selection in Evolution* (New York: Oxford University Press, 1993).

1. Population of entities (units of evolution) exist with three properties:

   multiplication (one can give rise to two),
   variation (not all entities are alike), and
   heredity (like usually begets like during multiplication).

2. Differences between entities will influence the likelihood of surviving and reproducing. That is, the differences will influence their fitness.
3. The population will change over time (evolve) in the presence of selective forces.
4. The entities will come to possess traits that increase their fitness.[8]

These four principles are often summarized as 'descent with modification.' Notice that this model stresses the reproductive fitness of the variants, such that they are more likely to pass on their characteristics to offspring.

### 7.4.2. *The Modern Synthesis Model*

By the middle of the twentieth century, Julian Huxley could boast of the modern synthesis, also called the neo-Darwinian synthesis.[9] In this welding together of the sciences of Darwinian evolution, Mendelian genetics, and biochemistry, scientists found a series of tools that have resulted in the hierarchical paradigm I discussed earlier. The features of this synthesis include:

- Genes: information in the form of the linear array of bases that make up the DNA molecules of chromosomes.
- The traits of an organism (phenotype): direct expression of the information found in the genes (genotype).
- Variations: result of subtle differences in this information (changes in base pairs).

[8] John Maynard Smith, *Symbiosis as a Source of Evolutionary Innovation: Speciation and Morphogenesis*, eds., L. Margulis and R. Fester (Cambridge, MA: MIT Press, 1991), 26–39.

[9] Julian Huxley, *Evolution, The Modern Synthesis* (London: Allen and Unwin, 1942).

- Changes in genes: mutational events that occur in a 'random' way.
- A population of entities: will have variations in traits that are the result of mutational events (genetic drift).

The force of natural selection operates on this pool of genetic variants, allowing those with greater reproductive fitness to be represented in succeeding generations. The model emphasizes strict gradualism whereby mutations occur in small steps, usually one base pair at a time, over long geological periods.

The force of this synthesis is such that it is difficult to think of any biological problem with reference to its principles. It is important to note, however, that the synthesis retains much of the nineteenth-century philosophical underpinnings of both evolution and genetics. As a result, modern biology tends toward reductionism and determinism, conditioned as it is by the Newtonian framework that was assumed by both Darwin and Mendel.

This orientation can be seen in the work of many contemporary evolutionary biologists, including Richard Dawkins and Edward O. Wilson. In *The Blind Watchmaker*, Dawkins argues for gradual change and natural selection leading to all of the existing forms of life, focusing on the central role of DNA in the scheme. He writes that 'living organisms exist for the benefit of DNA rather than the other way around.'[10] In a similar vein, Wilson proposes that genes play a pre-eminent role, even in the development of human behavior and culture. He argues that the environment is important in the selection of genetic variants and in their expression. In the end, however, it is the gene and its information that determines who we are and what we do as humans.[11]

Both of these positions assume a reductionistic and deterministic foundation. Even though allowances are made for some higher-level patternings, these are really epiphenomena of the structures that make up the whole. Dawkins, in fact, defends this assumption by renaming it 'hierarchical reductionism.' He says that 'reductionism, in this sense, is just another name for an honest desire to understand

[10] Richard Dawkins, *The Blind Watchmaker* (New York: Norton, 1986), 126.

[11] Edward O. Wilson, *Consilience: The Unity of Knowledge* (New York: Vintage Books, 1998).

how things work.'[12] His statement does not reveal any understanding of the philosophical issues at stake, conflating as it does the methodological approach of science with the epistemic and ontological conclusions he then makes.

Is the Darwinian model, then, irretrievably embedded in this philosophical system? Some researchers in modern biology, such as those working on the Human Genome Project, would argue yes. On their website, one can find a slide that touts 'from DNA to humans' as a kind of working model.[13] On the other hand, no one working in biology disputes the complex nature of life and the levels of organization observed.

The late Stephen Jay Gould was a most prolific writer concerning issues in evolutionary biology. Gould made two contributions that continue to be controversial among the strict gradualists such as Dawkins and his interpreters. These are punctuated equilibrium and the notion of spandrels.

### 7.4.3. *The Punctuated Equilibrium Model*

As Gould states in his final majestic work on evolution, he and Niles Eldredge were led to their proposal of punctuated equilibrium by the 'species problem.'[14] Using the gradualistic approach (also called anagenesis), one assumes a continuum of change over deep time. The problem arises when one needs to assign boundaries between populations. Is this done arbitrarily or are there some defining features that allow such an assignment? As an alternative explanation, Gould and Eldredge proposed the concept of punctuated equilibrium.[15] In Gould's most recent description of this theory he writes: 'Eldredge and I argued that the vast majority of species originate by splitting, and that the standard tempo of speciation, when expressed

[12] Dawkins, *Blind Watchmaker*, 13.

[13] Found on the Department of Energy Human Genome Project Site at http://www.ornl.gov/TechResources/Human_Genome/graphics/slides/images1.html.

[14] Stephen Jay Gould, *The Structure of Evolutionary Theory* (Cambridge, MA: Belknap, Harvard University Press, 2002), 775.

[15] Stephen Jay Gould and Niles Eldredge, 'Punctuated Equilibria: An Alternative to Phyletic Gradualism,' in T. J. M. Schopf, ed., *Models in Paleobiology* (San Francisco: Freeman, Cooper, and Co., 1972), 82–115.

in geological time, features origin in a geological moment followed by long persistence in stasis.'[16]

The critics of this model are champions of gradualism as the principal, if not only, mechanism by which evolution takes place. Dawkins, for instance, caricatures punctuated equilibrium as anti-Darwinian. He argues that, since the suddenness in the theory is really over geological time scales, there is no difference here. However, since Dawkins is a strict gradualist, he also argues against the rapidity of change that would be required for punctuated equilibrium to be true.

Nevertheless, Gould and Eldredge have had a significant effect on evolutionary thinking, if in no other way than to decrease the reliance on the gradualist approach. Does punctuated equilibrium alter the reductionistic setting in which modern biology finds itself? In general, the answer would be no, since there is no change here in the emphasis on the gene as the determiner of traits. However, since Gould and Eldredge rely on speciation rather than gradualism, there is a consideration of natural selection acting on the organism level. Their theory proposes that animals can be reproductively isolated by environment and geography, thus creating a population that can rapidly evolve into a different species. This model leans away from the single steps of the gradualist toward a macroevolutionary framework. Dawkins discounts the importance of macromutational events, but I will return to these below.

### 7.4.4. *The Exaptation Model*

If punctuated equilibrium was not enough to set his critics on edge, certainly Gould's next proposal more than drove them over that edge. In 1979, Gould and Richard Lewontin published the paper that introduced the concept of spandrels into the literature of evolutionary biology.[17] Gould and Lewontin attack the standard adaptationist approach of evolutionary biology as 'panglossian,' or what Gould had come to call 'just so stories.' They used the architectural model of spandrels to explain their reasoning. A spandrel (or pendentive)

[16] Gould, *Structure*, 776.

[17] Stephen Jay Gould and Richard Lewontin, 'The Spandrels of San Marco and the Panglossian Paradigm: A Critique of the Adaptationist Programme,' *Proceedings of the Royal Society of London, Series B* 205 (1979): 581–598.

is a somewhat triangular space that is found between two adjacent arches in medieval cathedrals. It is actually best thought of as a space left over. Gould and Lewontin refer to the wonderful mosaics of the four evangelists adorning the spandrels in the great cathedral of St. Mark in Venice, Italy. They point out that you could assume the spandrels were put there to contain the art, when, in fact, they are really necessary features required by the use of arches to support a dome. In the same way, they argue, certain features of organisms are dictated by the ways in which living systems develop in accordance with physical principles. Such structures are then used by evolution because of the advantages they confer. As they wrote in their paper:

> organisms must be analyzed as integrated wholes, with *baupläne* so constrained by phyletic heritage, pathways of development, and general architecture that the constraints themselves become more interesting and more important in delimiting pathways of change than the selective force that may mediate change when it occurs.[18]

The reference here to organisms as 'integrated wholes' is in clear distinction to the reductionistic approach favored by the gradualists. Gould by no means intended spandrels to be seen in the same light as those structures touted by intelligent design theorists as 'irreducibly complex.' Rather, Gould envisioned spandrels as nonadaptive structures that set constraints on the ultimate structure of an organism and that can be co-opted for use in an evolutionary pathway. He coined the term 'exaptation' for this kind of evolutionary use in order to distinguish it from adaptation.[19]

Do these biological models imply that complexification is an emergent property of life? Certainly a strictly neo-Darwinian model would point strongly in the direction of causal reductionism. As we move toward the models presented by Gould, however, biology appears to be more amenable to thinking holistically and allowing space for emergence. The idea of nonadaptive structures such as spandrels setting boundary conditions for living systems is at least tending towards emergentism.

[18] Ibid., 581. [19] Gould, *Structure*, 43.

## 7.5. New Models and New Directions

How are these models changing? More importantly, what kinds of evidence are accumulating to press for change? We can ask these questions in the Kuhnian sense, in that explanatory problems arising from the current paradigms in the field are what drive the need for such re-examination and change, if warranted.

### 7.5.1. *Sources of Mutational Change*

First, it is important to realize that the gradualist ideal of single-step changes, championed by Dawkins and others, is only one kind of genomic change that is observed. Lynn Margulis has summarized the array of hereditary alterations that are observed.[20] Table 7.2 is adapted from her paper.

**Table 7.2.**

| Mutations ('micro' hereditary alterations) | Karyotypic Alterations ('macro' hereditary alterations) | Genomic Acquisitions ('mega' hereditary alterations) |
|---|---|---|
| Base pair changes (AT ' GC) | Polyploidy (2N = 4N) | Transformations (DNA uptake) |
| Deletions/Insertions (ACTG ' ATG) | Polyteny (2N = 2N) | Transduction (phage, virus, or replicon acquisition) |
| Duplications (ATCG ' ATCGATCGT | Polyenergids (2N = ×N) | Bacterial conjugation |
| Transpositions (CGCCCATG ' GCGATCCG | Robertsonian fusions (2N = 2N − 1) | Meiotic sex |
| | Karyotypic fissions (2N = 2N') | Symbioses |

The first column includes those kinds of genetic changes that would be called gradual, single-step alterations, occurring in one or only a few bases at a time. These micro-hereditary alterations are what Dawkins believes to be the principal source of variation upon which evolution acts. However, the macro- and mega-hereditary changes referred to in the table play significant roles in observed variation. In

[20] Lynn Margulis, in Margulis and Fester, *Symbiosis*, 1–14.

some cases, such changes appear to have led to alterations that have had quite sudden consequences.

### 7.5.2. *The Origin of Eukaryotes*

Margulis proposed that eukaryotic cells, that is, cells with internal, membrane-bound organelles, originated by an endosymbiotic relationship between two or more ancestral cells. While precursors of this model can be found in earlier literature, it was Margulis who most clearly articulated and defended this endosymbiotic model. Her model is an explanation for the existence of the energy-generating organelles, the mitochondria and the chloroplasts.

Recently, Langton and I have been concerned with this model and its predictive value for the existence of other eukaryotic intracellular structures.[21] In particular, we considered the endomembrane system of the cell, involved in the synthesis and movement of proteins throughout the cell. We modeled the origin of this membrane system as a network of interacting cells, living in a parallel symbiotic relationship and sharing functions by export through the extracellular space surrounding the network. Over time, the relationship of this network or colony of cooperating cells becomes a selective advantage. Note that, in this case, the selection is for an emergent property of the system—the relationship in the network itself. In our model, the cells involved in the cooperative colony come to interact even more closely until, finally, the colony fuses into one cell. In this model, what was the extracellular space between the cells becomes the intracellular endomembrane system involved in moving proteins from one location to another. Our model explains the observation that the topology of the endomembrane system in the eukaryotic cell is such that the interior (lumen) of the endomembrane system is topologically equivalent to the outside of the cell. Thus, using a strictly Darwinian notion of selection for reproductive fitness, our network evolves into the eukaryote, able to expand into new ecological niches because of its increased metabolic potential. Selection here has been for an emergent property of the colony, the interactions between the cells.

[21] Christopher Langton and Martinez Hewlett, manuscript in preparation.

### 7.5.3. *Scale-Free Networks and Biological Systems*

Of course, networks have been known for a long time. However, it is only recently that the nature of the interactions in a network has begun to be explored. Several recent texts of both scientific and general interests have been published about what is being called the new science of scale-free or small world networks.[22] Most interestingly, the kind of network termed 'scale-free' has properties such that the network is more than the sum of its parts. When the number of connections between nodes in a regular network is plotted versus the number of nodes, a typical distribution over a particular range is obtained. However, for scale-free networks such is not the case. In these networks a few nodes are highly connected, leading to a plot that, on a log-log scale, is linear for a large range of values. These networks are also called 'small world' in recognition that one of the first observations of this kind of behavior was in human social connections, published in 1967 by Stanley Milgram.[23]

Scale-free networks exist at all levels of organization, from the network of interacting proteins within the cells of an organism,[24] to the colonial organization of social insects,[25] to the World Wide Web,[26] and finally, to the network of human sexual interactions.[27]

This latter example is quite interesting. In epidemiology, viral infections are characterized by a certain threshold of infection. Below the threshold, it is unlikely that the virus will spread in the population, while above the threshold the virus will spread. Influenza viruses are typical of this pattern. In contrast, human immunodeficiency virus (HIV), the causative agent of AIDS, spreads in the human sexual contact network. In this kind of scale-free network, HIV has a zero threshold for spread. This means that once the virus enters the

[22] Albert-Lászlό Barabási, *Linked: The New Science of Networks* (Cambridge, MA: Perseus, 2002); and Duncan Watts, *Six Degrees: The Science of a Connected Age* (New York: Norton, 2003).

[23] Stanley Milgram, 'The Small World Problem,' *Psychology Today* 2 (1967): 60–67.

[24] L. Giot *et al.*, 'A Protein Interaction Map of Drosophila Melanogaster,' *Science* 302 (2003): 1727–1736.

[25] Jennifer Fewell, 'Social Insect Networks,' *Science* 301 (2003): 1867–1870.

[26] Réka Albert, Hawoong Jeong, and Albert-Lászlό Barabási, 'Diameter of the World Wide Web,' *Nature* 401 (1999): 130–131.

[27] Fredrik Liljeros *et al.*, 'The Web of Human Sexual Contacts,' *Nature* 411 (2001): 907–908.

network the chances of spread are 100%. This is a startling result that has major implications for the control of AIDS in populations.

The scale-free properties of these networks are emergent. That is, it is not the nodes that produce these properties, but the relationships and connections between the nodes. Therefore, the network as a whole cannot be understood by simply summing the parts. Such networks are not methodologically or philosophically resolvable by reductionism. In fact, reductive approaches lead in the opposite direction. Albert-László Barabási, in his recent text on this topic, humorously points out the limits of the reductive approach in this case:

> Have you ever seen a child take apart a favorite toy? Did you then see the little one cry after realizing he could not put all the pieces back together? Well, here is a secret that never makes the headlines: We have taken apart the universe and have no idea how to put it back together.[28]

Barabási continues with a critique of reductionism and a defense of emergence. To place this in perspective, it must be said that Barabási is a physicist. It would seem, from these examples, that the science of scale-free networks presents a potential shift in the philosophical perspective of the sciences, including biology. This shift is being driven, I maintain, by the failure of the explanatory power of the reductionist approach to complex systems.

## 7.6. Conclusion

Science is about building models of reality. These models are based upon paradigms that underlie the discipline at any given time. The strength of these models depends upon their explanatory and predictive power. Here, I mean predictive in the sense of generating additional experimental approaches.

The models that are used in biology rely both on the observations being made and on the philosophical setting in which these observations take place. Until recently, models based on a strictly reductive methodological and epistemological approach have been extremely

[28] Barabási, *Linked*, 6.

productive. However, as more and more complex systems are analyzed, it has become clear that the value of such models is limited. As the concept of emergent phenomena gains greater acceptance, models that incorporate higher-order levels of organization will become increasingly important.

What does this mean for the science and theology discussion? On the one hand, as science begins to be less oriented philosophically toward the strictly reductive view, a pathway can be seen that might allow intellectual acceptance of such non-material issues as consciousness and spirituality. On the other hand, there remains the divide between physicality and spirituality that may continue to limit investigative exchange between these two fields.

# 8

## Nonlinear Science and the Cognitive Hierarchy

*Alwyn Scott*

### 8.1. Introduction

The activity of a human brain is organized into many levels, from the dynamics of individual membrane proteins, through the switching of isolated patches of membrane and collisions among propagating nerve impulses, to the complex interactions of the neocortex with human culture.[1] A key concept in understanding these dynamics is that of a *cell assembly* of neurons, which was defined by Donald Hebb as follows:[2]

Any frequently repeated, particular stimulation will lead to the slow development of a 'cell-assembly,' a diffuse structure comprising cells . . . capable of acting briefly as a closed system, delivering facilitation to other such systems and usually having a specific motor facilitation. A series of such events constitutes a 'phase sequence'—the thought process.[3]

[1] A. C. Scott, *Neuroscience: A Mathematical Primer* (New York: Springer-Verlag, 2002).

[2] D. O. Hebb, *Organization of Behavior: A Neuropsychological Theory* (New York: John Wiley and Sons, 1949); *idem*, 'The Structure of Thought,' in P. W. Jusczyk and R. M. Klein, eds., *The Nature of Thought* (Hillsdale, NJ: Lawrence Erlbaum Associates, 1980), 19–35; and *idem*, *Essay on Mind* (Hillsdale, NJ: Lawrence Erlbaum Associates, 1980).

[3] P. B. Andersen, C. Emmeche, N. O. Finnemann, and P. V. Christiansen, eds., *Downward Causation: Minds, Bodies and Matter* (Aarhus, Denmark: Aarhus University Press, 2000), 23.

This concept leads naturally to a *cognitive hierarchy* of distinct dynamic levels in which each level of description is built upon—or emerges from—those below.[4] Although much of the brain's hierarchical structure is a matter of observation, implications of Hebb's perspective for the studies of the mind are not fully appreciated. This paper describes the cognitive hierarchy, paying particular attention to two aspects: the role played by nonlinear phenomena and the claims of reductive materialism.

## 8.2. What Are Nonlinear Phenomena?

The short answer to this question is that nonlinear phenomena are those for which the whole differs from the sum of its parts. Beyond this simple slogan, one can point out an impressive array of dynamic effects currently studied under the aegis of nonlinear science, including (but not limited to) the following: *emergent structures* (cell assemblies, tornadoes, tsunamis, lynch mobs, optical solitons, black holes, schools of fish, cities, Jupiter's Great Red Spot, nerve impulses), *filamentation* (rivers, bolts of lightning, woodland paths, optical filaments), *chaotic effects* (low-order chaos, the 'butterfly effect,' strange attractors, Julia sets, turbulence), *threshold phenomena* (an electric wall switch, the trigger of a pistol, flip-flop circuits, tipping points), *spontaneous pattern formation* (fairy rings of mushrooms, the Gulf Stream, fibrillation of heart muscle, ecological domains), *harmonic generation* (digital tuning of radio receivers, conversion of laser light from red to blue), *synchronization* (Huygens's pendulum clocks, electric power generators connected to a common grid, circadian rhythms, hybernation of bears, flashing of Indonesian fireflies), and *shock waves* (sonic booms from jet airplanes, the sound of a cannon, bow waves of a boat, sudden pile-ups in smoothly-flowing automobile traffic), to name but a few. All of these phenomena and more can play roles in the nonlinear dynamics of hierarchical systems.[5]

A yet deeper answer points out that nonlinearity implies a statement about the nature of causality. This task is undertaken below after consideration of the hierarchical nature of living organisms.

[4] Scott, *Neuroscience.*

[5] A. C. Scott, ed., *Encyclopedia of Nonlinear Science* (New York: Routledge, 2004).

## 8.3. The Biological Hierarchy

Before taking up the cognitive hierarchy, let us fix ideas by considering a related structure, the *biological hierarchy:*

**Biosphere**
**Species**
**Organisms**
**Organs**
**Cells**
**Processes of replication**
**Genetic transcription**
**Biochemical cycles**
**Biomolecules**
**Molecules**

Several comments with respect to this hierarchy are in order. First, it is only the general nature of this hierarchy that is of interest to us here, not the details. We might include fewer or more levels in the diagram or account for branchings into (say) flora and fauna or various phyla. Although such refinements may be useful in particular discussions, the present aim is to become acquainted with the general nature of a nonlinear dynamic hierarchy, so a relatively simple diagram is appropriate. Second, the nonlinear dynamics at each level of description generate *emergent structures*, and nonlinear interactions among these structures provide a basis for the dynamics at the next higher level.[6] Third, the emergence of new dynamic entities stems from the presence of closed causal loops, in which positive feedback leads to exponential growth that is ultimately limited by nonlinear effects.

Finally, it must be noted that philosophers disagree about the ontological nature of emergent levels. Do they differ merely by their labels, convenient for academic organization, or are these levels qualitatively different aspects of reality? In attempting to answer this question, it is important to understand how the upper levels are related to lower levels, which brings us face to face with the doctrine of *reductionism.*

[6] Scott, *Nonlinear Science.*

## 8.4. Biological Reductionism

Since the days of Galileo and Newton, the reductive program has been surprisingly successful in prising out explanations for the behavior of the natural world. Thus, this perspective is now widely accepted by the scientific community as the fundamental way to pose and answer questions. Basically, the reductive approach to understanding proceeds in three steps.

- *Analysis.* Assuming some higher-level phenomenon is to be explained, separate the dynamics of that phenomenon into components, the behaviors of which are individually investigated.
- *Theoretical formulation.* Guided by empirical studies and imagination, develop a theoretical formulation of how the components interact.
- *Synthesis.* In the context of this theory, derive the higher-level phenomenon.

Among the many aspects of nature that have fallen to this approach, one can mention planetary motion (based on the concepts of mass and gravity and on Newton's laws of motion), electromagnetic radiation (based on the concepts of electric charge, electric fields, and magnetic fields related through Maxwell's electromagnetic equations), atomic and molecular structures (based on the concepts of mass, electric charge, Planck's constant, and Schrödinger's equation for the dynamics of quantum probability amplitudes), and nerve impulse propagation (based on the concepts of voltage, membrane permeability, ionic current, and the Hodgkin–Huxley equations for current flow through a voltage-sensitive membrane).

Generalizing from such specific examples, some believe that all natural phenomena can be understood in this way.[7] Others believe there exist natural phenomena that cannot be completely described in terms of lower-level entities—life and the human mind being outstanding examples. In its more extreme form, this latter position is called *substance dualism*: the view that important aspects of the biological realm do not have a physical basis. A less salient position is

[7] S. Weinberg, *Dreams of a Final Theory: The Search for the Fundamental Laws of Nature* (New York: Pantheon Books, 1992).

*property dualism*, which asserts that there are aspects of biology that cannot be explained in terms of atomic or molecular dynamics. To a statement of belief there is not a scientific response, but if we can agree on the physical basis of life, the scope of the discussion narrows. Let us agree, therefore, that all biological phenomena *supervene* on the physical in the following sense. If the constituent matter is removed, the phenomenon in question disappears, or as philosopher Jaegwon Kim puts it in the context of cognitive phenomena: 'Any two things that are exact physical duplicates are exact psychological duplicates as well.'[8] This position is called *physicalism*, and among biologists it is now widely accepted for the phenomenon of life. In other words, there is no Bergsonian 'life force' or *élan vital* that exists independent of the atoms comprising a living organism. Most neuroscientists also believe that a person's mind (or consciousness) would not survive removal of the molecules of his or her brain. With this assumption, two questions arise: (1) Does reductionism follow from physicalism? and (2) Does physicalism allow property dualism?

Over the past two decades, these questions have been considered by Kim, who reluctantly concludes that physicalism does indeed imply reductionism and sits uneasily with property dualism.[9] Let us review his argument with reference to Figure 8.1.

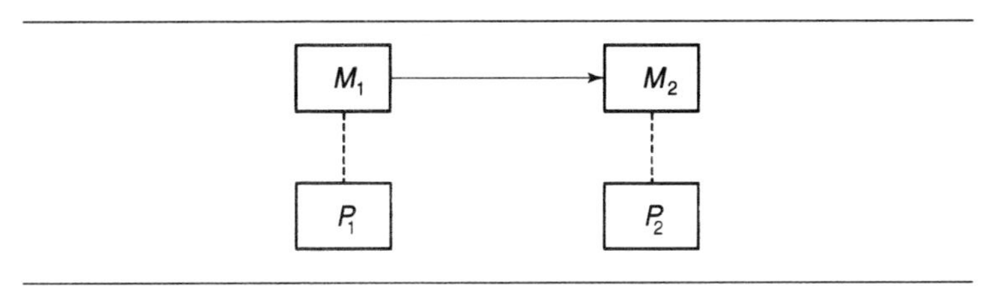


**Figure 8.1.** The causal interaction of higher-level phenomena ($M_1$ and $M_2$) that supervene on lower-level properties ($P_1$ and $P_2$).

This figure represents higher-level mental phenomena ($M_1$ and $M_2$) that supervene on lower-level physical properties ($P_1$ and $P_2$), where supervenience is indicated by the vertical dashed line. In other words, if the properties $P_1$ are removed, then the phenomenon $M_1$ will disappear, with a similar relationship between $P_2$ and $M_2$.

[8] Jaegwon Kim, *Mind in a Physical World* (Cambridge, MA: MIT Press, 2000), 12.
[9] Ibid.

Now suppose that there is observed to be a *causal relationship* between $M_1$ and $M_2$ (indicated by the horizontal arrow in Figure 8.1), under which the initial upper-level observation of $M_1$ always leads to a corresponding upper-level observation of $M_2$. Because under the assumption of physicalism $P_1$ ($P_2$) must be present to provide a basis for $M_1$ ($M_2$), we could as well say that $P_1$ causes $P_2$, which is a formulation of the upper-level causality in terms of the lower-level properties. In other words, one could interpret the relation between phenomena $M_1$ and $M_2$ in terms of a corresponding relation between $P_1$ and $P_2$, thereby undercutting a position of property dualism.

Finally, there is a practical argument. Even if reductionism turns out not to hold in all aspects of biological or mental organization, it is still a prudent strategy for the majority of biologists and cognitive scientists to take as a working hypothesis. Why? Often the riddles of one generation become standard knowledge of the next; thus, the dualist (substance or property) is ever in danger of giving up too soon on the search for reductive formulations.

## 8.5. Objections to Reductionism

Thus, reductionism based on physicalism is a serious philosophical position meriting careful response. Those who object to the reductionist position must offer substantial objections. Let us consider some.

### 8.5.1. *Constructionism versus Reductionism*

Condensed-matter physicists—who study aggregates of atoms and molecules—tend to question reductionist claims. Thus, Philip Anderson asserts:

> [T]he reductionist hypothesis does not by any means imply a 'constructionist' one: The ability to reduce everything to simple fundamental laws does not imply the ability to start from those laws and reconstruct the universe. In fact the more the elementary-particle physicists tell us about the nature of the fundamental laws, the less relevance they seem to have to

the very real problems of the rest of science, much less to those of society. The constructionist hypothesis breaks down when confronted with the twin difficulties of scale and complexity.[10]

What is it about 'scale and complexity' that creates problems for the constructionist hypothesis?

### 8.5.2. *Immense Numbers of Possibilities*

Severe computational difficulties arise because the number of possible emergent structures at each level of the biological hierarchy is too large to be counted. To sharpen ideas in theoretical biology, physicist Walter Elsasser introduced the term *immense* to characterize a number that is finite but greater than a *googol* ($10^{100}$) and thus inconveniently large for numerical studies.[11]

To grasp this concept, consider the proteins. Because there are 20 different amino acids and a typical protein comprises some 200 of them, the number of possible proteins is greater than $20^{200}$, which is greater than a googol. The number of possible protein molecules is therefore immense, meaning that all the matter in the universe falls far short of that required to construct but one example of each possible protein molecule.[12] Throughout the eons of life on earth, most of the possible protein molecules have never been constructed and never will be. Those particular proteins that are presently known and used by living creatures were selected in the course of evolution through a succession of historical accidents that are consistent with but not governed by the laws of physics and chemistry.

So it goes at all levels of the biological hierarchy. The number of possible entities that might emerge from each hierarchical level—to form a basis for the dynamics of the next level—is immense, suggesting that happenstance guides the evolutionary process.[13]

[10] P. W. Anderson, 'More Is Different: Broken Symmetry and the Nature of the Hierarchical Structure of Science,' *Science* 177 (1972): 393–396.

[11] R. E. Crandall, 'The Challenge of Large Numbers,' *Scientific American* (February 1997): 72–78; W. M. Elsasser, *Reflections on a Theory of Organisms: Holism in Biology* (Baltimore: Johns Hopkins University Press, 1998) (first published in 1987).

[12] Elsasser, *Reflections*; See also A. C. Scott, *Stairway to the Mind* (New York: Springer-Verlag, 1995).

[13] S. J. Gould, *Wonderful Life: The Burgess Shale and the Nature of History* (New York: Norton, 1989).

It follows that biological science differs fundamentally from physical science, which deals with *homogeneous* sets having identical elements. Thus, a physical chemist has the luxury of performing as many experiments as are needed to establish *laws* governing the interactions among atoms as they form molecules. In the biological and social sciences, on the other hand, the number of possible members in most interesting sets is typically immense, so experiments are necessarily performed on *heterogeneous subsets* of the classes of interest. Because the elements of heterogeneous subsets are never exactly the same, it follows that experiments cannot be precisely repeated. Thus, causal laws cannot be determined with the same degree of certainty in the biological and social sciences as in the physical sciences.

In other words, psychologists establish *rules* rather than laws for interpersonal interactions. At the levels of biology, neuroscience, and social science, therefore, the horizontal arrow in Figure 8.1 should be drawn fuzzy or labeled with an estimate of its reliability.

### 8.5.3. *The Nature of Causality*

Whether one is concerned with establishing dynamic laws in the physical sciences or seeking rules in the biological and social sciences, the notion of *causality* requires careful consideration.[14] As was noted above, this study is essential for appreciating nonlinear phenomena, but it is not new. Some 25 centuries ago, Aristotle noted that 'We have to consider in how many senses *because* may answer the question *why*.[15] As a 'rough classification of the causal determinants of things,' he suggested four types of cause:

*Material cause.* Material cause stems from the presence of some physical substance that is needed for a particular outcome. Aristotle suggested that bronze is an essential factor in the making of a statue, but the concept is more general. The epidemic of gunshot wounds in the United States, for example, is materially caused by the large number of handguns in private homes.

*Formal cause.* The material necessary for some particular outcome must be available in the appropriate form. The blueprints of a house

[14] M. Bunge, *Causality and Modern Science*, 3rd edn. (New York: Dover, 1979).

[15] Aristotle, *The Physics*, trans. P. H. Wicksteed and F. M. Cornford (Cambridge, MA: Harvard University Press, 1953).

are necessary for its construction, the DNA sequence of a particular gene is required for synthesis of the corresponding protein, and a violinist needs the score to play a concerto.

*Efficient cause.* For something to happen, according to Aristotle, there must be an '*agent* that produces the effect and starts the material on its way.' Thus, a golf ball moves through the air in a certain trajectory because it was struck at a particular instant of time by the head of a club.

*Final cause.* Events may come about because they are desired by some intentional organism; thus a house is built because someone wishes to have shelter from the elements.

For those familiar with the jargon of mathematics, the following paraphrasing of Aristotle's definitions may be helpful.

1. At a particular level of the biological hierarchy, a *material cause* might be a time or space average over dynamic variables at lower levels of description and enter a hierarchical formulation as a slowly varying *parameter* at the level of interest.
2. Again, at a particular level of the biological hierarchy, a *formal cause* might arise from values of dynamic variables at higher levels of description, which enter as a *boundary condition* at the level of interest.
3. An *efficient cause* is represented by a *stimulation–response* relationship. Following Galileo, this is the standard sense in which physical scientists use the term causality.[16]
4. In mathematical terms, it is not clear (to me, at least) how one might formulate a *final cause*.

Although this classification may seem tidy, reality is often more intricate. Thus, Aristotle noted that causes might be difficult to sort out in particular cases, with several of them often 'coalescing as joint factors in the production of a single effect.'[17] Such interaction among causes is the key property of nonlinear phenomena.

Distinctions among 'joint factors' are not always easy to make. A subtle difference between formal and efficient causes appears in the metaphor for Norbert Wiener's *cybernetics*: the steering mechanism

[16] Bunge, *Causality*. [17] Aristotle, *Physics*.

of a ship.[18] If the wheel is connected directly to the rudder (via cables), then the forces exerted by the helmsman's arms are the efficient cause of the ship executing a change of direction. For larger vessels, however, control is established through a servomechanism in which the position of the wheel merely sets a pointer that indicates the desired position of the rudder. The forces that move the rudder are generated by a feedback control system that minimizes the difference between the actual and desired positions. In this case, one might say that the position of the pointer is a formal cause of the ship's turning, with the servomotor of the control system acting as the efficient cause.

Finally, when a particular protein molecule is constructed within a living cell, sufficient amounts of appropriate amino acids must be available to the messenger RNA as material causes. The DNA code, controlling which amino acids are to be arranged in what order, is a formal cause, and the chemical (electrostatic and valence) forces acting among the constituent atoms are efficient causes.

For mathematicians, it is not surprising for several different types of causes to be involved in a single event. We expect that parameter values, boundary conditions, and forcing functions will all combine to influence the outcome of a given dynamics. What other complications of causality are anticipated?

### 8.5.4. *Nonlinear Causality*

In mathematics, the term 'nonlinear' is defined in the context of relationships between causes and effects. Suppose that a series of experiments on a certain system have shown that cause $C_1$ gives rise to effect $E_1$; thus,

$$C_1 \to E_1$$

and similarly,

$$C_2 \to E_2$$

expresses the relationship between cause $C_2$ and effect $E_2$. This relation is *linear* if

$$C_1 + C_2 \to E_{12} = E_1 + E_2 \tag{1}$$

[18] N. Wiener, *Cybernetics* (New York: John Wiley & Sons, 1961).

If, on the other hand, $E_{12}$ is *not* equal to $E_1 + E_2$, the effect is said to be a *nonlinear* response to the cause.

Equation (1) indicates that for a linear system the cause can be arbitrarily divided into convenient components $(C_1, C_2, \ldots C_n)$, whereupon the effect will be correspondingly divided into $(E_1, E_2, \ldots E_n)$. Although convenient for analysis—providing a basis for Fourier analysis and Green function methods—this property is not found in the realms of biological, cognitive, and social sciences.[19]

Far more common is the nonlinear situation, where the effect from the sum of two causes is not equal to the sum of the individual effects. The whole is not equal to the sum of its parts. Nonlinearity is less convenient for the analyst because multiple causes interact among themselves, allowing possibilities for many more outcomes, confounding the constructionist. For just this reason, however, nonlinearity plays a key role in the course of biological evolution and the organization of the human mind.

### 8.5.5. *The Nature of Time*

Causality is intimately connected with the way we view time—thus, the statement '$C$ causes $E$' implies that $E$ does not precede $C$ in time[20]—yet the properties of time may depend on the level of description.[21] Thus, the dynamics underlying molecular vibrations are based on Newton's laws of motion, in which time is *bidirectional*. In other words, the direction of time in Newton's theoretical formulation can be changed without altering the qualitative behavior of the system. At the level of a nerve impulse, on the other hand, time is unidirectional, with a change in its direction making an unstable nerve impulse stable and vice-versa. In appealing to Figure 8.1, therefore, the reductionist must recognize that the nature of the time used in formulating the causal relationship between $P_1$ and $P_2$ may differ from that relating $M_1$ and $M_2$.

[19] Scott, *Nonlinear*; *idem*, *Encyclopedia*. [20] Bunge, *Causality*.

[21] J. T. Fraser, *The Genesis and Evolution of Time* (Brighton, UK: Harvester Press, 1982); *idem*, *Of Time, Passion, and Knowledge: Reflections on the Strategy of Existence*, 2nd edn. (Princeton: Princeton University Press, 1990); A. T. Winfree, *When Time Breaks Down: The Three-Dimensional Dynamics of Electrochemical Waves and Cardiac Arrhythmias* (Princeton: Princeton University Press, 1987); *idem*, *The Geometry of Biological Time* (New York: Springer-Verlag, 2001).

### 8.5.6. *Downward Causation*

Reductionism assumes that causality acts upward through the biological hierarchy, where the causality can be interpreted as both efficient and material. Formal causes, on the other hand, can also act *downward* because variables at the upper levels of a hierarchy can place constraints on the dynamics at lower levels.[22] Although such examples provide convincing evidence of downward causation, the means through which it acts are not well understood. To sort things out, Claus Emmeche and his colleagues have recently defined three types of downward causation.[23]

*Strong downward causation* (SDC). Under SDC, it is supposed that upper-level phenomena can act as efficient causal agents in the dynamics of lower levels. In other words, upper-level organisms can modify the physical and chemical laws governing their molecular constituents. Presently, there is no empirical evidence for the downward action on efficient causation, so SDC is almost universally rejected by biologists.

*Weak downward causation* (WDC). WDC assumes that the molecules comprising an organism are governed by some nonlinear dynamics in a phase space, having attractors (which include the living organism) each with a corresponding basin of attraction. Under WDC, a higher-level influence might move certain lower-level variables from one basin of attraction to another. Because many examples of such nonlinear systems have been studied both experimentally and theoretically,[24] there is little doubt about the scientific credibility of this means for downward causation. Building on the seminal suggestions of Alan Turing,[25] biologists Stuart Kauffman[26] and Brian Goodwin,[27] among others, have presented detailed discussions of ways that WDC influences the development and behavior of living organisms.

[22] Andersen *et al.*, eds., *Downward Causation.*

[23] C. Emmeche, S. Køppe, and F. Stjernfelt, 'Levels, Emergence, and Three Versions of Downward Causation,' in Andersen *et al.*, eds., *Downward Causation.*

[24] Scott, *Nonlinear*; *idem*, *Encyclopedia.*

[25] A. M. Turing, 'The Chemical Basis of Morphogenesis,' *Philosophical Transactions of the Royal Society of London* B237 (1952): 37–72.

[26] Kauffman, *The Origins of Order.*

[27] B. Goodwin, *How the Leopard Changed its Spots: The Evolution of Complexity* (New York: Scribner's, 1994).

*Medium downward causation* (MDC). Accepting WDC, proponents of MDC go further in supposing that higher-level dynamics (e.g. the emergence of a higher-level structure) can modify the local features of an organism's lower-level phase space through the downward actions of formal causes. In biology, MDC opens the possibility of closed causal loops spanning several layers of the hierarchy. In this picture, an organism emerges from the underlying phase space, which it in turn modifies. Over two decades ago, biochemists Manfred Eigen and Peter Schuster suggested that closed causal loops around at least three layers of dynamic description were necessary for the emergence of living organisms from the oily foam of the Hadean oceans.[28]

### 8.5.7. *Open Systems*

In contrast to most formulations of classical physics, biological organisms are *open systems*, requiring a steady input of energy (sunlight or food) to maintain their metabolic activities. A familiar example of an open system is provided by the flame of a candle. From the size and composition of the flame and the candle, it is possible to compute the (downward) propagation velocity of the flame ($v$), thereby establishing a rule for where the flame will be located at a particular time.[29] Corresponding to

$$M_1 \rightarrow M_2,$$

in Figure 8.1, such a rule is the following. If the flame is at position $x_1$ at time $t_1$, *then* it will be at position

$$x_2 = x_1 + v(t_2 - t_1)$$

at time $t_2 > t_1$. Because the flame is an open system, it follows that a corresponding relation

$$P_1 \rightarrow P_2$$

cannot be written—not even 'in principle'—for the physical substrate. Why not? Because the atoms comprising the physical substrate

[28] M. Eigen and P. Schuster, *The Hypercycle: A Principle of Natural Self-Organization* (Berlin: Springer-Verlag, 1979).
[29] Scott, *Nonlinear*.

are *continually changing*.[30] The flame's heated molecules of air and wax vapor at time $t_2$ are entirely different from those at time $t_1$. Thus, knowledge of the detailed positions and speeds of the molecules present in the flame at time $t_1$ tells nothing about those at time $t_2$. What remains constant is the flame itself—a higher-level *process*.

### 8.5.8. *Closed Causal Loops*

In his analysis of reductionism, Kim misses the concept of a closed causal loop, asking: 'How is it possible for the whole to causally affect its constituent parts on which its very existence and nature depend?' Causal circularity, he claims, is unacceptable because it violates the following 'causal-power actuality principle.'

> For an object, $x$, to exercise, at time $t$, the causal/determinative powers it has in virtue of having property $P$, $x$ must already possess $P$ at $t$. When $x$ is being caused to acquire $P$ at $t$, it does not already possess $P$ at $t$ and is not capable of exercising the causal/determinative powers inherent in $P$.[31]

There are two replies to this assertion, one theoretical and the other empirical: (1) From a theoretical perspective, Kim errs in supposing that a coherent structure somehow pops into existence at time $t$, which would indeed be surprising. An emergent entity (or coherent structure), however, begins to form through a process of exponential growth initiated by an instability that appears at some lower level of description. Eventually, this growth is limited by nonlinear effects, and a stable entity emerges. Upon being barely lit, for example, the flame of a candle grows rapidly in size before settling down to its natural size.

Similarly, in Kim's notation, both $x$ and $P$ should be viewed as functions of time ($t$), which may be related as

$$\frac{dx}{dt} = F(x, P),$$

$$\frac{dP}{dt} = G(x, P),$$

[30] M. H. Bickhard and D. T. Campbell, 'Emergence,' in Andersen *et al.*, eds., *Downward Causation*.

[31] Jaegwon Kim, 'Making Sense of Downward Causation,' in *Downward Causation*, eds. Andersen *et al.*

where $F$ and $G$ are general nonlinear functions of both $x$ and $P$. (The time scales of $F$ and $G$ can be very different, allowing $P$ to remain approximately constant during the dynamics of $x$.) The emergent structure is not represented by $x(t)$ and $P(t)$ (which are functions of time), but by $x_0$ and $P_0$, satisfying

$$0 = F(x_0, P_0),$$
$$0 = G(x_0, P_0).$$

Assuming that $x_0$ and $P_0$ are an asymptotically stable solution of this system,

$$x(t) \to x_0,$$
$$P(t) \to P_0.$$

as $t \to \infty$, exemplifying the establishment of a dynamic balance between downward and upward causations.

Thus, Kim's causal-power actuality principle is recognized as an artifact of his static analysis of an essentially dynamic situation.

(2) Going back to James Watt in the eighteenth century, engineers have used negative feedback to 'govern' the speed of engines. Since the 1920s, negative feedback loops are invariably used to stabilize the performance of amplifiers, making long distance telephone communications possible. These designs routinely employ closed causal loops in which a signal from the output terminals is brought back to the input. Occasionally, this feedback signal becomes positive rather than negative and leads to oscillations (called 'singing') that are viewed as unwanted emergent structures. For oscillators, on the other hand, positive feedback is an essential element of the design.

Nonlinear science offers many examples of positive feedback and subsequent emergence of coherent structures.[32] In the physical sciences, structures emergent from positive feedback loops include tornadoes, tsunamis, optical solitons, and Jupiter's Great Red Spot, among many others. Biological examples include the nerve impulse, cellular reproduction, flocks of birds and schools of fishes, and the development of new species. In the social sciences, there are lynch mobs, natural languages, and the founding of a new city.

[32] Scott, *Nonlinear*; *idem*, *Encyclopedia*; Wiener, *Cybernetics*.

Such emergent structures are essential elements in the *cognitive hierarchy*, which shares many features of the biological hierarchy.

## 8.6. The Cognitive Hierarchy

The preceding discussion of the hierarchical nature of biological science was presented in some detail as an introduction to the main subject of this paper: the cognitive hierarchy of the human brain, which is the most complex dynamic entity known to exist in the universe. In considering the brain's hierarchical structure from a reductionist perspective, one must be at least as careful as the biologist, if not more so.

With corresponding caveats about including fewer or more levels and allowing for branchings, a convenient version of the cognitive hierarchy takes the following form.

**Human culture**
**Phase sequences**
**Complex assemblies**
...
...
**Assemblies of assemblies of assemblies**
**Assemblies of assemblies**
**Assemblies of neurons**
**Neurons**
**Nerve impulses**
**Nerve membranes**
**Membrane proteins**
**Molecules**

Although this diagram differs from the biological hierarchy in important ways, many of the previous comments carry over into the present discussion. In particular, each cognitive level has its own nonlinear dynamics, involving closed causal loops of positive feedback, out of which can emerge an immense number of possible entities. A necessarily small subset of these possibilities does in fact emerge, providing a basis for the nonlinear dynamics of the next higher level.

Perhaps the most significant difference between the biological and cognitive hierarchies stems from the *internal levels,* which involve Hebb's cell assemblies.[33] Extracted from the cognitive hierarchy, these levels are

**Complex assemblies**

**...**

**...**

**Assemblies of assemblies of assemblies**
**Assemblies of assemblies**
**Assemblies of neurons**

the existence of which is deduced from theoretical speculation and circumstantial evidence rather than direct observation.[34]

Because individual assemblies share the basic dynamic properties of a neuron (threshold behavior and all-or-nothing response), Hebb proposed that they can organize themselves into higher-level assemblies of assemblies (called 'second-order assemblies'), which in turn become components of third-order assemblies and so on up to the complex assemblies that form the basis for normal thought.[35] Because it is not known how many internal cell-assembly levels there are or how they are organized, this region of the brain is presently a *terra incognita* of science—one of the more interesting unexplored frontiers.

Just as with the biological hierarchy, we expect to find formal causation acting downward in the cognitive hierarchy. The phenomenon of learning—whereupon (possibly synaptic) interconnection strengths among neurons change in response to experiences of the organism—provides an example of medium downward causation (MDC) because the local character of the underlying phase space is altered by global dynamics. As with the biological hierarchy, strong downward causation (SDC) is almost universally rejected by neuroscientists as both unproven and implausible.

Reductionism, in the context of neuroscience, is often interpreted as the view that all of the brain's behavior can be formulated in terms

[33] Hebb, *Organization of Behavior.* [34] Scott, *Neuroscience.*

[35] D. O. Hebb, 'The Structure of Thought,' in *The Nature of Thought,* eds. P. W. Jusczyk and R. M. Klein (Hillsdale, NJ: Lawrence Erlbaum Associates, 1980), 19–35; *idem, Essay on Mind* (Hillsdale, NJ: Lawrence Erlbaum Associates, 1980).

of local membrane dynamics. Referring back to Figure 8.1, $P_1$ ($P_2$) now represents the membrane states upon which a particular mental phenomenon $M_1$ ($M_2$) supervenes. Take away a $P$ and the corresponding $M$ disappears, according to the doctrine of physicalism, to which most neuroscientists subscribe.

The doctrine of *cognitive reductionism*, therefore, holds that any causal relationship between $M_1$ and $M_2$ (which is indicated by the horizontal arrow in Figure 8.1) can 'in principle' be formulated in terms of the underlying membrane states, $P_1$ and $P_2$. Proponents of this view should formulate responses to all of the objections raised above against biological reductionism.

As with the biological hierarchy, downward causation (WDC, MDC, or both) leads to additional opportunities for closed causal loops. These multilevel loops are far more intricate than those that can be represented by a diagram like the following:

This simple picture implies that **A** causes **B**, which in turn causes **A**; it is appropriate for describing the emergence of coherent structures at a particular dynamic level—the flame of a candle and a nerve impulse being clear examples. In the context of modern nonlinear science, each such diagram would correspond to the presence of an attractor in the phase space describing the system dynamics.[36] Complex cell assemblies, on the other hand, comprise subassemblies or attractors emerging at many different levels of the cognitive hierarchy, which can in turn become interconnected in an immense number of different ways.

Figure 8.2 suggests how various subassemblies of a complex cell assembly might be distributed over the principal lobes of the left hemisphere of the human neocortex. The *occipital* lobes, at the rearmost tip of each hemisphere, are related to vision because they accept signals from the eyes. The *parietal* lobes, on the upper rear of each hemisphere, handle judgments of weight, size, shape, and

[36] Scott, *Nonlinear*.

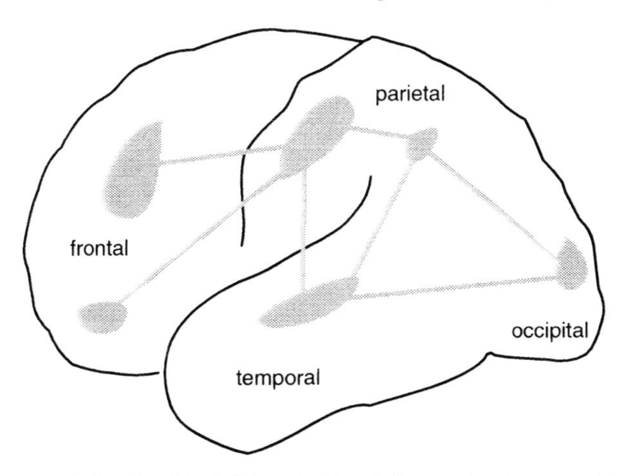


**Figure 8.2.** A sketch of the left-hand side of a human brain suggesting how the subassemblies (shaded areas) of a complex cell assembly might be distributed over various lobes of the neocortex.

feel. The *temporal* lobes, near the temples, deal with language and the perception of sound, among other things, which is not surprising given their proximity to the ears. Finally, the *frontal* lobes, immediately behind the forehead, govern voluntary movements and some logical processes, which is why a frontal lobotomy tends to alter personality.

The shaded areas in the figure indicate where neocortical subassembly neurons (described by Hebb as a 'three-dimensional fishnet') might be sparsely located. Each of these subassemblies, under the theory, would in turn be composed of further subassemblies representing more basic perceptual or conceptual elements.

Thus, the type of dynamic object that may emerge in the hidden internal levels is extremely intricate, but that's not the whole story. Although the lower levels

**Neurons**
**Nerve impulses**
**Nerve membranes**
**Membrane proteins**
**Molecules**

are open to direct empirical investigation (via X-ray studies, electron and optical microscopy and the techniques of electrophysiology), puzzles remain.[37]

At the higher levels of

**Human culture**
**Phase sequences**

it is again possible to directly observe the dynamics, but new difficulties appear. By the term 'phase sequence,' Hebb implied a 'thought process' in which:

> Each assembly action may be aroused by a preceding assembly, by a sensory event, or—normally—by both. The central facilitation from one of these activities on the next is the prototype of 'attention.'[38]

We all experience ongoing trains of thought in every waking moment, but these processes do not occur in a social vacuum.

Throughout such trains, our individual thoughts both comprise and are formed by the particular human culture in which we are immersed. In other words, our minds are molded by levels of cultural reality of which we are often unaware. Thus, corresponding levels of social reality—described by anthropologist Ruth Benedict as 'patterns of culture'—must be included in realistic models of the human brain.[39]

Finally, the possibility of causal interactions among the various levels of the cognitive and biological hierarchies should be included in the overall theoretical perspective. At lower levels, this is obvious because the physiological condition of a neuron clearly affects the manner in which it relates incoming and outgoing streams of information, but higher cognitive levels also have causal biological effects. Cultural imperatives to ingest a psychoactive substance, for example, can alter the dynamics of membrane proteins and lead to mental changes that influence bodily health with subsequent psychological effects in a wending path of branching causes and effects that daunts analysis.

[37] Scott, *Neuroscience.* [38] Hebb, *Organization of Behavior*, xix.
[39] R. Benedict, *Patterns of Culture* (Boston: Houghton Mifflin, 1934, 1989); Scott, *Stairway*.

Presently, the types of phenomena that could emerge from such intricate causal networks—spanning several levels of both the biological and cognitive hierarchies—are only dimly imagined.

## 8.7. Directions of Future Research

Although the foregoing comments may seem pessimistic about the future of research on brain dynamics, that is not my intent. On the contrary, the awesome intricacy of the human brain presents unusual challenges, making the twenty-first century a most exciting time. Awaiting the assaults of vigorous young students are several interesting problem areas, including the following.

### 8.7.1. *Empirical Confirmation of Hebb's Cell Assemblies*

In the half-century since Hebb's theory of cell assemblies was first proposed, the experimental techniques of electrophysiology have greatly improved. Classical methods have been refined and new techniques introduced, leading Nicolelis, Fanselow, and Ghazanfar to comment in 1997:

> What we are witnessing in modern neurophysiology is increasing empirical support for Hebb's views on the neural basis of behavior. While there is much more to be learned about the nature of distributed processing in the nervous system, it is safe to say that the observations made in the last 5 years are likely to change the focus of systems neuroscience from the single neuron to neural ensembles. Fundamental to this shift will be the development of powerful analytical tools that allow the characterization of encoding algorithms employed by distinct neural populations. Currently, this is an area of research that is rapidly evolving.[40]

In assessing this perspective, it is important to remember that observing the dynamic behavior of a 'three-dimensional fishnet' comprising several thousand neurons (each receiving several thousand synaptic inputs) and spread over much of the brain is a daunting task, yet not hopeless. Although there is presently no possibility

[40] M. A. L. Nicolelis, E. E. Fanselow, and A. A. Ghazanfar, 'Hebb's Dream: The Resurgence of Cell Assemblies,' *Neuron* 19 (1997): 219–21.

of taking microelectrode readings from most of the neurons in an assembly, records from as few as two may offer interesting opportunities for research because the experimenter can ask whether the recorded voltages are *correlated* and observe how the degree of correlation depends upon the global behavior of the organism.

### 8.7.2. *Modeling of Cell Assemblies*

Using currently available computers, neurologically realistic simulations of cell assemblies have recently been reported by Fransén and Lansner, which confirm the dynamic properties originally proposed by Hebb.[41] As the numerical power available to computational neuroscience continues to grow, this activity should become an increasingly important window into the dynamic nature of the human brain.

At higher levels of description, Haken has shown how the dynamics of 'order parameters' (which correspond to activity levels of assemblies) can be related to psychological experiments.[42] At the level of subjective perception, switchings between assemblies are experienced as one stares at the Necker cube shown in Figure 8.3. (Constructed by a Swiss geologist in the mid-1800s, this image seems to jump from one metastable orientation to another, like the flip-flop circuit of a computer engineer.)

Defining order parameters ($\xi_1$ and $\xi_2$) to represent neural activities related to the two perceptions of the Necker cube suggests the equations

$$\frac{d\xi_1}{dt} = \xi_1\left[1 - \xi_1^2 - B\xi_2^2\right],$$
$$\frac{d\xi_2}{dt} = \xi_2\left[1 - \xi_2^2 - B\xi_1^2\right],$$

[41] E. Fransén and A. Lansner, 'Low Spiking Rates in a Population of Mutually Exciting Pyramidal Cells,' *Network* 6 (1995): 271–288; *idem*, 'A Model of Cortical Associative Memory Based on a Horizontal Network of Connected Columns,' *Network* 9 (1998): 235–264; *idem*, 'Modelling Hebbian Cell Assemblies Comprised of Cortical Neurons,' *Network* 3 (1992): 105–119.

[42] H. Haken, *Principles of Brain Functioning: A Synergetic Approach to Brain Activity, Behavior and Cognition* (Berlin: Springer-Verlag, 1996).

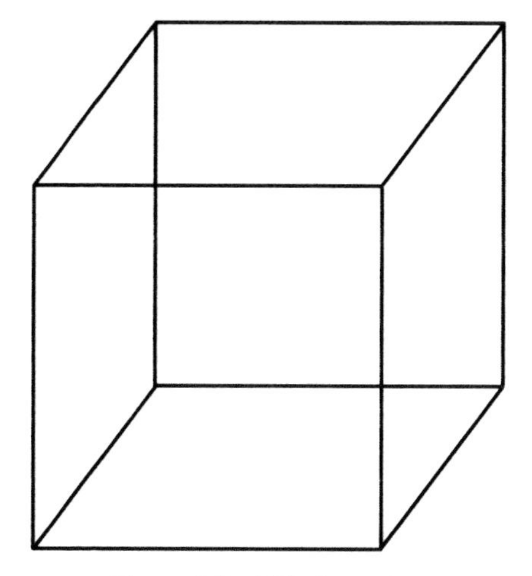

**Figure 8.3.** A Necker cube.

as a nonlinear dynamic model.[43] For $B > 1$, this system has two stable states:

$$(\xi_1, \xi_2) = (0, 1) \quad \text{and} \quad (\xi_1, \xi_2) = (1, 0)$$

which represent our two perceptions of the orientations of the cube. Thus can the data of Gestalt psychology be related to cell assembly models of the neocortex.

### 8.7.3. *General Hierarchical Formulations*

In this paper, the hierarchical structures of biological and cognitive systems have merely been described, leaving several unresolved issues. Building on the work of Eigen and Schuster,[44] Voorhees,[45] Fontana and Buss,[46] Baas,[47] and Nicolis,[48] can one formulate the various

[43] Ibid.; Scott, *Nonlinear.* [44] Eigen and Schuster, *The Hypercycle.*

[45] B. H. Voorhees, 'Axiomatic Theory of Hierarchical Systems,' *Behavioral Sciences* 28 (1983): 24–34.

[46] W. Fontana and L. W. Buss, '"The Arrival of the Fittest": Toward a Theory of Biological Organization,' *Bulletin of Mathematical Biology* 56 (1994): 1–64.

[47] N. A. Baas, 'Emergence, Hierarchies, and Hyperstructures,' in *Artificial Life III*, ed. C. G. Langton (Reading: Addison-Wesley, 1994).

[48] J. S. Nicolis, *Dynamics of Hierarchical Systems: An Evolutionary Approach* (Berlin: Springer-Verlag, 1986).

causal relationships among levels of a nonlinear dynamic hierarchy (weak and medium downward causation, for example) in a manner that is suitable for mathematical analyses?

This is not at all a trivial matter because the time and space scales for models of living creatures differ by many orders of magnitude as one goes from the biochemical levels to the whole organism, creating a challenge for the numerical modeler. Not only the scale but—as we have seen—the nature of time differs at biomolecular and cellular levels of description.[49] Thus, the laws governing the dynamics of the atoms within a molecule are bidirectional in time, meaning that the direction of time's arrow can be reversed in the mathematical formulation without changing the qualitative behavior. In living systems, this is not the case.

### 8.7.4. *Meaningful Information*

Several possibilities for information processing in the dendritic and axonal trees of real neurons have been proposed, and means for higher-level information processing have been suggested.[50] How can these various models be incorporated into a mathematical formulation for the global activity of a real brain? Indeed, what should we intend by 'information' in the context of living brain?

In engineering science, this term derives from the need to store and transmit a series of zeros and ones representing a photograph, computer code, musical recording, or whatever.[51] (A system with $p$ different stable configurations is said to store an amount of information equal to $\log_2 (p)$ bits.) Living organisms, however, are *intentional*, with self-determined programs of activity. Thus, the importance of a particular fact is related to what the creature is concerned about,[52]

[49] Fraser, *Genesis*; *idem*, *Of Time, Passion, and Knowledge: Reflections on the Strategy of Existence*, 2nd edn. (Princeton: Princeton University Press, 1990); Winfree, *When Time*; *idem*, *The Geometry*.

[50] Scott, *Neuroscience*.

[51] L. Brillouin, *Science and Information Theory* (New York: Academic Press, 1956); C. E. Shannon and W. Weaver, 'Recent Contributions to the Mathematical Theory of Communication,' in *The Mathematical Theory of Communication*, eds. Shannon and Weaver (Urbana, IL: University of Illinois Press, 1963).

[52] W. J. Freeman, *How Brains Make Up Their Minds* (New York, Columbia University Press, 2000).

and an item of meaningful information can involve many details of the type that are captured by Hebb's cell assemblies.

After a learning experience, for example, a person might recognize a geometrical figure, and this recognition could be viewed as one item of information that might or might not be meaningful according to his or her current interests. In more intricate situations, a person might recognize that those red berries are poisonous, it is going to rain, the car he just hit is driven by a policeman, his spouse is angry, and so on. The vast collection of such assemblies that each of us carries about were crafted moment by moment, day by day, and year by year as we became adults, providing the spectrum of contexts in which we interpret experience.

Considering the little we know of internal assemblies and the ways they interact, how might the concept of meaningful information be mathematically defined for an intentional organism? Might a sharpened concept of meaningful information lead to a formulation for Aristotle's final cause?

# 9

## The Emergence of Causally Efficacious Mental Function

*Warren S. Brown*

### 9.1. Introduction

One objective of this volume is to consider whether reductionism is a necessary outcome of physicalism. In this chapter, I will explore a physicalist, but non-reductive, understanding of human mental function that accounts for robust forms of human agency. I will be attempting to demonstrate how mental properties with causal roles might emerge from the physical activity of hyper-complex human brains.

I first offer a brief critique of the 'problem of mental causation' as based on a holdover from a Cartesian view of mind as essentially internal to the body and largely passive. This critique sets forth the view of mind that will be central to this chapter: mind is embodied and embedded in action in the world.

In section 9.3, I argue that the mental must first be seen within the context of a phylogenetic progression of adaptive capacities. Section 9.4 deals with the mental as present in modes of action (i.e. action-loops). Mental processing involves the re-expression in action of beliefs embodied in organisms' complex memory structures. This view of the mental as present in action begins to approach an explanation of higher mental capacities when one imagines hierarchies in levels of behavioral control and in levels of evaluation of behavioral feedback (section 9.5). Not all that is intelligent about human mental function is present within the nervous systems of organisms, as I emphasize in a brief discussion of external scaffolding and language (section 9.6). I then describe the *coup de maître* of mental efficacy, imagination and mental modeling. Finally, I take up the issue of consciousness—both how it might be embodied in brain physiology, and what difference it makes for behavioral adaptability and human agency.

## 9.2. The Illusion of a Problem

There is vast literature in philosophy on the problem of mental causation. Nancey Murphy expresses what is at stake as follows: 'If mental events are intrinsically related to . . . neural events, how can it *not* be the case that the contents of mental events are ultimately governed by the laws of neurobiology? If neurobiological determinism is true then it would appear that there is no freedom of the will, that moral responsibility is in jeopardy, and . . . that our talk about the role of reasons in any intellectual discipline is misguided.'[1]

Fred Dretske expresses it this way: 'The project is to see how reasons—our beliefs, desires, purposes, and plans—operate in a world of causes, and to exhibit the role of reasons in the *causal* explanation of human behavior. In a broader sense, the project is to understand the relationship between the psychological and the biological—between, on the one hand, the *reasons* people have for moving their bodies and, on the other, the *causes* of their bodies' consequent movements.'[2]

I speculate that the *problem* of mental causation is, in some respects, an illusion—a residual of a Cartesian view of mind where the mind is interior and effectively disembodied. Mind is implicitly thought to be a characteristic or property of something not unlike an inner homunculus—as in the 'Cartesian theatre.' I myself find it difficult to shed this model of mental function—it is so deeply ingrained in my thinking.

[1] Nancey Murphy, 'Supervenience and the Downward Efficacy of the Mental: A Nonreductive Physicalist Account of Human Action,' in *NP*, 147.

[2] Fred Dretske, *Explaining Behavior: Reasons in a World of Causes* (Cambridge, MA: Bradford Books, 1988), x.

An alternative possibility is that 'mind' is something that occurs in acting. At their core, mental processes are neither prior to, nor apart from, doing. As Andy Clark expresses it, 'minds make motions.'[3] Mind is always 'on the hoof'—contextualized in action. Mind is not the activity of the brain, nor is it a non-material emergent in some numinous and dualist sense. Rather, mind is a description of the brain and body operating as one in solving real problems in the field of action. One might say that 'mind' should always be a verb—we 'mind' we do not 'have a mind.' Therefore, I will defend an embodied (i.e. physicalist), embedded (i.e. contextualized in action), and active view of mind, and explore the implications of this view for mental causation.

So what I meant when I said that the problem of mental causation is illusory is that mind is, in this definition, causes-in-process. Perhaps it would be more accurate to propose that mental causation poses a different sort of problem than many previous discussions have imagined, but it is the sort of problem that is comprehensible within a nonreductionist psychobiology.

Of course, there *is* the issue of the 'internalness' of thinking that I will approach in a brief discussion of symbolic and off-line processes, as well as in the section on consciousness. I suggest that the capacity for off-line symbolic thought can be understood as piggy-backing on processes involved in ongoing adaptive action.

## 9.3. Mental Constructs and Lower Animals

A combination of the concepts of physicalism, emergence, and evolutionary biology imply that we should find a phylogenetic progression of mentalist properties. What we humans experience as thinking, planning, deciding, and acting for reasons have their roots in similar but simpler forms of mental activity in lower (or less complex) organisms. Beginning with the richness and complexity of human mental capacities often seems to leave us with incomprehensible mysteries that incline us toward forms of dualism. But if we can appreciate the expanding capacity for mental transactions as one moves up levels

[3] Andy Clark, *Being There: Putting Brain, Body, and World Together Again* (Cambridge, MA: Bradford Books, 1997), 1.

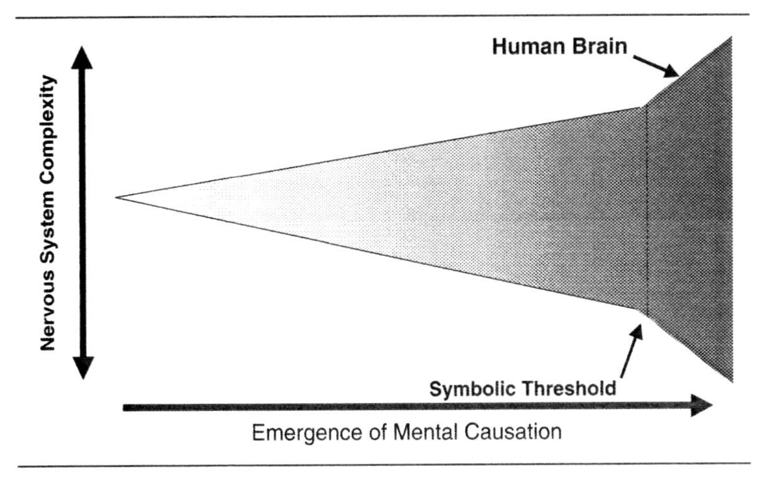


**Figure 9.1.** A Phylogenetic Model of Mental Causation.

of complexity among animal species, human mental function can begin to appear less mysterious. Thus, in exploring the implications of nonreductive physicalism and emergence, one needs to take evolutionary biology seriously.

Although helpful in dissolving mysteries, studying animals creates a problem in the application to lower animals of abstract concepts derived from an understanding of human mental processes—concepts such as reasons, beliefs, decision, and consciousness. One cannot decide precisely where in the continuum of complexification of species such terms apply and where they do not. In fact, the idea of emergence seems to imply that these questions can seldom be answered 'yes' or 'no' with respect to any particular behavior in lower animals. The human quality under consideration will typically have some richer or poorer (thicker or thinner, saturated or pale) meaning when applied to organisms at various levels of the phylogenetic hierarchy of nervous system complexity.

Figure 9.1 is a visual model suggesting the distribution of mental characteristics along a continuum, without clear boundaries or sudden transitions. The presence or absence of a mental dimension is indicated by the increased density of shading. Complexity of the nervous systems of various species is represented by the width of the cone at any particular point. Thus, this figure provides graphic representation of the idea that mental characteristics emerge with the increasing complexity of the nervous system, and, most importantly, that there are no clear lines to be drawn that separate the presence and absence of most mental properties. Figure 9.1 shows one exception: one line (i.e. one discrete transition point) at the human end of the scale to suggest what Terrence Deacon has described as a 'symbolic threshold' in both human evolution and child development.[4] I find Deacon's arguments compelling. However, Deacon also suggests that, while there is a symbolic threshold, symbolic ability is nested in prior iconic and indexical systems clearly present in lower, less complex animals.

## 9.4. What Is the Mental?

I now return to the issue of the concept of mind as *embodied* (that is, constituted by physical processes involving the body as well as the brain) and as *embedded* (that is, present in contextualized action in the world).

### 9.4.1. *Is the 'Mental' in the Brain?*

The embodied nature of mind is illustrated by Donald MacKay in a *thought experiment* in which you imagine that you have your own operating brain sitting on a table in front of you (still functionally connected to your body and nervous system).[5] In this experiment, you electrically stimulate your own brain in the primary visual area, and, of course, you will see flashes of light. MacKay believes that several questions can be answered from this simple thought experiment:

1. *Where are the physical events that are creating your experience of lights flashing?* They are in the part of the brain being stimulated. Thus, brain activity is necessary (but not sufficient) for your subjective mental experience of lights flashing. However, the mental event of seeing-lights-flashing is a phenomenon attributable to you as a whole person.

[4] Terrence Deacon, *The Symbolic Species: The Co-evolution of Language and the Human Brain* (London: Penguin, 1997), 73ff.

[5] Donald M. MacKay, *Behind the Eye* (Oxford: Basil Blackwell, 1991), 5–8.

2. *Where are the lights as far as you are concerned?* They are out in your immediate action-space. They are part of your map of the external world—the field with which you are currently interacting. Mental events, even if dependent on internal physiological events, are about the field-of-action (or about you-in-the-field-of-action).
3. *But where are you? You* are sitting in front of your brain. *You* are not on the table in front of you. Therefore, all descriptions of human mental experience are about the whole embodied person, not about the brain as a separate organ. In this respect, the mental has to be something resident in the whole person.

MacKay makes an even more general point concerning our use of any mentalist terms with respect to human behavior: we should attempt to exercise 'semantic hygiene,' that is, we should only use mentalist language to describe what the *whole* person (or animal) is doing or experiencing, not to refer to the activity of some *part* of the person (such as the brain).[6] The brain is necessary for mind, but not sufficient.

### 9.4.2. *The Mental as Embedded in Action*

Based on the above, it becomes clear that 'mental states' are best conceptualized as 'contextualized brain states.'[7] They are not internalized in the sense of existing in a form that is uncoupled from contexts of interaction with the real world. In its root form, a 'mental state' is the cognitive resources being brought by the organism to an ongoing interaction with the world. It is not a description of a separate, distant, abstract planner and initiator of such interactions. Mental activity is the entire embodied complex organism embedded in a process of interacting with the current field of action (real or imagined).

[6] Ibid., 8–10.

[7] Nancey Murphy and Waren S. Brown, *Did My Neurons Make Me Do It?: Philosophical and Neurobiological Perspectives on Moral Responsibility and Free Will* (Oxford: Oxford University Press, 2007).

Consider an example from human behavior. The following is an account of an experiment regarding procedural knowledge done by D. C. Berry and D. E. Broadbent:

> Berry and Broadbent . . . asked subjects to try to control the output of a hypothetical sugar factory (which was simulated by a computer program) by manipulating the size of the workforce. Subjects would see the month's output . . . and then choose the next month's workforce. [The rule relating workforce and output was a complex mathematical formula.] . . . . Oxford undergraduates were given sixty trials at trying to control the output of the sugar factory. Over the sixty trials, they got quite good at controlling the output of the sugar factory. However, they were unable to state what the rule was and claimed they made their responses on the basis of 'some sort of intuition' or because it 'felt right.' Thus, subjects were able to acquire implicit knowledge of how to operate such a factory without corresponding explicit knowledge.[8]

Interacting with the sugar factory program is clearly a form of very high-level mental activity even though the students had no *conscious* awareness of exactly what they had learned and why they were eventually able to perform the task so well. Research on automaticity suggests that only a small portion of moment-to-moment human behavior is consciously controlled.[9] More importantly, knowledge regarding how to solve this problem only appeared in the interaction between the person (body and brain) and the computer program. While experience with the program obviously modified their brains in some manner, this brain modification was inert and only became 'mental' in the process of interacting with the computer program. In this particular case, no conscious access to the knowledge was possible outside of engagement in the task. Mental causation was the elicitation of the procedural memory within ongoing action-in-the-world.

The critical role of consciousness in human agency, and its relationship to such automatic behaviors, will be explored in section 9.8. My point here is that some very high-level cognitive operations are realized in action that may never become fully conscious,

[8] Described in John R. Anderson, *Cognitive Psychology and Its Implications* (New York: Worth, 2000), 236–237.

[9] J. A. Bargh and T. L. Chartrand, 'The Unbearable Automaticity of Being,' *American Psychologist* 54(1999): 462–479.

yet they should not be excluded from considerations of mental causation.

### 9.4.3. *The Role of Memory in Mental Processes*

The previous example of procedural learning and the incorporation of such knowledge into behavior also illustrates the importance of memory as a major contributor to the efficacy of mental processing. Fred Dretske considers associative learning to be central to his analysis of how it can be said that animals have beliefs and act for reasons. Learning creates what Dretske refers to as 'structuring causes' of behavior, as contrasted with 'triggering causes.'[10] Learning structures the nervous system in a way that provides the reason why, for this organism, stimulus A causes behavior X. Further, learning explains how stimulus B can come to predict the likelihood of A and thus also trigger behavior X. To turn this around, learning is the explanation of why behavior X is known by the animal as an effective behavior in the context A. Thus, an animal can be said to act (behavior X) for the *reason* of the learned *belief* that this behavior in the context of A (or B that predicts A) will have an outcome relevant to current needs or goals. In Dretske's formulation, mental causation is rooted in this sort of acting for *reasons* on the basis of *beliefs*, and he argues cogently that animals do this to the degree that associative learning reshapes the nervous system, amounting to new 'structuring causes' of behavior.

One of the fundamental properties of even simple nervous systems is rapid pattern completion in the face of incomplete information. Andy Clark points out that neural networks also do this well.[11] The demands of rapid ongoing behavior typically do not allow organisms either to await further information, or to sample the entire sensory environment. Based on a minimal amount of information sampled, the brain fills in the remainder in order to make a rapid response. This process of pattern completion is based on memory records of previous experience. Extrapolation based on incomplete information is one of the important rudiments of mental activity.

[10] Dretske, *Explaining Behavior*, 42. [11] Clark, *Being There*, 59–62.

We know from cognitive psychology that memories are held in associative networks which become activated by current external or internal contexts. There is a process, typically called 'spreading activation,' by which wider networks of more or less distantly associated memories become pre-activated in ways that guide the directions of future behavior (or, subjectively in the case of humans, further thought). Thus, efficacious mental activity is not only what is contributed by learning to immediate responses, but also contributions from complex networks of learned associations that can prime future thought and behavior in ways related to previous experiences.

This emphasis on the importance of memory is also consistent with Terrence Deacon's definition of third-order emergent systems.[12] While first- and second-order emergence involve amplification due to complex and recurrent interactions between elements, third-order emergence involves, in addition, some form of information and memory. Previously occurring higher-order neural patterns (i.e. memories) exert a cumulative influence over the future of the system by being repeatedly re-elicited by new, but similar situations, and therefore are re-entered into the lower-order neural dynamics. The result of the addition of memory is that 'specific historical moments of higher-order regularity or of unique micro-causal configurations can . . . exert a cumulative influence over the entire causal future of the system.'[13] Thus, Deacon's third-order emergence, which is clearly the form of emergence that is important in the efficacy of the mental in more complex organisms, is dependent on learning and memory.

### 9.4.4. *The Essence of the Mental in Action Loops*

Let me back up a bit to suggest how we need to view the behavior of organisms. As mentioned previously, Dretske talks of 'triggering causes,' which suggests that organisms are passive and become active only at the point of some triggering event. Many of the causal diagrams used by philosophers of mind give the same impression. I do

[12] T. W. Deacon, 'The Hierarchical Logic of Emergence: Untangling the Interdependence of Evolution and Self-Organization, in B. Weber and D. Depew, eds., *Evolution and Learning: The Baldwin Effect Reconsidered* (Cambridge, MA: MIT Press, 2003).

[13] Deacon, this vol., 105–106.

not think that 'triggering' represents a correct and informative way to view causes in animal behavior. It is more accurate biologically to view all organisms as characterized by three important properties.[14]

1. *All organisms are continuously active.* Behavior is emitted from inside the organism, not triggered by external stimuli. Thus, behavior is always (or nearly always) voluntary, rather than elicited in a passive organism by a particular stimulus. Neuroscientist Martin Heisenberg has demonstrated in his experiments on the behavior of fruit flies (drosophila) that their flying behavior is a matter of trying out potential directions of flight in order to satisfy current internal needs. No external stimuli can be detected that cause or trigger attempted changes in direction.[15] In this sense, the behavior is voluntary—that is, emitted, not triggered.

2. *All behavior of organisms is goal-directed.* Behavior is tried out in order to meet some internal goal (whether these goals are represented simply or complexly within particular organisms). Even protozoa have goals to find nutrients and avoid toxic substances. Like the fruit fly, they emit swimming behavior in order to try out directions of movement to determine if they will result in more nutrients or less toxicity. In more complex organisms capable of associative learning, Dretske suggests that there exist both *pure desires* (based on biological needs) and *derived desires* that are acquired due to some association with pure desires.[16] Thus, 'goal-directed' can be relative to immediate biological needs, or to derived desires that are not immediately relevant to such needs.

3. *All organisms have the ability to evaluate the outcome of behavior and modify their ongoing behavior in relationship to evaluations.* As we can already appreciate in the behavior of protozoa and drosophila (as well as laboratory rats, monkeys, and human beings), the activity of all organisms is a constantly recurring loop of emitting behaviors, evaluating the behaviors based on a comparison of sensory feedback and internal criteria for desired outcomes, and adjusting ongoing behavior based on the evaluation of feedback.

[14] See Murphy and Brown, *Did My Neurons Make Me Do It?*

[15] Martin Heisenberg, 'Voluntariness (Willkürfähigkeit) and the General Organization of Behavior,' in R. J. Greenspan and C. P. Kyriacou, *Flexibility and Constraint in Behavioral Systems* (New York: John Wiley & Sons, 1994).

[16] Dretske, *Explaining Behavior*, 111.

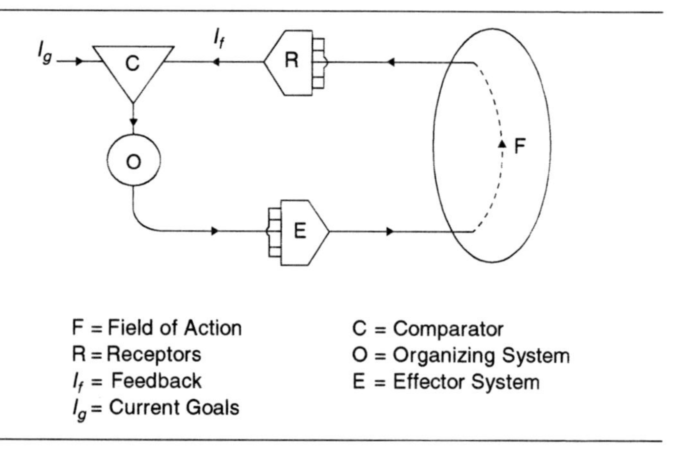


**Figure 9.2.** MacKay's conceptualization of the minimum fuctional architecture of an agent. Behavioral effector systems (E) operate within the field-of-action (F). The ongoing actions and outcomes are monitored by sensory receptors (R) and fed into a comparator system (C) that compares this input ($I_f$) to the desired goal ($I_g$). Based on this comparison, behavior (E) is continued or reorganized by the organizing system (O).

Donald MacKay used an information-engineering diagram of the minimal functional architecture of an agent that captures these aspects of the behavior of all organisms.[17]

Figure 9.2 expresses the fact that behavior always involves a process of continuous loops of action–feedback–evaluation–action ('action loops'). Behavior has no beginning or end, and is not typically triggered by external stimuli. Rather, feedback from the field of action is used to constantly monitor the success of the current goal-directed behavior of the organism with respect to desired outcomes, and necessary adjustments in behavior are made. This view of the behavior of organisms puts a different spin on the idea of a 'cause'—that is, 'causation' is not triggering of action in an otherwise inert organism. It is not necessary to find the cause that set a particular behavior in motion. Rather, causes are the processes by which a continuously active organism evaluates and modulates its action. For example, genetic causes (which we would not typically consider to be

[17] MacKay, *Behind the Eye*, 43.

'mental') are genetically programmed ways of modulating ongoing action based on feedback. Similarly, the causal role of the mental is evident in *modulations* of ongoing goal-related behavior that is based on more complex, learned forms of evaluation, or on evaluative criteria established in conscious, verbal thought.

I would grant that modulations of ongoing goal-directed behavior based on memories of past experiences seem a rather pale form of mental causation. However, there are still important points that need to be made about off-line simulations of action-loops (i.e. running imaginary mental scenarios), about symbolic/linguistic processing, and about consciousness in order to understand more sophisticated forms of human mental causation.

### 9.4.5. *Representations for Action Loops*

The central point of this essay is that mental activity is rooted in action loops, and is, in this sense, both embodied and embedded. Mental activity is not dissociated from action at the moment (or stimulations of such action). In this view of mental processing, what sorts of representations are most useful? It would seem that a mode of representation that is very close in form to action loops would be most efficient.

Very much in this vein, Donald MacKay has argued that the nature of perception is the elicitation of a *conditional readiness to reckon with* the current elements of the sensory world.[18] Thus, perception is generating a matching response in the form of the activation within memory of the sensorimotor record of previous *actions* with respect to this stimulus pattern. This activated sensorimotor record provides predictive information about ways one might interact with what is currently being sensed. For example, to perceive a pen lying on the desk in front of me as a pen is for my nervous system to activate a neural network that consists of learned potentialities for actions involving pens.

Similarly, MacKay considers a *belief* to be a conditional readiness to reckon with that which is not immediately present in the sensory environment, but might well become present should circumstances

[18] Ibid., 109–113.

change in a particular way.[19] For example, if you were to say that George W. Bush is standing outside of the door, the consequence of me believing you would be to have activated a readiness to reckon with Mr. Bush under the condition that I decide to go out the door. My map of the field of potential actions has been altered by activation of new potentialities for the actions that would be required should I go out of the door. This new map of potentials for action-in-the-world is the representation of my belief.

The role of motor systems in representation is particularly clear in recent research on mirror neurons. The idea of mirror neurons comes from the demonstration that the neural systems of the supplementary motor cortex that would be involved in initiating and controlling a specific movement are active while a monkey passively observes another monkey making this particular movement. Giacomo Rizzolatti, the discoverer of mirror neurons, writes, 'The fundamental neurophysiological mechanism that underlies understanding of an action is a direct matching of the observed action with the motor representation of that action. This matching is made by the mirror neuron system.'[20] The perceptual recognition and representation of the action of another animal involves an implicit simulation of making a similar action. Rizzolatti argues for the important role of mirror neurons in facilitating imitation. Such perception and potential imitation based on activity in mirror neurons is reminiscent of MacKay's idea of a conditional-readiness-to-reckon.

## 9.5. Hierarchy and Complexity

I have been proposing a view of mind as resident in action loops, which, thus far, have been conceived in rather simple terms. However, a complete account of the causal role of the mental needs also to involve an understanding of more complex forms of mental processing. Yet even these more complex forms can be understood

[19] Ibid., 112.

[20] Giacomo Rizzolatti, 'The Mirror Neuron System and Imitation,' in Susan Hurley and Nick Charter, eds., *Perspectives on Imitation: From Neuroscience to Social Science, Volume 1: Mechanism of Imitation and Imitation in Animals* (Cambridge, MA: MIT Press, 2005), 56.

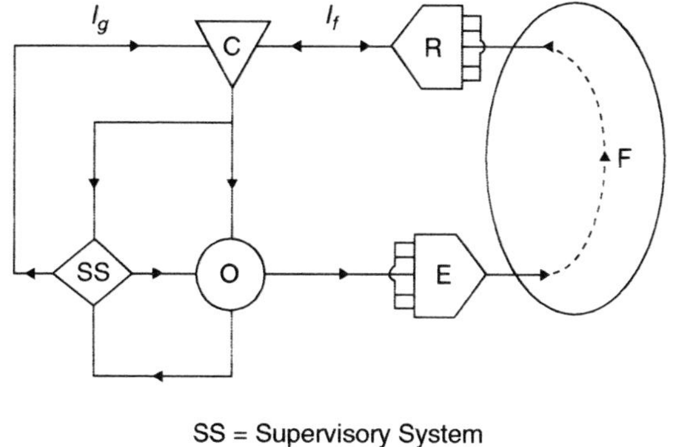


**Figure 9.3.** Functional architecture of an agent that sets its own goals.

as elaborations of action-loops. Thus, we need to consider the overlays of more complex forms of modulation of action loops. MacKay expands his functional architecture of an agent to suggest overlays of more complex and abstract forms of action modulation. He suggests that action loops are modulated by supervisory systems that function to set the goals of local action loops. Figure 9.3 incorporates the concept of a supervisory system (SS).[21]

MacKay makes it clear that this 'supervisor' is not some sort of homunculus, nor even a centralized place where, in Daniel Dennett's terms, 'it all comes together.' Rather, a supervisory system is a larger action loop within which the original loop is nested. Therefore, this diagram when elaborated to illustrate the nature of a supervisory system can be seen in Figure 9.4.[22]

In Figure 9.4, we see that the functional architecture of a supervisory system is a meta-comparator (MC) and meta-organizer (MO)—that is, the same sort of architecture that comprises the original action loop, involving action, feedback, evaluation, and modified action. MacKay points out that it is reasonable to consider increasingly more complex levels of nesting of such modulatory feedback loops, as in Figure 9.5.[23]

The supervisory system of Figure 9.5 can also be expanded to a meta-meta-comparator and meta-meta-organizer . . . and so on. Only

[21] MacKay, *Behind the Eye*, 51. [22] Ibid., 141.
[23] Modification of MacKay, *Behind the Eye*, 141.

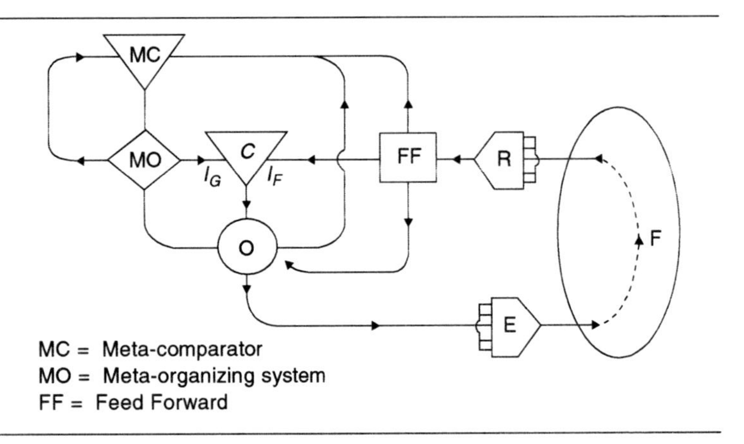


**Figure 9.4.** Functional architecture of a supervisory system: Hierarchy of organization and recurrent loops.

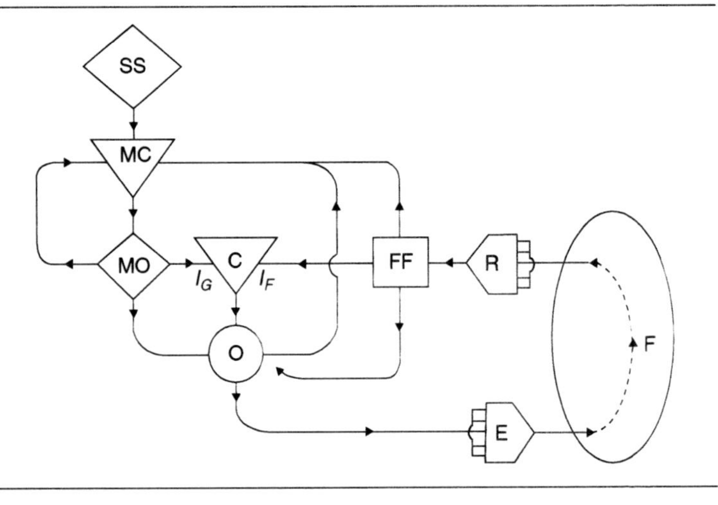


**Figure 9.5.** Continued expansion of the hierarchy of supervisory systems.

the limits of imagination constrain the possibilities for nesting, particularly given the incredibly complex network that is the human brain. MacKay does not speculate as to what might be the various forms of comparisons and evaluations in the higher-level loops, but one can imagine them to involve more complex memories, information distributed over longer epochs of time, more abstract forms of representation eventually involving symbolic systems, and more extended and elaborate forms of behavioral organization.

## 9.6. The Mental 'Out and About'

We have seen that the core of mental causation is embodied in modulations of action loops, as well as how such processes can be marshaled into the emergence of higher-level cognitive processes. There are many additional factors to consider in order to give a reasonable account for the richness of human mental function. Some of these additional elements are forms of representation and memory that are not entirely within the brain or body.

### 9.6.1. *External Scaffolding*

Clark argues for the importance of 'external scaffolding' in the emergence of higher mental processing.[24] 'Scaffolding' refers to all of the ways that an organism relies on external supports for augmenting mental processing. This process is present in animals, but particularly critical in human cognition. He writes, 'We use intelligence to structure our environment so that we can succeed with less intelligence. Our brains make the world smart so that we can be dumb in peace! ... It is the human brain *plus* these chunks of external scaffolding that finally constitutes the smart, rational inference engine we call mind.'[25]

Thus, external scaffolding plays a critical role in the emergence of efficacious mental processes. Some of the most important elements in mental causation are things we build into our environments for use in later cognitive processes, and ways in which our mental activity involves direct manipulation of the environment such as to keep some of the calculations and representations external to the brain. Human culture involves a vast array of artifacts that scaffold cognitive processing, the most remarkable of which is language.

The emergence of nonreducible properties of mind and the causal role of the mental becomes much less mysterious if we take into

[24] Clark, *Being There*, 179–192. [25] Ibid., 180.

account the continued interactions between the person and the physical and social environments—interactions that involve creation and use of external structures to scaffold cognition. Scaffolding allows us successfully to solve problems that would exceed the capacity for immediate, on-line brain processing.

### 9.6.2. *Language*

Human mental causation cannot be adequately discussed without some consideration of the contribution of language. It is the primary form of external scaffolding of higher human mental abilities. However, symbolic representation and language also make an important contribution to the highest forms of internal (off-line) processing within the human brain. It is not possible to do justice to this topic in the remainder of this chapter. However, several general observations are worth making in order to link language to our current view of mental causation.

E. Hutchins and B. Hazelhurst performed a neural network experiment that illustrates both the emergence of more and more useful symbols and the contribution of symbol systems to group problem-solving.[26] The simulation involved a group of 'citizens,' each a connectionist neural network. Input to each citizen involved both information about events themselves, and symbols from other citizens representing these events. The task was to predict the tides based on the phases of the moon. The simulation also involved sequential generations of groups of citizens that did not inherit the knowledge of the previous generation, but did inherit the symbols with their current meanings. There were two outcomes of this simulation that are important in considering the role of language in the emergence of mental causation: (1) there was a gradual evolution of better and better symbols, and (2) the improved symbols allowed later generations to learn the environmental regularities involved in the task that earlier generations could not learn. Thus, the emergence of better symbols allowed for the emergence of better group problem-solving.

This illustration fits nicely with Clark's emphasis on the social scaffolding of cognition as previously described. Linguistic symbols

[26] E. Hutchins and B. Hazelhurst, 'Learning in the Cultural Process,' in C. Langton *et al.*, eds., *Artificial Life II* (Redwood City, CA: Addison-Wesley, 1991), as reported in Clark, *Being There*, 189–190.

pre-structured individual learning and group problem-solving such as to allow later generations to accomplish a task that could not be mastered by previous generations. Language thus provides scaffolding for both internal and external problem-solving. Language provides 'structuring causes' (to use Dretske's terminology) for forms of problem-solving not otherwise possible. Language is, in Clark's words, 'a computational transformer that allows a pattern-completing brain to tackle otherwise intractable classes of cognitive problems.'[27]

Some of the contributions of language to cognition specifically described by Clark are: (1) offloading of memory into the socially maintained language environment; (2) use of labels as perceptually simple cues regarding a complex environment; and (3) coordination of the action of individuals and groups via internal self-talk, dialog, or in written plans, schedules, and so on.[28]

Deacon has also provided an important analysis of the role of language in the emergence of human thought.[29] Some of these contributions are:

1. *Distancing of action from the demands of immediate motivations and needs*: Language facilitates behavior involving delayed need-gratification by augmenting the ability to consider alternative actions through entertaining various symbolic 'what if' scenarios.
2. *The ability to form a self-concept*: The symbolized self can become the object of evaluation. Clark also describes the importance of language in allowing for second-order cognition (or meta-cognition) in the form of self-representation and self-evaluation.
3. *Expanded empathy*: We can enter into the experiences of others through emotional engagement created by stories. Such empathy also enhances the development of a Theory of Mind—the ability to model and predict the mental life of other individuals.[30]

[27] Clark, *Being There*, 194. [28] Ibid., 200–201.

[29] Deacon, *Symbolic Species*.

[30] S. Baron-Cohen, A. Leslie, and Uta Frith, 'Does the Autistic Child Have a "Theory of Mind"?' *Cognition* 21 (1985): 37–46.

4. *A virtual common mind among groups of people*: Common semantics, metaphors, and stories create cultural groups with similar worldviews.
5. *Ethics*: Language encodes communal values for judging between 'what if' scenarios.

## 9.7. Imagination and Mental Modeling

Finally, we come to an extension of embedded and embodied cognition that feels intuitively more like the robust 'mental causation' sought by philosophers. This is the mental modeling of behavior. One of the capacities that develops with a more complex nervous system is an ability to simulate action loops off-line.

An example from the work of psychologist Wolfgang Köhler strongly suggests that problem-solving solutions can be tried out in the form of internal mental modeling prior to behavioral execution even in apes.[31] Köhler describes the attempts of a chimpanzee named Sultan to obtain some bananas outside his cage. Some sticks were available, but none of them was long enough for Sultan to reach the bananas. After many unsuccessful attempts at solving the problem, Sultan quit trying and sat in the corner, apparently ruminating. He suddenly went back to the problem and immediately executed the proper solution of putting two sticks together end-to-end, allowing him to reach the bananas. It appears that Sultan was able to solve the problem via off-line, imaginative, mental modeling prior to executing the correct solution. Of course, this experience of off-line problem-solving is common to human beings, but the point of this illustration is that such off-line behavioral scenarios occur in the absence of language and self-talk.

The idea of simulations of behavior is becoming increasingly important not only in the understanding of complex cognition, but also at lower levels of nervous system operation. There is a phenomenon in accounts of visual perception that is called a 'corollary discharge.'[32] A command to make a saccadic eye-movement is

[31] W. Köhler, *The Mentality of Apes* (New York: Harcourt, Brace, 1927).

[32] B. Kolb and I. Q. Whishaw, *Fundamentals of Human Neuropsychology*, 5th edn. (New York: Worth, 2003), 403–404.

communicated both to the lower brain nuclei controlling eye movements and to the visual cortex, the latter signal being the corollary discharge. Based on this corollary discharge, the visual input created by the eye movement is not interpreted by the visual cortex as a movement of the external world. The visual world appears stationary as we move our point of eye fixation from place to place. The corollary discharge is presumed to be a predictive simulation of the sensory impact of the anticipated eye movement.

Finally, recent work on mirror neurons, described earlier, is also informative.[33] Mirror neurons are neurons in the supplementary motor cortex that would be involved in initiating and controlling a specific movement, but that also become active while a monkey passively observes another monkey making the same movement. The perception of the action of another animal involves an implicit simulation of making a similar action—a simulation rich enough to serve as the basis for imitation.

Given these examples, it appears that the brain is adept at sensorimotor simulation. Such simulations undergird the notion of the ability of the brain to run off-line simulations of problem-solving action loops. Based on memories of past experiences, action–feedback–evaluation–action loops can be simulated off-line in order to predict the likely consequences of any particular action.

When combined with the ability to use language, off-line scenarios become a form of inner problem-solving using self-talk. A number of cognitive psychologists believe that all thinking is done with inner speech. Subjectively, inner speech seems to be the predominant mode of conscious thought. It is, however, not likely to be the only form of off-line mental modeling available to the human brain. The main point is that internal, imaginative problem-solving and thinking *piggyback on action-loops* via the ability to run sophisticated sensorimotor simulations off-line.

## 9.8. Consciousness

In the background of much of what has been discussed thus far has been the phenomenon of consciousness. What is consciousness?

[33] Rizzolatti, op. cit. (n. 19).

What role does it play in mental causation? How is it physically embodied? These are complex questions about which volumes have been written. I will briefly describe some theories and neuropsychological studies of consciousness as they relate to the concepts of action loops and causes-in-process.

### 9.8.1. *Models of Consciousness*

The term 'consciousness' covers several forms of animal and human functioning. Merlin Donald suggests that there are three forms of consciousness (or 'awareness,' as he calls them).[34] *Level 1 Awareness* involves sensory binding and selective attention such as to achieve perceptual unity. Thus, action loops would be modulated by formed images, rather than disconnected sensory properties. According to Donald, *Level 1 Awareness* occurs as the brain's sensory cortices interact with the thalamus, and may be dependent upon the establishment of coherence of electrical activity between various brain areas. *Level 2 Awareness* involves short-term or working memory. A working memory allows larger amounts of information to be bound, related, and held over seconds to minutes. This, in turn, permits more active mental processing, modulation of action based on more temporally extensive information, and greater autonomy from immediate environmental influences. Recurrent interactions between the frontal lobes and posterior sensory cortex are critical for this active form of mental processing in what is called working memory.[35] This second form of awareness also includes body awareness and a stable model of the environment that when coupled with information from a historical memory in animals with more complex brains (i.e. in higher primates) begins to allow for a rudimentary self-system. Finally, *Level 3 Awareness* involves even longer-term awareness and governance and even more intense voluntary control of action. Here the evaluation and modulation of action can occur with respect to a very long time frame (minutes to hours or even days), enhanced

[34] Merlin Donald, *A Mind So Rare: The Evolution of Human Consciousness* (New York: Norton, 2001).

[35] Joaquin M. Fuster, 'Prefrontal Neurons in Networks of Executive Memory,' *Brain Research Bulletin* 42 (2000): 331–336.

by symbolic processes and enculturated criteria for the evaluation of behavioral outcomes (i.e. values).[36]

Philosopher Thomas Metzinger provides an account of consciousness and the sense of self, expressed in terms of representational systems.[37] Consciousness is embodied in currently activated 'world models.' He suggests that consciousness can be broken down into four increasingly rich variants, each adding additional characteristics and constraints to the world model. First, there is *minimal consciousness* that merely involves a representational state constituted by a world model within a window representing presence (or 'now'). Second, Metzinger proposes *differential consciousness* in which the world model is meaningfully and hierarchically segregated and there is a temporal (sequential) structure. In differential consciousness, the world map provides information about a dynamic and segmented scene. *Subjective consciousness* begins to approach that which characterizes human consciousness. Here the world model includes a self-model, allowing for a consciously experienced first-person perspective. The world map is now centered continuously around this self-representation. Finally, there is *cognitive subjective consciousness.* What is added here is the ability to elicit and process *off-line* various world models (including a representation of the self). 'These systems would be able to engage in the activation of globally available representational structures *independently of current external input*.'[38] Thus, this sort of conscious system could 'engage in future planning, enjoy explicit, episodic memories, or start genuinely cognitive processes like the mental formation of concepts.'[39] At this level of consciousness, cognitive systems would be able to 'represent themselves as representational systems.... They would be thinkers of thoughts.' Thus, Metzinger describes consciousness as an activated model of the world, where the dynamic complexity of the model allows for richer forms of consciousness.

Metzinger's theory of consciousness can be fitted into our framework where the fundamental basis of agency is the modulation of action-loops simply by presuming that, at the most basic level, the

[36] Ibid., 178–204.

[37] Thomas Metzinger, *Being No One: The Self-Model Theory of Subjectivity* (Cambridge, MA: MIT Press, 2003).

[38] Ibid., 205. Author's italics.

[39] Ibid.

organism is modeling (or mapping) the world in terms of immediate past, current, and potential action–feedback–evaluation–action loops. Particularly critical for a notion of the highest forms of mental causation is Metzinger's suggestion that the most sophisticated form of consciousness (cognitive subjective consciousness) allows for off-line elicitation and manipulation of world models involving implications for action, that is, simulated action loops.

### 9.8.2. *A Plausible Neuroscience of Consciousness*

Gerald Edelman and Giulio Tononi present a neurobiologically plausible model of consciousness anchored in neuroanatomy, neurophysiology, and clinical neurology.[40] With respect to a phenomenological description of consciousness, they suggest a two-part model roughly consistent with both the three-level typology developed by Merlin Donald and the four-level theory of Metzinger. Edelman and Tononi speak of *Primary Consciousness* that is evident in the ability of animals to 'construct a mental scene,' but with limited semantic or symbolic content, versus *Higher-order Consciousness* that is 'accompanied by a sense of self and the ability in the waking state explicitly to construct past and future scenes. It requires, at minimum, a semantic capacity and, in its most developed form, a linguistic capacity.'[41] Both forms of consciousness include body awareness and some degree of reference in space. For Edelman and Tononi, the fundamental characteristics of consciousness are *unity* (i.e. it is experienced at any moment as an undivided whole); *informativeness* (i.e. one conscious state is experienced out of billions of possibilities); and *privateness* (i.e. the experience is only accessible to the conscious agent, although reflections of the state are available both in brain physiology and in the behavior and verbal descriptions of the agent).

Edelman and Tononi argue further that a state of consciousness and its content (whether Primary or Higher-order Consciousness) is a temporary and dynamically changing process within the cerebral cortex that is characterized by a high degree of functional interconnectedness between widespread areas created by rapid, two-way

[40] Gerald M. Edelman and Giulio Tononi, *A Universe of Consciousness: How Matter Becomes Imagination* (New York: Basic Books, 2000).

[41] Ibid., 103.

recurrent (or reentrant) neural interactions. They call such a state of widespread functional integration a 'dynamic core.' A dynamic core is a complex and highly differentiated neural state that includes different subsets of neurons or neural groups as the core changes from moment to moment. It is the specific neural groupings involved, and the functional relations between the groupings, that define the nature and content of consciousness at any particular moment.

Dynamic cores (and thus consciousness) are characteristic of the mental life of all animals to the degree that the cerebral cortex has sufficiently rich recurrent connections. As they express it, consciousness becomes possible by a

> transcendent leap from simple nervous systems, in which signals are exchanged in a relatively insulated manner within separate neural subsystems, to complex nervous systems based on reentrant dynamics, in which an enormous number of signals are rapidly integrated within a single neural process constituting the dynamic core. Such integration leads to the construction of a scene relating signals from many different modalities with memory based on an entire evolutionary history and an individual's experience—it is a remembered present. This scene integrates and generates an extraordinary amount of information within less than a second.[42]

The higher-order consciousness characteristic of humans comes into play when symbolic representations and language are incorporated into dynamic cores, including the ability to represent the self, and to use symbols to note time (past, present, and future). Thus, higher-order consciousness is a developmental achievement dependent on social interactions and social scaffolding.

Edelman and Tononi point out that the anatomy of the cerebral cortex is unusual compared to the rest of the brain in having a high degree of two-way reentrant interconnections. Most of the intercellular interconnections in other parts of the brain are essentially one-way, or if there are feedback paths, they are indirect and multicellular paths. However, the interconnections within the cerebral cortex, as well as the loops between the thalamus and the cortex, are reentrant pathways—that is, there are immediate and direct return connections.

[42] Ibid., 211.

Regarding the neurophysiology of consciousness, a dynamic core forms by establishment of a network of temporary functional linkages that are marked by spatially differentiated, coherent, high frequency brain waves (i.e. EEG signals indicative of patterns of nerve cell responding). Consciousness is incompatible with undifferentiated electrical signals characteristic of the pervasive, very slow, high amplitude EEG patterns of non-dreaming sleep and coma.

Since a dynamic core is a functionally interactive subset of the cortex, not all of the cerebral cortex is admitted to a dynamic core at any particular moment. The contextual sensitivity of a dynamic core is created by inputs from sensory and memory systems that are not directly involved in dynamic cores themselves. Similarly, direct motor output is influenced by the current functional state of the core without necessarily being directly included in the recurrently interactive networks. Dynamic cores are also influenced by and exert influences on other brain systems, including the basal ganglia (motor regulation), cerebellum (learned automatic behavioral sequences), hippocampus (memory), amygdala (negative emotions), and hypothalamus (vegetative, autonomic, and hormonal regulation). However, these brain areas also do not become directly involved in a dynamic core because they do not have the reentrant form of reciprocal connections that are characteristic of interneuronal cortical connections. Other important inputs to a dynamic core are contributed by the evaluative systems of the locus coeruleus (norephinephrine), the raphé nucleus (serotonin), and the ventral tegmental area (dopamine) that signal aspects of the significance of information—but, again, these structures cannot directly participate in a dynamic core.

The Dynamic Core Hypothesis gives a reasonable account of the relationship between conscious control of behavior and automaticity. In the early learning stages of difficult tasks or behaviors, the performance must be incorporated in and regulated by the dynamic core. However, once the performance is well learned (automated) it can go forward efficiently based on the activity of a smaller subgroup of cortical or subcortical neurons that do not need to be incorporated into the current dynamic core for adequate functioning. For example, under normal circumstances the basic lexical and syntactic aspects of language processing can go on in the background while the dynamic core embodies the ideas that one is attempting to express. Unconscious thought processes that influence the core are those cortical and subcortical processes that are not directly incorporated into the dynamic core, but exert some level of influence on the current shape of the core. Explicit memory, then, is the re-elicitation of a particular form of dynamic core that is based on previous experience. Unconscious, implicit memories are unconscious precisely because they do not become bound to the dynamic core.

The Dynamic Core Hypothesis is consistent with what one would expect if the brain were an extremely rich version of a complex dynamic system. The massively recurrently connected cerebral cortex is beautifully suited for emergence of the properties described by complex dynamic systems analysis.[43] Thus, the dynamic core that embodies a particular moment of consciousness can be described as a point in a very high-dimensional space of potential patterns of neural activity. The probabilities of occurrence of various dynamic cores (points within this high-dimensional neural space) can be represented by a probability topology, with the probabilities of occurrence of the most likely dynamic cores representing attractors within the dynamic system.

### 9.8.3. *Disturbances of Consciousness*

The role of consciousness in behavioral flexibility and adaptability can be readily understood by considering disabilities and residual abilities of individuals with disturbances of consciousness due to forms of brain damage.

Much has been written in philosophy of mind about the phenomenon of 'blind sight.' This occurs in persons with damage to the visual cortex on one side of the brain, who consequently lose the ability to see anything in the opposite side of the visual world (i.e. loss of conscious visual perception). 'Blind sight' refers to the ability of some of these individuals to reach out and intercept a target that is moving within the area of blindness that they nevertheless report that they cannot see. Some would suggest that this proves that phenomenal

[43] Alicia Juarrero, *Dynamics in Action: Intentional Behavior as a Complex System* (Cambridge, MA: Bradford Books, MIT Press, 1999).

awareness is unnecessary for action. The correct interpretation is that phenomenal awareness is unnecessary for some primitive forms of action (intercepting a moving target), but necessary for other forms of action. Thus, when the target moves from the area of blindness into the area of visual perception and awareness, the person can tell specifically what the moving object is. Phenomenal awareness of stimulus meaning allows the person to consider different possibilities for action with respect to the flying object based on conscious perception of the nature of the object, and to name the object. Thus, phenomenal awareness opens the possibilities for a wider and more flexible range of adaptive actions.

The contribution of consciousness to behavior is also apparent in the contrast between individuals with anosognosia versus individuals with other forms of agnosia. Anosognosia is the clinical term for a disorder (usually from damage to the right parietal lobe) that involves paralysis of one side of the body, with *adamant denial* of the disability. Somehow the body representation of these individuals cannot be updated to provide information to conscious awareness regarding their unilateral paralysis. Thus, they are not able to adjust their behavior to take the paralysis into account. In contrast, other forms of agnosia (meaning 'absence of knowledge of'), such as the inability to recognize faces, create deficits regarding which the patient is acutely aware, and this awareness allows them to adjust their behavior to accommodate the deficit.

One extreme case of the impact of awareness in learning to compensate for agnosia was described by the famous Russian neuropsychologist A. R. Luria. Luria reproduced (with his own accompanying commentary) portions of the painstakingly written 3,000-page diary of a Russian soldier called Zasetsky who had suffered a brain injury during World War II that left him with very severe deficits in visual perception (agnosia) and a very limited working memory. He had great difficulty reading or writing because he both struggled to recognize letters and could not hold the letters in mind long enough to read words or sentences. What is remarkable about Zasetsky's story was his acute consciousness of his limitations and his persistence over a twenty-year period in learning to overcome his deficits sufficiently to write his extensive diary. Zasetsky's conscious awareness of his deficiencies allowed him to marshal adaptive strategies that are not brought to bear by persons with anosognosia who are unable to consciously recognize their disability. Since conscious awareness is necessary for such complex forms of behavioral flexibility, it cannot be epiphenomenal.

Finally, the fractioning of the content of consciousness created by severing of the connections between the right and left cerebral hemispheres (resulting in individuals with a 'split-brain') is also revealing of the nature of consciousness. The surgical procedure (done for the relief of otherwise intractable epilepsy) severs all neural connections between the cerebral cortices, including the 200 million axons of the corpus callosum. The result is that sensory information, cognitive processing, and motor control within each hemisphere are isolated from processing in the other hemisphere. For example, information occurring only in the left side of the patient's visual world will be seen only by the right hemisphere and, therefore, can be responded to only by the patient's left hand (controlled by the right hemisphere that is privy to the visual information). What has been suggested is that the split-brain patient now has split consciousness. Each hemisphere knows and processes different information without the benefit of the sharing of information from the opposite hemisphere. There are now two separate domains of consciousness, at least at the level of higher-order information and behavioral control. Expressed using Edelman's and Tononi's model, the absence of the large body of interhemispheric reentrant neural connections of the corpus callosum means that a dynamic core becomes isolated within one or the other hemisphere and will not be able to include processes and information available in the opposite hemisphere. However, the continued unity of the lower brain (not split by this surgery) means that the lower-level evaluative and regulative systems of the person remain integrated, such that the person maintains a unitary sense of being.

### 9.8.4. *Consciousness and Mental Efficacy*

The models of consciousness described here help to sort out the variety of phenomena that have been the referent of the term 'consciousness.' The Dynamic Core Theory of Edelman and Tononi provides a plausible neurophysiological model that is well supported by the research literature and by clinical observations. This model both suggests how phenomenal consciousness might be embodied and enlightens us regarding many of the broader cognitive implications of a state of consciousness. Observations of neurological patients with various disturbances of consciousness make it clear that disruption of consciousness has major implications for both thought and behavior. The presence or absence of phenomenal awareness is consistently linked to the presence or absence of higher forms of behavioral flexibility, adaptability, and agency. All of this suggests that consciousness opens persons up to wider behavioral options. Consciousness makes behavioral alternatives explicitly available to awareness, allowing for modulation of real or simulated action-loops with respect to a wide variety of information and memories, all bound together in the temporary workspace that is a dynamic core.

## 9.9. Summary

This chapter has attempted to outline the bare biological bones of the emergence of efficacious mental function, from lower animals to humans. The emergence has come about by evolutionary development that has allowed: (1) more and more elaborately nested hierarchies of action-loops; (2) increasing capacity for learning, leading to the embodiment of new criteria for evaluation of action; (3) the capacity to run action-loops off-line, in sensorimotor simulations; (4) development of cortical systems capable of supporting more and more complex dynamic cores; and (5) symbolic representation (language) and the enormous amount of external scaffolding inherent in human culture. From these capacities, which are elaborations of basic biological processes, have emerged increasingly efficacious forms of mental processing.

In this essay, I have been motivated by the hope that the examination of these bare bones might reduce some of the mysteries of the emergence of mental causation. I believe these biological bones can provide a skeletal framework for understanding more complex and robust forms of human mental causation, including those forms of mental causation necessary for presuming human beings to be responsible moral agents and capable of meaningful religiousness.[44]

[44] Figures from Mackay, *Behind the Eye*, are reprinted with permission.

# Part III

# Theology

# 10

## Reductionism and Emergence: Implications for the Interaction of Theology with the Natural Sciences

*William R. Stoeger, SJ*

### 10.1. Introduction

Nancey Murphy and Robert Van Gulick have critically surveyed and presented various types of both reductionist and emergentist theses, indicating implications for various focal problems in philosophy and philosophy of science. In this essay, I shall presume those discussions and concentrate on setting out key relationships between reductionism and emergence on the one hand and issues central to relating theology with the natural sciences on the other.

To set the stage, I shall first relate 'theology' to 'faith,' and then compare and contrast theology and natural sciences as sources of knowledge. Along the way I shall provide a characterization of philosophical knowledge. After describing in summary form some of the important issues in relating theology and the natural sciences. I shall, thirdly, discuss the contributions a refined and careful treatment of reductionism and emergence would make to their coherence. Fourthly, I shall indicate the important philosophical and theological resources available to guide further work in this area. Finally, I shall suggest a portrait of divine action which is consistent with Christian revelation and enriched by our knowledge of the natural sciences. As we shall see, a careful treatment of reductionism and emergence is essential for a rich and proper understanding of divine creative action in both nature and history which is consonant with our scientific knowledge of nature, its dynamisms, and its relative autonomy. It will also help us to understand who we are as human beings, capable of cognition and free action, open to the transcendent, and yet immersed in the world. Furthermore, such an account of divine action has indirect bearing on the origin of value and meaning, the incompleteness of creation, cosmic and personal destiny, and our appreciation of the roles of transience, suffering, and both natural and moral evil.

## 10.2. Theology, Science, and Philosophy

In referring in this volume to the interaction between theology and the natural sciences, we have intentionally avoided using the words 'religion,' 'faith,' and 'spirituality,' though obviously the natural sciences have a significant impact on all of these. This is to signal our primary focus on the cognitive core, the theological dimensions of faith, religion, and spirituality and its relationship to the cognitive contents of the natural sciences. Both science and theology represent ways of knowing, disciplines directed towards enhanced understanding. As such, we want to probe the connections, similarities, and differences between them. Here we focus on the reducibility of complex phenomena to simpler underlying processes and relationships, the emergence of qualitatively new phenomena, and the enrichment our understanding of them brings both to the natural sciences and to theology. At the same time, we strive to recognize the meaning theology brings to these pervasive fundamental features of reality.

The most fundamental realities underlying theology are revelation and faith. In fact, following Avery Dulles, we may say that faith is the positive ongoing response in discernment and commitment to perceived divine revelation.[1] Theology, according to the often quoted

[1] Avery Dulles, SJ, 'The Meaning of Faith Considered in Relationship to Justice,' in *The Faith That Does Justice*, ed. John C. Haughey (New York: Paulist Press, 1977), 10–46.

classical definition, is 'faith seeking understanding,'[2] the faithful person or community struggling to understand experience and the world that it reveals in light of his, her, their experience of faith—in light of the perspectives and knowledge, the discernment and commitment out of which they live. Theology's object is 'revealed truth.' Thus, as already indicated, theology expresses, however inadequately and provisionally, the cognitive content of faith. It is this which is to be related to and integrated with the knowledge originating from the natural sciences.

If we now compare theology and the natural sciences as ways of knowledge, we can fill out our very brief characterization of theology above by saying that theology is the discipline directed towards understanding God, the presence, action, and revelation of God in our world, ourselves, and our response to that transcendent reality. The natural sciences, in contrast, are disciplines oriented towards a detailed qualitative and quantitative understanding and modeling of the regularities, processes, structures, and interrelationships ('the laws of nature') which characterize reality—relying on rigorous repeatable analysis and experiment (observation). In reflecting on these two very different disciplines, it is very helpful to articulate the general similarities and differences between them.

As is obvious from our discussion so far, both the natural sciences and theology are oriented towards truth—towards knowledge and understanding—although they are each concerned with different types of truth and have different motivations for attaining that truth. Thus, they each presuppose an absolute commitment to truth and to understanding. Other common features are: (1) both use carefully developed methods and procedures in pursuing knowledge and understanding, and definite criteria to validate their conclusions; (2) both are practiced and experienced in communities; (3) both seek to go beyond where they are now; and (4) both share common personal and cultural fields—that is, they automatically influence and interact with one another in individual thinkers and believers and in society at large.

[2] Anselm of Canterbury, *Proslogion*, Preface (Prooemium), in *St. Anselm's 'Proslogion' with 'A Reply on Behalf of the Fool' by Gaunilo and 'The Author's Reply to Gaunilo,'* trans. M. J. Charlesworth (Oxford: Clarendon Press, 1965), 101–105; see also the translator's accompanying commentary, p. 53.

We can now easily see differences between the sciences and theology; these are characterized by: (1) different foci—that is, different interests and different types of fundamental questions and perspectives; (2) different fundamental assumptions or presuppositions; (3) different bodies of relevant evidence and methods, and different criteria for validation; (4) different knowledge products and different application of those products; (5) different communities which produce and use scientific and theological knowledge; and (6) the very different social and cultural impact of each discipline. We have mentioned these similarities and differences in order to provide a context for focusing on the issues at hand—the implications of reductionism and emergence for the interaction of theology and the natural sciences. Further discussion can be found in the recent theology–science literature.[3]

Before leaving these general considerations, it is worthwhile to complete our disciplinary summary by locating philosophy with respect to the natural sciences and to theology. Philosophical knowledge, though it does not rely upon divine revelation, is oriented toward the pervasive aspects and the ultimate issues of reality and the possibility of our knowledge of it—and not toward well-defined, easily isolated phenomena, as the natural sciences are. Philosophy often deals with questions which are either presupposed by or neglected by the sciences.

## 10.3. General Issues Relating Theology and the Natural Sciences

There is a fairly large number of issues which either must be considered when studying the relationships of theology with the sciences as disciplines oriented towards knowledge and understanding or provide common ground for their interaction. First, in order to relate theology and the sciences in any meaningful way, it is crucial to recognize the competencies and limitations of each, which are closely

[3] See especially William R. Stoeger, SJ, 'Contemporary Cosmology and Its Implications for the Science–Religion Dialogue,' in *PPT*, 219–247; *idem*, 'What Contemporary Cosmology and Theology Have to Say to One Another,' *CTNS Bulletin* 9/2 (Spring 1989): 1–15.

related to some of the central points summarized in the previous section. Obviously, theology is not equipped to, nor even interested in, describing and modeling the laws of nature more accurately at their various levels of operation. This is the realm of the sciences. Nor are the sciences capable of dealing with questions of God and of ultimate value and meaning. These are the objects of theology and philosophy. Furthermore, the languages used by the natural sciences and by theology are very different not only in what their terms signify, but also in the context within which they are to be understood. We cannot provide any treatment of these issues here, but it is helpful to point them out as part of the background against which to consider reductionism and emergence.

Once we have recognized the complementary competencies and limitations of theology and the natural sciences and negotiated their differing languages, we are faced with the epistemological issue of critical realism. Since both theology and the natural sciences are oriented towards knowledge and understanding, they claim to possess cognitive content. In fact, one of the key issues concerning theology and science is the relationship of the cognitive content of one to that of the other. Critical realism is the philosophical stance which affirms—with claimed justification—that knowledge is capable of reaching, or describing, what really exists outside of ourselves, of specifying to what extent and under what conditions it is reliable. There are ways in which critical realism with regard to the natural sciences can be supported over against idealism and instrumentalism. It is much less clear that critical realism can be supported in theology, though much recent work in theology and science has argued that it can.[4] Among the possible obstructions to doing so are the severe inadequacy of theological statements' ability to describe God and God's relationships with creation—even if they can disclose God and God's relationships, it is unclear that they can do more than that—and the very different way in which validation of theological

[4] For a summary of the approaches taken and the issues involved, Ian G. Barbour, *Religion and Science: Historical and Contemporary Issues* (San Francisco: Harper SanFrancisco, 1997), 110–121; Robert John Russell, 'Ian Barbour's Methodological Breakthrough: Creating the "Bridge" between Science and Theology,' in *50 Years in Science and Religion: Ian G. Barbour and his Legacy*, ed. R. J. Russell (Aldershot: Ashgate, 2004), 27–37.

assertions is achieved compared to the sciences. I suspect, along with others,[5] that critical realism can be extended to theology as long as it is based on a broad and adequate account of human rationality, a careful approach to analogy and a recognition of the different ways truth can be validated in theology and in science. The most promising basis for the first part of this support is the operation of well-informed imagination and Peircean retroduction.[6]

Practically all our knowledge and understanding of the laws of nature—the regularities, processes, structures, and relationships which characterize reality at each level of organization—derives from the natural sciences and our philosophical reflection on their findings. This includes our understanding of the potential reducibility of more complex phenomena to simpler ones and the pervasive emergence of radically new systems and entities from simpler and more fundamental objects and interrelationships. The reliability of our scientific knowledge is gradually established, as theory and controlled observation and experiment improve our descriptions and our models of the phenomena. We come to recognize that the natural sciences are giving us significant, but still imperfect and provisional, purchase on reality as it actually functions. This is the critical realist perspective.

The important question with regard to theology then is: Does the reliably modeled reducibility and emergence revealed by the natural sciences and critically accepted by philosophy of science enable, support, or at least allow the realities and relationships which theology claims to discover? If not, then there is a conflict between the conclusions of the two disciplines. In this case, either one or the other is in error—either with respect to its conclusions or with respect to its interpretation of its conclusions.

Thus, the key connection between reducibility and emergence and the science–theology interaction is: How do the forms of reducibility and emergence which can be reliably attributed to the natural world affect key theological issues—interpretations of our experience of our relationship to the divine and the understandings which flow from

[5] Barbour, *Religion and Science*, 117–121; Russell, 'Barbour's Methodological Breakthrough.'

[6] Ernan McMullin, *The Inference that Makes Science* (Milwaukee, WI: Marquette University Press, 1991), 112.

that? Is there coherence or consonance between the findings of the natural sciences and the conclusions of theology? Or is there, rather, dissonance and deep incompatibility? And how are such coherence and consonance, or dissonance and incompatibility, to be interpreted and resolved?

Indirectly supporting this critical realist approach, which has been so prominent in work in theology and science, is the ongoing moderately successful quest for understanding, coherence, and unity among the sciences, philosophy, and theology by many scholars. They have experienced a deep and growing awareness of the similarities and differences among these disciplines as well as the common personal and cultural fields they inhabit, and have begun to establish fruitful connections between them. The very fact that they are achieving some success validates the quest.

At this point, it is useful to discuss some of the specific points at which the reducibility and emergence of novelty described by the sciences will significantly impact our understanding and interpretation of philosophical or theological conclusions. It is also useful to discuss where philosophical or theological perspectives may, in turn, force a re-examination or reinterpretation of scientific models and conclusions. Obviously, the important issue of the limits and competencies of the natural sciences, of philosophy, and of theology arises here. These may not be known precisely to begin with, but in the course of the critical confrontation among disciplines they become better recognized and articulated.

Perhaps the most important and fundamental topic where this interaction will be significant and far-reaching is that of divine action—both God's universal creative action in nature and God's special action in history in virtue of God's special relationship to individuals or to peoples.[7] First, how can we characterize 'the transcendent' (God)? Secondly, how does God interact with the universe? With all of nature? With conscious and self-reflective (self-transcendent) nature? The strong tradition of all three major monotheistic religions is that

[7] There has recently been a great deal of discussion of divine action from the perspective of the natural sciences (but not explicitly with regard to emergence). This research is well represented by the Vatican/CTNS (Center for Theology and the Natural Sciences, Berkeley, California) series of volumes on divine action: *QCLN*, *CC*, *EMB*, *NP*, and *QM*.

God is the Creator—the source of all being and order, through whom and in whom all things exist. As Creator, God would be the ultimate author of 'the laws of nature' as they actually function.[8] This is God's universal creative action. The traditional model used to articulate this is that of *creatio ex nihilo*, creation from absolutely nothing. In using this term, the essential content is the absolute dependence of reality on God the Creator at every moment, not some transcendent process which rivals or replaces the processes of nature. Insofar as we can analogically attribute agency or causality to God as Creator, God is then considered 'the primary cause,' the cause unlike and more fundamental than any other cause. All other causes are 'secondary' and ultimately rely on, though are not completely determined by, the primary cause.

In addition, there is God's special action in history, on behalf of God's chosen people, through apparent orchestration of historical events such as the Exodus and the deliverance of Israel from captivity in Babylon, through the prophets, and in New Testament times through the Incarnation, the Resurrection of Jesus, the sending of the Spirit, the answering of prayer, and so forth. Are these two categories of divine action consonant with what we know from the natural sciences concerning reducibility and emergence? If not, can the dissonance or conflict be resolved—to the satisfaction of the canons of both disciplines? By all accounts, there *seems* to be acceptable consonance or coherence between scientific descriptions and accounts of the laws of nature and a proper understanding of God's universal creative action. This is due, primarily, to the fact that the natural sciences are mute with regard to the ultimate sources of existence and order. There is a considerable challenge, however, in reconciling special divine action with the natural sciences.

Closely related to this issue of special divine action—in fact, an essential part of it—is both an understanding, vis-à-vis a critical interpretation of scientific findings and constraints, of our pervasive experiences of the transcendent and the openness of the personal and the social to divine revelation. Among our experiences of transcendence are abstract thinking; the construction of meaning;

[8] William R. Stoeger, SJ, 'Contemporary Physics and the Ontological Status of the Laws of Nature,' in *QCLN*, 209–234.

poetic and religious experience; the apparently absolute claims of truth, beauty, goodness, and authentic love; the witness of lives of radical authenticity, holiness, heroism, and self-sacrifice. Can these phenomena, as they are traditionally interpreted by philosophers and theologians, be reconciled with our scientific and philosophical understanding of reducibility and emergence? It is not my intention to answer these questions, but merely to pose them with some emphasis. Towards the end of this chapter, I shall briefly outline an approach for answering them.

## 10.4. Contributions of Accounts of Reducibility and Emergence to Science and Theology

As an introductory reflection on the implications of reductionism and emergence for relating theology with the natural sciences, we must include the actual or anticipated benefits that precise, insightful, and well-supported descriptions of reducibility and emergence are making, or can make, to the natural sciences, to philosophy, and to theology.

First, arriving at a carefully differentiated, well-conceived, well-tested, and *properly limited* attribution of reducibility to the full range of physical, chemical, biological, neurological, psychological, sociological, and religious phenomena is helping us, and will help us further, to describe more accurately and with greater nuance some of the key relationships between the many different levels of organizational complexity. This includes coming to a more precise appreciation of the limits of the various types of reducibility and the importance of genuine modes of emergence. At any stage of our investigations, however, our conclusions are provisional. Furthermore, we need carefully to avoid extending our conclusions beyond those levels where they can be validated. This increasingly nuanced description of the possibilities and limits of reducibility also helps the natural sciences recognize and appreciate more fully the power and competence of their methods and procedures as well as their own limits.

If it turns out that causal reducibility really holds, which also implies determinism, then there can be no real human freedom, and the avenues of divine revelation are limited to those which are bottom-up. If so, the transcendental aspects of our experience are simply manifestations of basic causal structures and influences. So, in fact, would be all emergent phenomena! We would still have to understand these manifestations and how their disguised appearances surface at higher levels from lower-level causes. There are strong indications that causal reducibility—and rigid causal determinism—do not hold. But it is important here to be clear about what the consequences would be—for scientific conclusions regarding emergent phenomena, and for the theological and philosophical issues we have broached. It is worth noting that, even if causal reducibility and determinism span all levels of complexity, there would still be room for *creatio ex nihilo* and a need for primary causality. This is simply because, however tightly the phenomena and levels of organization in nature are causally linked from the bottom up, an ultimate explanation for its existence and order is required.

Secondly, a more complete and nuanced account of various types of emergent phenomena, as presented by Deacon[9] and summarized by Murphy,[10] greatly enriches our scientific understanding of complexity and, indirectly, our philosophies of nature and of science and our theology of creation. All these disciplines reflect upon what the natural sciences reveal. In particular, such increased understanding of emergent phenomena—of the ways in which new and more complex entities and behaviors emerge from simpler and more fundamental entities and behavior—would lead to: (1) a more complete and adequate account of 'the laws of nature,' particularly to those pertaining to self-organization and information generation; (2) more precision about the various types of reducibility and their limits; (3) ongoing refinement of our understanding of causality and relationality—bottom-up, top-down, whole–part, intentional—as well as of the significance of the development of functionality, teleonomy and teleology within systems; (4) a better understanding of the relationships between inanimate and animate, pre-sentient and sentient, pre-conscious and conscious capacities and behaviors; (5) a better understanding of what 'spirit,' or 'soul,' is, and of the relationships

[9] Deacon, Chapter 4, this vol.

[10] Murphy, Chapter 1, this vol.

between matter and 'spirit,' body and 'soul,' brain and mind (this directly impacts the important issue of dualism, and the mystery that is matter); (6) an account of—or at least help in understanding—the openness of conscious beings to the transcendent, and of the immanence of the transcendent; (7) the origin of value and meaning.

As an example, we focus briefly on contribution (5). When we consider the higher reaches of emergent phenomena, particularly consciousness, understanding, and goal-directed action, we encounter a key philosophical/theological issue—the relationship between matter and 'spirit,' brain and mind, body and 'soul.' How are we to understand these relationships? How should we conceive the soul or the spirit? We shall not enter into extensive discussion here,[11] but simply point out that we really need to avoid any type of substance dualism and to stress the importance of constitutive relationships, both internal and external, in realizing the tremendous potentialities of matter. Elsewhere we have suggested that, at least at the higher levels, it is this network of constitutive relationships, including the metaphysical ones, that we should identify with 'soul' or 'spirit.'[12]

Related to the issue of dualism is the question of human freedom and intentionality (purpose). Are we really free to act within the obvious constraints our bodies and environment impose? Or is this an illusion? How can human freedom of choice be reconciled with reducibility? It is already implicit in the questions regarding special divine action, which I shall treat more fully in section 6.

In turn, this increased understanding of the natural world contributes: (1) key understandings and perspectives necessary for a more thorough account of how God acts in and through the laws of nature at every level; (2) confirmation of the exclusion of God acting as an intervener, as a secondary cause, within creation, thus reinforcing the formational and functional integrity of creation itself, and of nature—its relative autonomy (at the level of secondary causes, there is nothing missing that must be supplied by God to complete

[11] See the various essays in *NP*.

[12] William R. Stoeger, SJ, 'The Mind–Brain Problem, the Laws of Nature, and Constitutive Relationships,' in *NP*, 129–146, especially 144–146. Constitutive relationships are those which contribute to making an entity or organism what it is. Practically all of them are dynamic.

creation, or to enable new things to emerge); (3) further elaboration of divine immanence and transcendence, as well as of divine hiddenness and vulnerability within creation; (4) clarity concerning the possible need for 'the windows of opportunity' provided by quantum indeterminacy to enable God's special action in history; (5) increased understanding of the meaning and role of transience, suffering, natural and moral evil in nature; (6) possible light on special divine action in history in terms of the openness of self-reflective consciousness to the 'immanently transcendent' and in terms of personal relationship; (7) improved appreciation of the incompleteness of creation, and of personal, communal and cosmic destiny—how can theology, and certain brands of philosophy, coherently affirm a purpose and a directionality in the cosmos, along with a positive destiny for it and its creatures, when the sciences are unable to do so? Although it is far beyond the scope of this paper to discuss these contributions, I shall briefly reflect more fully on each of them in section 10.6.

## 10.5. Resources Available

Outside the natural sciences and the philosophy of science themselves, there are a variety of philosophical and theological resources available for helping us appropriate, assess, and interpret conclusions about reducibility and emergence. We have already briefly discussed two of these: critical realism and the notions of primary and secondary causality associated with *creatio ex nihilo*. Both of these resources are being used extensively and flexibly in research on the relations between science and theology.

A third resource that has been very helpful and promises to continue to be important is critical work on the meaning of 'the laws of nature,' in particular the distinction between the laws of nature as the regularities, processes, and relationships which actually function in nature, and the laws of nature as our imperfect, provisional descriptions of those regularities, processes, and relationships. Additionally, it seems crucial to recognize that some of the actual laws of nature—those that deal with the personal and the transcendental aspects of experience—may be beyond the capabilities of the natural sciences to investigate and describe.[13] And yet they are essential for understanding certain aspects of our world.

Other resources which are being used and can be used more fully in these investigations and discussions are: (1) process theology (Alfred North Whitehead, Charles Hartshorne, John Cobb, David Griffin, Ian Barbour, Joseph Bracken);[14] (2) the work Peacocke and others have done on top-down and whole–part causality, complementing the usual bottom-up causal analyses;[15] (3) the opportunities provided by quantum indeterminacy for an explicitly science-based, non-interventionist approach to special divine action, as pioneered by Robert J. Russell;[16] (4) the arguments and perspectives provided by work and reflection on the Anthropic Principle.[17]

## 10.6. Emergence and Divine Action

I conclude this paper with a brief sketch of how we may conceive both universal creative and special divine action in light of what we know from the sciences. This treatment relies on our best descriptions of the limits of reducibility and the possibility for emergence, as supported by careful analysis and interpretation of the data from science.

What are the consequences of our discoveries about reducibility, determinism, and emergence for divine action, philosophically and theologically understood? More complete and adequate accounts of emergent phenomena of all kinds, along with the precisions we have

[13] Stoeger, 'Contemporary Physics,' *QCLN*.

[14] Barbour, *Religion and Science*, 281–304; Joseph A. Bracken, *Society and Spirit: A Trinitarian Cosmology* (Selingsgrove: Susquehanna University Press, 1991); *idem*, *The Divine Matrix: Creativity as Link between East and West* (New York: Orbis Books, 1995), 52–69, 128–140.

[15] Peacocke, *Theology for a Scientific Age*, 2nd edn. (Minneapolis: Fortress, 1993) chap. 11; *idem*, 'The Sound of Sheer Silence,' in *NP*, 215–247.

[16] Robert John Russell, 'Divine Action and Quantum Mechanics: A Fresh Assessment,' in *QM*, 293–328; Wesley J. Wildman, 'The Divine Action Project, 1988–2003,' *Theology and Science* 2/1 (April 2004): 31–76, and references therein.

[17] See, for instance, Ernan McMullin, 'Indifference Principle and Anthropic Principle in Cosmology,' *Studies in the History and the Philosophy of Science* 24 (1993): 359–389; and William R. Stoeger, SJ, 'The Anthropic Principle Revisited,' in *Philosophy in Science*, Vol. 10, eds. William R. Stoeger, SJ, Michael Heller, and Józef M. Życiński (Tucson, AZ: Pachart, 2003), 9–33.

discussed regarding reducibility, determinism, and emergence, are deepening our understanding of 'the laws of nature,' the regularities, processes, structures, and relationships, especially those pertaining to self-organization, or autopoiesis. The most important category among these is constitutive relationships, which may be internal or external, synchronic or diachronic, and which may involve bottom-up, horizontal (same-level), and/or top-down causal influences.[18] As we consider the different orders of emergence Murphy and Deacon have described in their chapters, we can see that many of these causal influences are radically new and irreducible to causal influences at lower levels of organization. They emerge as matter becomes organized in more and more complex ways. For instance, goals develop within systems which are not in any way determined by what occurs at lower levels of organization. Thus, there are these emerging irreducible factors as matter becomes more and more self-organized. Such emergent complex causal relationships, in turn, are leading to a much better, but not yet adequate, understanding of how the key transitions between inanimate and animate, pre-sentient and sentient, preconscious and conscious entities and organisms occurred. Though we shall probably never know exactly how these actually happened,[19] it is becoming very clear that the laws of nature at the proper level are capable of explaining these emergent phenomena. The direct intervention of God, or of some designer, is not needed. In fact, as has been emphasized often in theological discussion, there are strong theological reasons for not making God just another secondary cause (instead of the primary cause) in the universe.[20]

Thus, as the natural sciences develop, revealing how new levels of complexity and function emerge in our evolving universe, the formational and functional integrity of nature—its relative autonomy—is confirmed. In addition there is confirmation that God as Creator does not act as a secondary cause within God's creation, but always as the primary cause—the ultimate source of being and order. God, as Creator, endows nature from the beginning with existence and with

18 Stoeger, 'The Mind–Brain Problem,' 136–137.

19 William R. Stoeger, SJ, 'The Immanent Directionality of the Evolutionary Process, and its Relationship to Teleology,' in *EMB*, 163–190.

20 William R. Stoeger, SJ, 'Describing God's Action in the World in Light of Scientific Knowledge of Reality,' in *CC*, 239–261.

capacities and dynamisms to evolve the rich diversity of remarkable structures and organisms which have emerged in the course of cosmic history. Included with this endowment is relative freedom and autonomy—the course of evolution was not rigidly determined from the beginning, but the rich potentialities were there. Some of these were actualized and others were not. The processes of evolution rely on the harnessing of chance within a larger framework of order and regularity. In fact, what has happened in our universe is that each of the hundreds of billions of hundreds of billions of star systems in our observable universe has become a separate evolutionary experiment. How many of them—or even whether or not any of them besides our own—have yielded life and self-conscious social life, we shall probably never know.

What we have also discovered is that the advance of cosmology and the natural sciences indirectly reinforce the doctrine of *creatio ex nihilo*, the utter dependence of all creation on the Creator. They do this first, as we have already seen, by indicating that God's direct intervention in the evolutionary process as another secondary cause is not needed. Secondly, at the same time they strongly reveal the radical contingency of all things and all phenomena—everything depends on something else for its continuation in existence. And thirdly, as we have also seen, because of this universally radical contingency, nothing that cosmology or the natural sciences investigates provides an ultimate explanation for existence, order, purpose, and meaning. When we come to the very beginnings of our universe, the quest for further understanding and intelligibility continues. That quest can only be satisfied by an ultimate source of being and order, by a Creator. That takes us beyond cosmology and the natural sciences to philosophy and theology.

Then there is special divine action—God's active presence in history in God's personal relationship with individuals or with believing groups of people. Although God apparently does not directly intervene in the natural processes of evolution, it seems to many that God does, in some sense, intervene or reveal God's self in a special way in nature and in history in order to answer prayer, effect the Incarnation and the Resurrection, and so on. Is that what happens, or is there some other way of understanding these events? One way to deal with this question is to stress that the regularities, processes, structures, and relationships that constitute the laws of nature as they actually function in nature—and not simply as we imperfectly understand and model them—include all the relationships that God has with individuals and with believing communities. These relationships—these 'higher laws of nature'—almost certainly subsume, modify, and marshal the 'lower laws of nature,' of physics, chemistry, and biology, as top-down causality always does, in one way or another. Thus, according to this view, any 'violation' or 'intervention' by God is only relative to our limited understanding of the full laws of nature, which include God's personal relationships with self-conscious entities who are open to the transcendent.[21]

This perspective, together with our reflections on reducibility and emergence, leads us to the increasing realization of God's immanence—God's active presence—in all the regularities, processes, and relationships in the universe, even in the transience and fragility of all that emerges. God not only creates the universe from nothing, but also holds it in existence at each moment. And God not only holds it in existence at each moment, but is also working and struggling as Creator through the laws of nature, and the processes of cosmic, biological, and social evolution, to coax it towards the realization of its destiny. The natural sciences are not capable of discovering that destiny. For the theologies of the various religions it has always been conceived as something like communion with the divine. Certainly, this painful distance from our origin and our destiny is indirectly supported by the incompleteness—and even futility—which we discern in nature when we limit our perspectives to those of the sciences.

Linked with this, of course, is the pervasiveness and persistence of cultural cosmologies, or myths of origin and of destiny. All societies and cultures possess some type of mythical cosmology to express the values, meanings, and orientations that vitalize them. And as they develop, these myths are re-explored and re-articulated in more critical and more logical form—in literature, philosophy, theology—in order to understand and validate the meanings and perspectives

[21] William R. Stoeger, SJ, 'Science, the Laws of Nature, and Divine Action,' in *Interdisciplinary Perspectives on Cosmology and Biological Evolution*, eds. Hilary D. Regan and Mark Worthing (Adelaide: Australian Theological Forum, 2002), 117–127.

that are implicit in cultural and religious legacies. What is essential for the society or culture is not 'the facts' about origin and destiny, but rather the meaning and significance they secure for the people in the face of uncertainty, natural and moral evil, death, and suffering.[22] This might be rephrased in terms of a sense of the directionality or purpose of existence. Without some such sense of purpose there is no adequate meaning or orientation available to communities or to the individuals who belong to them. It has become clear that the knowledge and understanding attained by the natural sciences does not of itself yield such meaning and orientation. Detailed and comprehensive physical, chemical, and biological knowledge, rich and important as it otherwise is, does not deliver us from pointlessness and meaninglessness.

Returning to the immanence of God in creation, we might wonder how it is connected to God's transcendence. Divine transcendence not only means that God is above and beyond all that exists and can be conceived, but also that God's presence and action know no restriction or barrier. From this point of view, God's transcendence enables God's immanence in creation. God as transcendent Creator is intimately present, available, and active in all that is. But, at the same time, God is endowing each entity with its own dynamisms and its own autonomy to be what it is and to act according to its own potentialities. Thus, in a very real way, God is also hidden in creation and to some extent vulnerable within creation. However, God is not some person or object we find within creation. 'God' is more of a verb than a noun. The autonomy God has given to God's creation on different levels means that God, out of love and respect for creation, has 'surrendered some of God's control' over creation to the laws of nature and to the initiative of semi-autonomous agents.

One possible channel of special divine action which has received considerable attention in recent years is that which is opened up by quantum indeterminacy at the sub-microscopic levels of physical reality. God could conceivably act, either constantly or episodically, within this 'window of opportunity' without violating the laws of

[22] See, for instance, William R. Stoeger, SJ, 'Cultural Cosmology and the Impact of the Natural Sciences on Philosophy and Culture,' in *The End of the World and the Ends of God: Science and Theology on Eschatology*, eds. John Polkinghorne and Michael Welker (Harrisburg, PA: Trinity Press International, 2000), 65–77.

nature in order to effect divine intentions at macroscopic levels of reality, including those characteristic of human behavior and history.[23] Although these proposals are very attractive from some points of view, they are also controversial. A more thorough understanding of emergent phenomena will enable us to see whether or not this quantum-mechanical opening for special divine action is really needed.

Another important avenue to explore in attempting to answer this question of special divine action is to see how we can conceive it as a manifestation of God's universal creative action more adequately and more broadly considered.[24] The basic insight is to recognize that God's special actions within history (e.g. Incarnation, Resurrection, etc.) are in virtue of God's universal role as the Creator of history. As Thomas Tracy explicates this, the key component is that 'God's creative action includes the continuous "giving of being" to the created world in its entirety,' enabling each being and system of beings within creation to function and develop according to its own capacities and dynamisms, including those of conscious, freely deciding and acting persons and communities of persons.[25] It is through what material beings and systems of beings at all levels accomplish through the operation of their God-given potentialities that God continues to act creatively in the world—God acting in and through secondary causes. Thus, ultimately all that happens within the created world can be considered an act of God.[26]

Thus, according to this approach, God's special acts would be those which are freely chosen and executed by persons and communities in complete openness to the creative action of God in their lives, so that God's initiative is fully realized in them. What is crucial to recognize is that God's universal creative action is neither uniform nor indifferent to particularity. It is, instead, richly differentiated—that is, differently expressed in each entity, organism, and person,

[23] See Robert John Russell, 'Special Providence and Genetic Mutation: A New Defense of Theistic Evolution,' in *EMB*, 191–223, and references therein. For a detailed assessment of these possibilities, see Wesley J. Wildman, 'The Divine Action Project, 1988–2003,' *Theology and Science* 2/1 (April 2004), 31–76.

[24] Thomas F. Tracy, 'Creation, Providence, and Quantum Chance,' in *QM*, 235–258.

[25] Tracy, 'Creation, Providence,' 238–239.

[26] Tracy, 'Creation, Providence.'

endowing each with its own individuality—and actively engaged with and supportive of the emergent capacities (such as personhood) at each level. At this point we can integrate the insight from our first approach, recognizing that the creative action of God towards individuals and communities invites them into a personal relationship with God's self.[27]

## 10.7. Conclusion

Here I have sketched the primary scientific, philosophical, and theological consequences of taking our scientific understanding of emergent phenomena in the universe seriously, along with those of failing to do so and instead allowing a thoroughly reductionistic paradigm to dominate. In particular, I have discussed the positive impact this has in reinforcing, enriching, and deepening our understandings of the following: God's universal creative action in nature and God's special action in history; God's transcendence and immanence; natural and moral evil; the incompleteness of creation, and personal and cosmic destiny; body and soul, matter and spirit, and relationships between them, avoiding dualism; the openness of human beings to the transcendent; human freedom. I have only given a rough blueprint of the framework for effecting this, and briefly indicated how the more coherent understandings are developed. There is much more to be explored—these orientations show us where to go and how to proceed.

[27] William R. Stoeger, SJ, 'Conceiving Divine Action in a Dynamic Universe,' in *Scientific Perspectives on Divine Action: Problems and Progress*, eds. Robert John Russell, Nancey Murphy, and William R. Stoeger, SJ (Vatican Observatory Publications and Center for Theology and the Natural Sciences, forthcoming).

# 11

## Emergence, Scientific Naturalism, and Theology

*John F. Haught*

### 11.1. Naturalism and Emergence[1]

The renowned scientist and writer Carl Sagan began his popular *Cosmos* TV series by stating that 'the universe is all there is, all there ever was and all there ever will be.' Sagan, now deceased, was a devotee of naturalism, by which I mean the belief that 'nature is all there is.' According to naturalism, there is no need to look beyond the physical world for a divine creator or source of meaning. Nature is self-contained and self-sufficient. And according to *scientific* naturalism, which I shall be highlighting here, science is the best way to understand it. The universe alone is life's ultimate creator and designer. The Darwinian recipe for evolution, consisting of random genetic changes, natural selection and an enormous depth of time, is enough to account for all the diversity and complexity of life, including beings endowed with minds. Why, then, would rational persons want to look beyond nature itself in order to arrive at a deeper explanation for all the outcomes of evolution?

According to evolutionist Richard Dawkins, if life had available to it only a biblical span of several thousand years to bring about something as complex as the human brain, Darwinian explanations

[1] Many of the ideas set forth in this essay are developed at greater length in my book *Is Nature Enough? Meaning and Truth in the Age of Science* (Cambridge: Cambridge University Press, 2006).

might not be enough, and naturalism would be questionable. But if life has four billion years to evolve, this immensity of time—in combination with random variations and Darwinian selection—is all that is needed. Minute cumulative changes, taken in conjunction with the relentless weeding out of nonadaptive variations by natural selection over an unimaginably lengthy span of time, would be sufficient to account for all instances of living complexity. Nature alone would be quite enough.[2]

This, at least, is the point of view taken by naturalism. Naturalism comes in many flavors, but in all its variety it usually includes the following core doctrines:

1. Outside of nature, which includes humans and their cultural creations, there is nothing.
2. It follows then that nature is self-originating.
3. Since there is nothing beyond nature, there can be no purpose to the universe.
4. Since there is no divine efficient or final cause, all causes are purely natural causes. This means that every natural event is itself a product of other natural events.
5. The various traits of living beings, including those of human persons, can be accounted for ultimately in purely natural terms (today this usually means scientific, evolutionary, and specifically Darwinian, terms).
6. There is no possibility of conscious human survival or resurrection beyond death.[3]

Today naturalism typically goes hand in hand with scientism, the belief that science is the only reliable pathway to genuine knowledge. And among the many scientific ideas that seem to support the naturalist belief that 'nature is all there is,' that of *emergence* now occupies an increasingly prominent station. 'Emergence' is a term that scientists employ to characterize nature's tendency over time to bring about complex ordered systems endowed with properties and

[2] Richard Dawkins, *Climbing Mount Improbable* (New York: Norton, 1996).

[3] This list corresponds roughly to the description of naturalism given by Charley Hardwick, *Events of Grace: Naturalism, Existentialism, and Theology* (New York: Cambridge University Press, 1996).

functions that had not been operative at previous or less complex levels of evolution. In the history of the universe, according to biologist Harold Morowitz, there have been at least twenty-eight distinct stages of emergence.[4]

The most conspicuous of these, of course, are the appearance of life on earth out of lifeless physical precursors, and the emergence of self-awareness out of a previously unconscious cosmos. But the spontaneous self-organization of physical entities and processes into more and more complex systems has been going on in less dramatic ways from the time of cosmic origins. Furthermore, it would seem reckless to assume that the cosmos has now reached the end of its fascinating adventure of emergence.

The relatively new scientific excitement about emergence follows upon the scientific knowledge, accumulated over the last century, that matter is not essentially passive but inherently energetic, and that the physical universe does not have to be ignited by exceptional interventions of supernatural agencies in order to bring about creative and interesting outcomes. If a lot of time is available and sufficient numbers of component entities are involved, routine physical and chemical processes operating solely in accordance with elementary, lower-level rules of nature can at critical junctures be taken up into new and 'higher levels.' At these new levels more encompassing rules of activity, and complex patterns irreducible to their antecedents, suddenly come into play.

## 11.2. The Novelty in Emergence

But where do these new emergent rules, patterns, and principles come from? From below, from above, or both? Or is there perhaps an alternative way of accounting for the novelty in emergence? It is intuitively obvious that during the course of evolutionary history something *more* has emerged out of something *less*. Simply contrast the human mind with primordial cosmic radiation, for example. Yet, where does the content of the *more* come from? Is there a reservoir of 'moreness' that has always been built into nature in a concealed way? Or does the novelty in emergence flow in from a realm of possibilities

[4] Harold Morowitz, *The Emergence of Everything: How the World Became Complex* (New York: Oxford University Press, 2002).

that transcends nature? New possibilities, as Alfred North Whitehead has observed, do not just float into the universe from nowhere. They must reside somewhere before they are actualized.[5] But where?

What is incontestable, in any case, is that out of a monotonous primal sea of elementary particles prevalent shortly after the Big Bang there has indeed emerged in our universe a succession of complex systems and organisms, including life, mind, and culture. If these novel levels of being have emerged from within the universe, then they reveal something *essential* about the universe. And so, the emergent phenomena of life and mind in particular cannot simply be bracketed out if one wishes to understand in depth the character of physical reality itself.

And yet the natural sciences, in spite of the fact that they are now forced to attend to the fact of emergence, typically divert their attention away from the *novelty* in emergent outcomes in order to focus primarily on the series of material and efficient causes that historically led up to such phenomena as life and mind; or they may also focus on the 'lower levels' of constituent elements that now underlie emergent realities.[6] Science's firm conviction is that any explanation of emergent outcomes must be simpler than the outcomes themselves. Mathematically speaking, says Gregory Chaitin, a scientific explanation is comparable to an economical algorithm: 'For any given series of observations there are always several competing theories, and the scientist must choose among them. The model demands that the smallest algorithm, the one consisting of the fewest bits, be selected. Put another way, this rule is the familiar formulation of Occam's razor: Given differing theories of apparently equal merit, the simplest is to be preferred.'[7]

Accordingly, to account in the simplest way for complex emergent products of evolution such as life and mind, scientists seek to lay out as parsimoniously as possible the physical and chemical components,

[5] Whitehead refers to these possibilities as 'eternal objects,' and compares them to the Platonic forms that are also said to reside in God; see, e.g., *Process and Reality*, corrected edn. ed. David Ray Griffin and Donald W. Sherburne (New York: Free Press, 1968), 43–44.

[6] See the discussion of the difference between explicit and tacit knowing below.

[7] Gregory J. Chaitin, 'Randomness and Mathematical Proof,' in Niels Gregersen, ed., *From Complexity to Life: On the Emergence of Life and Meaning* (New York: Oxford University Press, 2003), 23.

networks and laws operating prior to and beneath the (seemingly) improbable emergent products. The methodological assumption is that any adequate explanation of emergent phenomena must always be a simplification or else it is not really explanatory. Mathematically speaking, unless the outcome of an emergent process is algorithmically compressible—that is, able to be made intelligible in terms of relatively simple mathematical operations—it has not yet been 'explained' but only described. In the cases of life and mind, explanation entails our going *back* into the causal past and *downward* into an underlying domain of relative physical simplicity. Explanation moves backward in time, and downward in degree of complexity, to a more historically remote, physically simpler (and hence more mathematically manageable) set of explanatory elements.

Thus, scientific accounts of life and mind are constrained to do what is *logically* impossible: explain life and mind in terms of what is lifeless and mindless. Recognizing the logical problem involved here, eliminative materialists attempt to bring coherence to their quest for understanding by, in effect, denying that life and mind are anything more than labels for what is objectively lifeless and mindless. In effect, there can be no real novelty in a materialist picture of the universe. Emergent complexity, as the physical chemist Peter Atkins puts it, is nothing more than a mask covering up an underlying physical simplicity.[8] Nothing really new, in other words, can possibly make its way into the physically hidebound universe. In reaction to this materialist brand of naturalism, vitalists often go along with the materialist understanding of physical reality, but then locate life and mind in a hypothesized domain apart from or only loosely connected to the physical. Neither the materialist nor the vitalist approach, however, is adequate to the obvious fact of emergence.

## 11.3. The Dimension of Futurity

The scientific quest for the origins of life and mind is typically one that embarks on a journey into the cosmic past by formally bracketing out the novel features that had allowed the scientist to recognize

[8] P. W. Atkins, *The 2nd Law: Energy, Chaos, and Form* (New York: Scientific American Books, 1994), 200; and *idem*, *Creation Revisited* (New York: W. H. Freeman, 1992), 11–17.

life and mind as distinctively emergent and as interesting enough to deserve further exploration, in the first place. For example, science typically omits any explicit consideration of the *striving* that makes living beings noticeably unique and distinct from the non-living. And scientific method also deliberately ignores the trait of *subjectivity* that each of us *experiences immediately* in our mental functioning. Our subjectivity, of course, is no less part of the emergent universe than rocks, radishes, and rodents. But scientists are instructed, as it were, to suppose that subjectivity is not part of the objective world.

Authentic objectivity should lead one to acknowledge that if humans are part of nature then their subjectivity is also part of nature, as is the subjective experience that other sentient beings possess. Of course, the scientific method of looking at what temporally precedes and physically constitutes life and mind is, in a limited way, fruitful and illuminating. But if serious human inquiry is truly interested in understanding in depth the *actual* universe, that is, the one that contains living and thinking beings, it needs to consider also what the existence of these emergent phenomena are telling us, more generally speaking, about the natural world. Even scientists must admit, after all, that there is something novel and unprecedented about emergent phenomena. They find emergence remarkable, even eerie. So, I ask here whether it is enough to assert that the novelty in emergence was already present, though concealed, in the cosmic past. Clearly, life and mind are something more than simplicity masquerading as complexity, as the physical chemist Peter Atkins claims.[9] In emergence, something *new* is added to the historically antecedent simplicity. But this additional dimension can appear only if the universe is continually gifted with a future.

The ultimate explanation of emergence is in some sense, therefore, 'the coming of the future.'[10] And it is especially in our striving to understand in depth this dimension of futurity that the scientific question of emergence allows for theological illumination.[11]

[9] Atkins, *The 2nd Law*, 200.

[10] Wolfhart Pannenberg, *Faith and Reality*, trans. John Maxwell (Philadelphia: Westminster Press, 1977), 58–59.

[11] I have developed this point more fully in my two books *God After Darwin: A Theology of Evolution* (Boulder, CO: Westview Press, 2000); and *Deeper than Darwin: The Prospect for Religion in the Age of Evolution* (Boulder, CO: Westview Press, 2003).

## 11.4. A Richer Empiricism

However, it is not only the dimension of futurity presupposed by emergence that calls for deeper explanation than science alone can provide. The facts of *striving* and *subjectivity* are also two of the more conspicuously novel emergent aspects of evolution that do not fall within the range of conventional scientific comprehension. Science has not yet given any deeply illuminating explanation of these interesting facets of nature, but instead has left them out of its proper sphere of concern. However, it is not a violation of the empirical spirit to ask what kind of universe it is that could have given rise to the kind of beings that experience the world *from the inside*. To take these 'inside' experiences into account may not be classically scientific, but a *richer empiricism*, by which I mean an attentiveness to *all* the data of experience, demands that they not be ignored. Of course science, by definition, is perfectly within its rights to focus only on what is objectifiable, and hence to pass over in silence all inklings of striving and inwardness. But, after making this concession, I would argue that science's abstraction from 'interiority' must be supplemented by a more wide-ranging empirical method that takes the latter into account.

A richer empiricism, as I shall call it, refuses to leave out the inner experience of striving and intelligent subjectivity. Each of us has immediate access to these phenomena, and so it seems arbitrary to leave them out of all formal representations of the world. Since subjectivity and striving have unquestionably emerged from within the cosmic process itself, the spirit of empiricism calls for a more penetrating way of approximating the world than scientific method has generally allowed. I believe it is one of the great virtues of an empirically based *metaphysics* such as that of Alfred North Whitehead to make sure that aspects of the real world that science leaves behind are not purged from our formal awareness altogether.

The notion of an empirically based metaphysics may initially seem self-contradictory. Traditional metaphysics appeared to be talking about things that are not part of our intra-worldly experience. But as Whitehead has shown, metaphysics can also be empirical in the sense that it is *based on the widest possible range of what we actually experience in the world.* Metaphysics looks for the most general categories, those that apply to *all* kinds of experience. Its data include not only the world 'out there,' but also the 'inner' experience of our own purposive striving and subjectivity. By opening itself to these realities, metaphysics will be much more radically empirical than is scientific method, which abstracts altogether from our immediate experience of the world's 'insideness.' Science, in other words, is not the only way of getting in touch with the real world. A generalized empirical metaphysics can do that also.

## 11.5. The Importance of Tacit Knowing

Objections to such a proposal, of course, are inevitable. Scientific naturalism cannot imagine how the human mind may reliably contact the natural world in any other way than by explicitly objectifying everything. So, the question will arise as to how one 'gets at' those aspects of the real world that science leaves out of its field of vision. But, this is the wrong question to pose. We don't get *at* them so much as *with* them. As Michael Polanyi points out (see below) we first encounter them by way of a *tacit* knowing that accompanies all explicit or objective knowing. We know them first not by focusing on them, in the explicit way that science objectifies the world, but by knowing *with* them and *from* them. At first we apprehend our subjectivity and striving only tacitly. An experientially based metaphysics refuses to pretend that what we know tacitly is not part of the real world.

A richly empirical metaphysics strives to leave nothing real out of its formal picture of the world. Science, on the other hand, leaves out not only the facts of striving and subjectivity but additionally, as I indicated earlier, the wider fact of emergent *novelty*. This is a shortcoming that renders scientific objectification at least partially ineffective in grasping what is going on in the universe. Today, scientists refer to emergent phenomena as 'self-organizing' or 'autopoietic.' But these are descriptive, not explanatory terms, and they still fail to reach into the novelty in emergent phenomena.

Science by nature reduces the unknown as much as possible to the already known. It transforms the remarkable into the routine. And so, it is constrained to explain life in terms of the non-living, and mind in terms of the mindless. Thus, as organisms evolve from less to more sentient and conscious varieties, science is able to show that the laws of physics and chemistry remain inviolate, but it does not account for the emergent novelty as such. Often it can only watch as novelty appears, for example in the phenomena associated with chaos and complexity. Hence, scientists are limited to *believing* or hoping that life will be fully explainable in terms of the subroutines familiar to the physical sciences. From the perspective of the physical sciences one cannot 'see' anything going on in life that is not reductively specifiable. The deliberately impersonal methods of science require from the start that 'the observer' look away from anything that a more richly empirical and *personal* kind of knowing may have identified as distinctively novel in living and thinking beings.[12]

Still, even among scientists there is at least a tacit acknowledgment that in emergent phenomena, novelty spills forth abundantly from the substratal uniformity identified by physics and chemistry. At times, the novelty erupts explosively, though generally it appears more gradually. A wondrous variety of unpredictable modes of being have 'emerged' from the undifferentiated stuff of the early universe. Scientists would never have noticed emergence if they had been methodologically and consistently rigorous in focusing on the cosmic past and subordinate particulars. As Polanyi has shown, it is only because scientists themselves are personal subjects that they notice emergent novelty at all.[13] They 'get at,' or become aware of, emergence as such only through a tacit kind of knowing.[14]

After the curtain went up on the Big Bang universe, the rules of physics and, later, chemistry have remained unchanged, and science rightly focuses on what stays constant in nature. But, this means that scientific method is not skilled at coping with novelty as such, even though our tacit personal experience indicates that new things keep happening all the time. To the physicist emergent phenomena will

[12] See Michael Polanyi, *Personal Knowledge* (New York: Harper Torchbooks, 1964); *idem, The Tacit Dimension* (Garden City, NY: Doubleday Anchor Books, 1967); *idem, Knowing and Being*, ed. Marjorie Grene (Chicago: University of Chicago Press, 1969), 22–39; and Michael Polanyi and Harry Prosch, *Meaning* (Chicago: University of Chicago Press, 1975).

[13] This is the central argument of Polanyi's book *Personal Knowledge.*

[14] Ibid. See also Polanyi's *The Tacit Dimension.*

always seem to be simplicity masquerading as complexity. This is because physics is not wired to pick up the tacit signals of emergence. Of course, in the real world of personal knowledge, scientists, including physicists, observe emergence as a fact of nature, but scientific method would have to be transformed into a wider kind of empiricism than it now professes if it is to account for novelty, subjectivity, and self-awareness in a *formal* way.

So, the question still remains for thinking *persons*: What accounts for all the novelty in emergence? The pure scientific naturalist can only reply that everything that emerges has always been there *in nuce*, waiting for the play of large numbers and the operation of simple local rules to bring about occasional abrupt leaps onto unprecedented levels of physical activity. But, once again, this is a feeble description, not a robust explanation. Where, we should like to know, do the new rules and corresponding operations come from? To scientific naturalism, the answer is that the novel aspects in emergence have always been latent in the material past, requiring only the passage of time to uncoil. Time itself, in league with chance and necessity, then becomes the demiurge that brings about all the novelty in emergence. But, can the passage of time alone, no matter how gradual it is, really be an explanation?

Classical science, allergic as it is to final causes, diachronically explains phenomena from the temporal past forward toward the present. And only on the basis of what has already occurred does it hazard to talk about future outcomes. Also, more synchronically speaking, it explains what is 'higher' in terms of what is algorithmically simpler or hierarchically 'lower.' In either case, reductive science is constrained methodologically to view the process of emergence formally as unremarkable, since everything that goes on in emergence can only be an exemplification of what science already knows.

I emphasize that this self-blinding to novelty does not as such make science blameworthy. There is no need on the part of scientific method at any point to appeal to factors beyond the limits of what it considers natural. So far, so good. It is only when the scientist is also a scientific *naturalist*, that problems arise. For, according to the naturalist, nature is 'all that is, all there ever was, and all there ever will be.' There is no place for a wider empiricism that would take into account the facts of novelty, striving, and subjectivity. Consequently, there can be no legitimate place at any level of inquiry for theology in the explanation of emergence. As far as scientific naturalism is concerned, to introduce theology in contemporary attempts to comprehend emergence would simply stifle the natural longing to understand.

## 11.6. Subjectivity as Part of Nature

However, a richer, wider, and deeper empiricism than the kind enshrined in scientific naturalism would refuse to leave out the obvious facts of novelty, striving, and subjectivity as though these were not also part of the real world. A broader way of 'seeing' the universe can be found in the writings of a number of philosophers and religious thinkers who have been largely ignored by today's philosophical and scientific naturalists. I am thinking here especially of Whitehead, Bernard Lonergan, Polanyi, and Pierre Teilhard de Chardin—all lovers of science themselves—who share the conviction that the universe cannot be understood even in a basic way without taking into account the irreducible facts of novelty, striving, and (as Lonergan would say) intelligent subjectivity.

Among these 'wider empiricists,' Whitehead stands out in his criticism of the way in which classical scientific method abstracts from—and thereby ignores—subjective experience. Science generally isolates certain aspects of the natural world mathematically in order to arrive at some degree of clarity about a complex and often confusing universe. Such abstracting is entirely appropriate as a first step in understanding. The ideal of clarity is a noble one, and modeling is one way of attaining it. Indeed, the search for clarity is essential in all understanding. But, what Whitehead finds problematic is the tendency on the part of scientific naturalists (whom he calls 'scientific materialists') to identify science's lifeless and mindless abstractions with concrete reality itself. This confusion of map with territory he refers to as 'the fallacy of misplaced concreteness.'[15]

It is nothing less than a *logical* error that has led to the naturalist's self-subverting program of explaining life only in terms of what is

[15] Alfred North Whitehead, *Science and the Modern World* (New York: Free Press, 1925), 51, 58.

essentially dead, and mind in terms of what is intrinsically mindless. As Whitehead insists, every mental event that occurs in human experience (and indeed in the experience of other beings as well) is a very real part of the universe, and so our own immediate subjectivity cannot be arbitrarily left out of a broadly experiential understanding of the world. 'Scientific reasoning,' Whitehead remarks, 'is completely dominated by the presupposition that mental functionings are not properly part of nature.' Then he goes on to say that, '[a]s a method this procedure is entirely justifiable, provided that we recognize the limitations involved.'[16] However, modern thought, and especially scientific naturalism, either are not aware of these limits or have decided to ignore them. As a result, the naturalist's world is one that appears *essentially* devoid of subjectivity. It is a world purged altogether of mind *a priori* that then becomes the starting point of naturalistic attempts to account for the emergence of mind.[17] The project, of course, is logically doomed from the outset.

Jesuit philosopher Bernard Lonergan agrees that one cannot understand the universe, at least in any depth, without at some point taking into account the emergent fact of subjectivity, especially our own experience of intelligent striving. A genuine understanding of the real world, he argues at length, can only begin with an understanding of understanding itself. 'Thoroughly understand what it is to understand, and not only will you understand the broad lines of all there is to be understood but also you will possess a fixed base, an invariant pattern, opening upon all further developments of understanding.'[18] A *deeply* empirical scanning of one's own intelligent subjectivity must supplement and contextualize any scientific survey of the universe.

Again, the naturalist will want to know how to 'get at' this elusive phenomenon of intelligent subjectivity, but there is no mystery to gaining access to it. Each of us can notice immediately, for example, certain imperatives that govern our thought processes, especially as we are doing science. These imperatives are as follows: Be attentive!

[16] Alfred North Whitehead, *Modes of Thought* (New York: Free Press, 1968), 156.

[17] See also Alan B. Wallace, *The Taboo of Subjectivity: Toward a New Science of Consciousness* (New York: Oxford University Press, 2000).

[18] Bernard Lonergan, SJ, *Insight: A Study of Human Understanding* (New York: Philosophical Library, 1957), xxviii.

Be intelligent! Be critical![19] If you find yourself wondering about the truth of what I just said, it is because your own mind, at this very moment, is issuing to you these very imperatives to be attentive, to be intelligent, and to be critical. At first, your awareness of the imperatives is immediate and tacit. As you engage in scientific work you are not focusing *on* the imperatives, but are looking at the world *from* them and *with* them. But an adequate picture of the world demands that at some point you explicitly and systematically include in it the intelligent subjectivity that has been quietly spurring on the process of your own inquiry.

It is undeniable that your intelligent subjectivity exists and that it exists *within* the context of the evolving universe. Hence, the question needs to be asked: what exactly does the fact of your own intelligent subjectivity tell you about the nature of the universe that gives it domicile? What kind of universe is it that could have given rise at all to your own intelligent subjectivity and to the cognitional imperatives that you are now obeying?

Lonergan's own answer to this question is too long and complex to lay out here. But one can say, at the very least, that any universe that gives birth to intelligent subjectivity cannot be an *essentially* lifeless and mindless one. The origin of intelligence cannot be finally or fully accounted for, without contradiction, in terms of the mindless processes and constituents that constitute the ultimate foundations of the scientific naturalist's universe. If you suppose that it could, then the imperatives of your own mind would carry no authority whatsoever. Even science, which is based on the mind's invariant imperatives, would be a completely untrustworthy enterprise. Science cannot merit our respect unless the imperatives that underlie its own activity and progress are acknowledged to have emerged from a ground much deeper than the dead and mindless strata identified by scientific naturalism. A *richly* empirical approach must look for the wellspring of intelligent subjectivity in a dimension that lies much deeper than the relatively shallow—because abstract—algorithmic flats to which science has easy access. It is in exposing the shallowness of this ground, I believe, that an empirical metaphysics—along with

[19] Bernard Lonergan, SJ, 'Cognitional Structure,' in F. E. Crowe, SJ, ed., *Collection* (New York: Herder and Herder, 1967), 221–239.

a theology that adheres to its method—has a legitimate role to play in the explanation of emergence.

Polanyi, in a manner not entirely unlike that of Lonergan, has also demonstrated that our acts of knowing are not alien to the universe, but instead the key to understanding emergence in depth. Influenced partly by Gestalt psychology, Polanyi has observed that most of our knowing consists of tacit, *personal* integration of atomic particulars into coherent wholes.[20] Our knowledge moves from an initially tacit awareness of particulars to a focal comprehension of emergent wholes. For example, if I see a frog on the bank of a pond I can recognize it as a living system only because I have first *tacitly* adverted to its color, size, and shape, to its striving, breathing, living, sensing, and other features. These particular features, known tacitly by my 'personal knowledge' that 'indwells' the particular features of the frog, serve as clues that guide me toward focal knowledge of the organic, emergent whole.

Knowledge, as Polanyi conceives it, has both a *from* character and a *to* character, and emergence has the same structure. Knowledge moves *from* silent awareness of atomic particulars *to* explicit knowledge of comprehensive systems.[21] And in natural history emergence moves, in a parallel way, from subordinate particulars to wholes that are irreducible to, though reliant upon, their atomic particulars and physical subsystems.[22] Scientists, Polanyi argues, would never have found it even *interesting* to ask about the chemical, physical, or evolutionary constituents of life's and mind's emergence in the first place unless they had already *implicitly* recognized—by way of tacit and personal acts of knowing—that there is something irreducible about living organisms, namely, that they are *subjects striving* to exist and accomplish certain goals; and about minds, that they are *striving* to understand.[23]

Novelty, striving, and subjectivity are clearly emergent aspects of the real world, but they are first known *as emergent* by way of tacit personal indwelling rather than by objectifying scientific observation of particulars. Tacit knowing engages the reality of emergent novelty long before, and much more intimately than, scientific objectification

[20] Polanyi, *Tacit Dimension*, 3–25. [21] Ibid. [22] Ibid., 29–52.
[23] Polanyi, *Personal Knowledge*, 327ff.

does. In fact, from a Polanyian perspective, scientific analysis of particulars actually leads the mind away from the initially tacit awareness of emergence.

Finally, Teilhard de Chardin also seeks to correct the one-sided direction of scientific knowing. Instead of looking only to the chronological past or to the subordinate constituents of emergent wholes in order to understand them, what is needed, he thinks, is a 'reversal of perspective' that takes into account what emergent evolution has actually brought about. Instead of viewing life and mind exclusively in terms of their antecedent lifeless and mindless elements, Teilhard looks back at the material spheres studied by physics and chemistry by deliberately starting inquiry into nature at the point of the phenomenon of consciousness. This richly empirical method, less abstract and analytical than what is typically called science, is based on the conviction that human inquiry cannot begin to understand the physical universe without taking into account what that universe has produced by way of emergence. Once again, in Teilhard we encounter a rich empiricism that refuses to remain oblivious to the amazing outcomes to which evolution has in fact led. Teilhard's perspective is, I believe, much more radically empirical than is that of normal science since it refuses to leave out the *actual experience* of consciousness that takes place *within* the context of the universe.

If sincere searchers begin delving into the universe by first considering the fact of their own exploratory consciousness, they will never arrive at a point in their survey of the cosmic past where the universe fades into complete mindlessness. This is because the 'insideness' they experience in their own consciousness is in some way connected to and continuous with the totality of cosmic history. To argue otherwise—by positing a world at one time completely stripped of every degree of feeling or what Teilhard calls 'consciousness'—would entail a dualistic universe split decisively into mind on the one side and mindlessness on the other. Such a severe divide would make the task of explaining how mind emerged all the more impossible, logically speaking. Conditioned as we are to looking at the world mechanistically, it may seem counterintuitive to attribute a vein of subjectivity all the way down and all the way back to elemental cosmic dust. But, it will seem strange only because of the peculiarly abstract and 'objectivist' angle of vision that science typically takes when it looks at the universe.

The point that Teilhard shares with Whitehead, Lonergan, and Polanyi is that a more sweeping empiricism than that endorsed by scientific naturalism must look at the universe more seamlessly and from a perspective that keeps its eyes fixed on the 'insideness' into which the cosmic process has actually blossomed. Scientific naturalism's 'restriction of the phenomenon of consciousness to higher forms of life,' Teilhard notes, has become an excuse for 'eliminating it from its constructions of the universe.' Consciousness 'has been classed as a bizarre exception, an aberrant function, an epiphenomenon.' However, 'to integrate consciousness into a system of the world requires us to envision the existence of a new face or dimension of the stuff of the universe.'[24] 'Indisputably, deep within ourselves,' Teilhard continues, 'through a rent or tear, an "interior" appears at the heart of beings. This is enough to establish the existence of this interior in some degree or other everywhere forever in nature. Since the stuff of the universe has an internal face at one point in itself, its structure is necessarily bifacial; that is, in every region of time and space, as well, for example, as being granular, *coextensive with its outside, everything has an inside*.'[25]

## 11.7. Implications of a Rich Empiricism

With Whitehead I find it puzzling, therefore, that in their attempts to understand emergence, contemporary philosophers have now abandoned their role as critics of abstraction, and instead cling tenaciously to the relatively more impoverished empiricism espoused by scientific naturalism. With Lonergan, I wonder why so much discussion of emergence fails to heed the mind's imperatives to be attentive, insightful, and critical, and fails to inquire into what the obvious fact of intelligent subjectivity is telling us about the nature of the *universe*. With Polanyi, I am puzzled that an impersonal ideal of knowing, one that refuses to notice the facts of personality, subjectivity, and

[24] Pierre Teilhard de Chardin, *The Human Phenomenon*, trans. Sarah Appleton-Weber (Portland, OR: Sussex Academic Press, 1999), 23–24.

[25] Ibid., 24.

centered striving, has been elevated by modern and contemporary philosophy to a position of such cognitional authority that the rich fact of emergence is eviscerated of the very features that allow persons to take notice of it at all. And with Teilhard, I am especially troubled that so few scientists and philosophers truly *see* subjectivity as an emergent dimension that tells us something new about the *universe*, something that analytical methods leave out.[26]

The result of these oversights by modern thought has been that the world as represented by scientific naturalism and most philosophy today is one that is *essentially* barren of life and subjectivity. Theologian Paul Tillich has accurately called this worldview an 'ontology of death.'[27] What is even more troubling is that scientific naturalism asks us, at least by implication, to take this ontology of death as the ultimate explanation of life, striving, and subjectivity.

What I would propose instead is a variation on Teilhard's reversal of vision. This perspective will be just as important to *seeing* the universe as is the reductive viewpoint of science. Indeed, it will prove to be more radically empirical than conventional science since it refuses to take its eyes off that to which the emergent cosmic process has *actually* led. The angle of vision that I am suggesting asks that we imaginatively stand in the past and look forward at the emerging universe with our eyes focused firmly on the horizon of the future outcomes toward which we now know the process has actually led. By focusing on these future outcomes (life, mind, and spirit) it may be that only through a kind of indirect advertence will we take into account the physical or atomic particulars (atoms, molecules, proteins and genes, subsystems) which, over the course of time, will have become gradually integrated into more complex emergent wholes. Only out of the corner of our eyes we shall be taking into account what science looks at focally and explicitly.

While we are gazing forward from our imaginary perch in the evolutionary past toward the living and conscious emergents up ahead, we shall see—only indirectly and tacitly—the material components that science has made the focal objects of its own attempts to 'explain'

[26] Ibid.

[27] Paul Tillich, *Systematic Theology*, 3 vols. (Chicago: University of Chicago Press, 1963), 3:19.

the fact of emergence. In Polanyi's terms, we shall be attending *from* the atomic particulars *to* the emergent wholes that evolution is in the process of bringing about. By taking note of what tacit knowing entails, we shall be leaving much less out of our representations of the actual universe than ordinary science is conditioned to doing.

We may then begin to sense dimly, at the far edge of the dawning cosmic future, the outlines of a power of attraction up ahead that lovingly and persuasively gathers the multiple cosmic particulars into a convergent unity. Instead of taking the cosmic process as driven forward entirely by a dead and inertial past, our new angle of vision will allow us to experience the emergent universe as drawn forward by an inexhaustibly resourceful Future.

The cosmos looks a lot different—but no less interesting—when we adopt this reversal of perspective. For example, the physics of the early universe will no longer be understood apart from the *promise* the cosmos has always had of blossoming eventually into life and mind. Today many scientists are themselves viewing the physics of the early universe in this way. If we look at the universe only according to the method prescribed by science, we may not apprehend adequately the tendency of the universe to gather initial fragments into emergent novelty and unity.

Of course, scientific naturalists will resist this way of looking at things, justifying their avoidance by pointing to the dangers of the pathetic fallacy, that is, the tendency to read human characteristics into nonhuman nature. But these anxieties are misplaced. The perspective I am advocating simply asks that we not arbitrarily leave the facts of novelty, subjectivity, or striving out of the picture when we seek to understand what sort of reality the universe is. This approach is in fact much more radically empirical than the analytical method of ordinary science, since it includes within the horizon of its vision the most immediately obvious datum of our experience, namely, the performance of our own minds engaged in the process of striving to understand and know the world in which it is embedded.

Analytical science, of course, is not obliged to adopt the kind of vision I am proposing here. It is permissible to avoid looking too closely at the fact of one's own subjective awareness in order to attend to atomic particulars. One need not object to such an evasion as long as it is acknowledged to be a useful methodological strategy. However, the more uncontaminated by contact with nature mind is thought to be, the more science and philosophy will be encouraged to keep alive the illusory idea of a universe essentially devoid of subjectivity. This is most ironic since evolutionary science itself shows that each subject is fully a part of nature. By embracing the emergent fact of subjectivity an empirically based metaphysics can retrieve what science has left out, and a theology open to an inexhaustibly resourceful future can provide a more comprehensive understanding of emergence.

# 12

# Emergent Realities with Causal Efficacy: Some Philosophical and Theological Applications

*Arthur Peacocke*

In this paper I attempt to extend the concepts necessary for understanding relationships between the various levels of organization, and so of description, that have developed in recent years from scientific analyses of complex physico-chemical and biological systems to wider issues in philosophy and theology.

## 12.1. Hierarchies of Complexity and Emergentist 'Monism'

I shall presuppose here the hierarchy of complexity with a correlative hierarchy of sciences that has been described in earlier chapters.[1] I shall also presume at least this with the 'physicalists': all concrete particulars in the world (including human beings), with all of their properties, are constituted only of fundamental physical entities of matter/energy at the lowest level and manifested in many layers of complexity—a 'layered' physicalism. This is indeed a *monistic* view (a constitutively-ontologically reductionist one), that everything can be broken down into whatever physicists deem to constitute matter/energy and that no extra *entities* or *forces*, other than the basic

[1] See also Arthur Peacocke, *Theology for a Scientific Age: Being and Becoming—Natural, Divine and Human*, 2nd enlarged edn. (Minneapolis: Fortress Press, 1993), henceforth TSA, 36–43, 214–218.

four forces of physics, are to be inserted at higher levels of complexity in order to account for their properties. However, what is significant about natural processes and about the relation of complex systems to their constituents is that the concepts needed to describe and understand—as indeed also the methods needed to investigate—each level in the hierarchy of complexity are specific to and distinctive of those levels. It is very often the case (but not always) that the properties, concepts, and explanations used to describe the higher-level wholes are not logically reducible to those used to describe their constituent parts, themselves often also constituted of yet smaller entities. This is an epistemological assertion of a non-reductionist kind.

When the epistemological non-reducibility of properties, concepts, and explanations applicable to higher levels of complexity is well-established, their employment in scientific discourse can often, *but not in all cases*, lead to a putative and then to an increasingly confident attribution of reality to that to which the higher-level terms refer. 'Reality' is not confined to the physico-chemical alone. One must accept a certain 'robustness' (Wimsatt[2]) of the entities postulated or, rather, discovered at different levels and resist any attempts to regard them as less real in comparison with some favored lower level of 'reality.' Each level has to be regarded as a cut through the totality of reality, if you like, in the sense that we have to take account of its mode of operation at that level. New and distinctive kinds of realities at the higher levels of complexity may properly be said to have *emerged*. This can occur with respect either to moving, synchronically, up the ladder of complexity or, diachronically, through cosmic and biological evolutionary history.

Much of the discussion of reductionism has concentrated upon the relation between already established theories pertinent to different levels. This way of examining the question of reductionism is less appropriate when the context is that of the biological and social sciences, for which knowledge hardly ever resides in theories

[2] W. C. Wimsatt has elaborated these criteria of 'robustness' for such attributions of reality to emergent properties at the higher levels in his 'Robustness, Reliability and Multiple-Determination in Science,' in *Knowing and Validating in the Social Sciences: A Tribute to Donald T. Campbell*, eds. M. Brewer and B. Collins (San Francisco, CA: Jossey-Bass, 1981).

with distinctive 'laws.' In these sciences, what is sought is more usually a *model* of a complex system which explicates how its components interact to produce the properties and behavior of the whole system—organelle, cell, multi-cellular organism, ecosystem, and so on. These models are not presented as sentences involving terms which might be translated into lower-level terms for reduction to be successful, but rather as visual systems, structures, or maps, representing multiple interactions and connecting pathways of causality and determinative influences between entities and processes. When the systems are not simply aggregates of similar units, then it can turn out that the behavior of the system is due principally, sometimes entirely, to the distinctive way its parts are put together—which is what models attempt to make clear. This incorporation into a system constrains the behavior of the parts and can lead to behavior of the systems as a whole, which is often unexpected and unpredicted. As W. Bechtel and R. C. Richardson have expressed it: 'They are *emergent* in that we did not anticipate the properties exhibited by the whole system given what we knew of the parts.'[3] They illustrate this from a historical examination of the controversies over yeast fermentation of glucose and oxidative phosphorylation.

What is crucial here is not so much the unpredictability, but the inadequacy of explanation if only the parts are focused upon, rather than the whole system. 'With emergent phenomena, it is the interactive organization, rather than the component behavior, that is the critical explanatory feature.'[4]

There are, therefore, good grounds for utilizing the concept of 'emergence' in our interpretation of naturally occurring, hierarchical, complex systems constituted of parts which themselves are, at the lowest level, made up of the basic units of the physical world. I shall denote this position as that of *emergentist 'monism.'*[5] *Emergentist*

[3] William Bechtel and Robert C. Richardson, 'Emergent Phenomena and Complex Systems,' in A. Beckermann *et al.*, eds., *Emergence or Reductionism? Essays on the Prospects of Nonreductive Physicalism* (New York: Walter de Gruyter, 1992), 266.

[4] Ibid., 285.

[5] As does Philip Clayton. The scare quotes (single inverted commas) indicate that I am not espousing those earlier forms of monism according to which reality is all of one *kind.* This position is intended to recognize that everything in nature can be broken down into whatever physicists deem to constitute matter/energy: It could indeed be called 'emergentist materialism' but for the fact that matter and energy are,

'*monism*' affirms that natural realities, although basically physical, evidence various levels of complexity with distinctive internal interrelationships among their components such that new properties, and also new realities, emerge in those complexes—in biology in an evolutionary sequence.

## 12.2. Whole–Part Influence (or Causation)

If we do make such an ontological commitment about the reality of the 'emergent' whole of a given total system, the question then arises of how one is to explicate the relation between the state of the whole and the behavior of parts at the micro-level. It transpires that extending and enriching the notion of causality now becomes necessary because of new insights into the way complex systems, in general, and biological ones, in particular, behave.

A more substantial ground for attributing reality to higher-level properties and the entities associated with them is the possession of any distinctive causal (I would say, rather 'determinative') efficacy of the complex wholes which has the effect of making the separated, constituent parts behave in ways they would not if they were not part of that particular system (that is, in the absence of the interactions that constitute that system). For *to be real is to have causal power.*[6] New causal powers and properties can then properly be said to have *emerged* when this is so.

Subtler understanding of how higher levels influence the lower levels allows application in this context of the notion of a determining ('causal') relation from whole to part (of system to constituent). This is not to ignore the 'bottom-up' effects of parts on the wholes, which depend on their properties for the parts being what they are, albeit now in the new, holistic, complex, interacting configurations

at the deepest level, interchangeable and that the fundamental building blocks of the natural world may have to be described in quite other terms, e.g. as superstrings. The term 'monism' is here also emphatically *not* intended (as is apparent from the non-reductive approach adopted here) in the sense in which it is taken to mean that physics will eventually explain everything.

[6] A dictum attributed to S. Alexander by J. Kim in 'Non-Reductivism and Mental Causation,' in *Mental Causation*, eds. J. Heil and A. Mele (Oxford: Clarendon Press, 1993), 204.

of that whole. A number of related concepts have in recent years been developed to describe these relations in both synchronic and diachronic systems—that is, both those in some kind of steady state with stable characteristic emergent features of the whole and those which display an emergence of new features in the course of time.

The term '*downward causation*' or '*top-down causation*' was employed by Donald Campbell to denote the way in which the network of an organism's relationships to its environment and its behavior patterns together determine in the course of time the DNA sequences at the molecular level in an evolved organism—even though, from a 'bottom-up' viewpoint, a molecular biologist would tend to describe its form and behavior as a consequence of those same DNA sequences.[7] Other systems could be cited, such as the Bénard phenomenon, the famous Zhabotinsky reaction and glycolysis in yeast extracts.[8] Indeed, Harold Morowitz has identified some 28 emergent levels in the natural world.[9]

Many examples are now known also of dissipative systems which, because they are open, far from equilibrium, and nonlinear in certain essential relationships between fluxes and forces, can display large-scale patterns in spite of random motions of the units—'order out of chaos,' as Ilya Prigogine and Isabel Stengers dubbed it.[10]

In these examples, the ordinary physico-chemical account of the interactions at the micro-level of description simply cannot account for these phenomena. It is clear that what the parts (molecules and ions, in the Bénard and Zhabotinsky cases) are doing and the patterns they form are what they are *because* of their incorporation into the system-as-a-whole; in fact, these are patterns *within* the systems in question. The parts would not behave as observed if they were not parts of that particular system. The state of the system-as-a-whole is influencing (i.e. acting like a 'cause' on) what the parts actually do.

[7] D. T. Campbell, '"Downward Causation" in Hierarchically Organised Biological Systems,' in *Studies in the Philosophy of Biology*, eds. F. J. Ayala and T. Dobzhansky (Berkeley and Los Angeles: University of California Press, 1974), 179–186.

[8] For a survey with references see A. R. Peacocke, *The Physical Chemistry of Biological Organization* (Oxford: Clarendon Press, 1983, 1989).

[9] Harold Morowitz, *The Emergence of Everything: How the World Became Complex* (New York: Oxford University Press, 2002).

[10] I. Prigogine and I. Stengers, *Order Out of Chaos* (London: Heinemann, 1984).

A wider use of 'causality' and 'causation' than Humean temporal, linear chains of causality as previously conceived (A→B→C...) is now needed to include the kind of whole–part, higher- to lower-level, relationships that the sciences have themselves recently been discovering in complex systems, especially the biological and neurological ones. One should perhaps better speak of 'determinative *influences*' rather than of 'causation,' as having misleading connotations. Where such determinative influences of the whole of a system on its parts occurs, one is justified in attributing reality to those emergent properties and features of the whole system which have those consequences. Real entities have influence and play irreducible roles in adequate explanations of the world.

Here the term '*whole–part influence*'[11] will be used to represent the net effect of all those ways in which a system-as-a-whole, operating from its 'higher' level, is a determining factor in what happens to its constituent parts, the 'lower' level.

With arrows representing such influences, the determining relations between the higher (H) and lower (L) levels in such systems and their succession of states (1, 2, 3...) may be represented as in Figure 12.1.

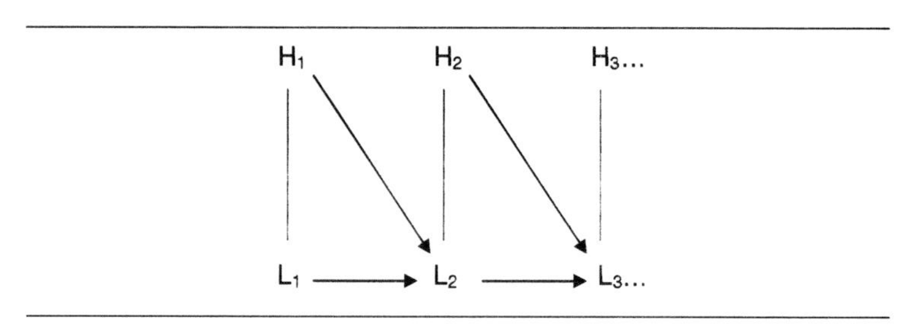


Figure 12.1

The vertical lines represent the mereological relation between the state of the whole system H and the entities of which it is constituted at the lower level L at particular times. The diagonal arrow is meant to indicate that the holistic state $H_2$, which is composed of constituents $L_2$, is determined by (is 'caused by,' is a consequence of) the holistic state $H_1$ *jointly* with $L_1$. How might this understanding of the relation

[11] It must be stressed that the 'whole–part' relation is *not* regarded here necessarily, or frequently, as a spatial one. 'Whole–part' is synonymous with 'system–constituent'.

of 'higher' to 'lower' levels in complex physical, physico-chemical and biological systems contribute to our understanding of the relationships in other complex realities? There are both philosophical and theological applications.

## 12.3. Philosophical Applications

### 12.3.1. *Mind–Brain–Body Relation*

Much of the discussion of the relation of higher to lower levels in hierarchically stratified systems has centered on the mind–brain–body relation, on how mental events are related to neurophysiological ones in the human-brain-in-the-human-body—in effect the whole question of human agency. In this context, a series of levels[12] can also be delineated, each of which is the focus of a corresponding scientific study, from neuroanatomy and neurophysiology to psychology.

The still intense philosophical discussion of the mind–brain–body relation has been, broadly, concerned with attempting to elucidate the relation between that which is colloquially regarded as the 'top' level of human mental experience and the 'lowest,' bodily physical levels. The question of what kind of 'causation,' if any, may be said to be operating from a 'top-down,' as well as the obvious and generally accepted 'bottom-up,' direction is still much debated in this context.[13]

Reality could, it was argued above, putatively be attributable to that to which non-reducible, higher-level predicates, concepts, laws, and so on, applied. These new realities, with their distinctive properties, could properly be called 'emergent' and be said to influence the behavior of their constituent parts. Mental properties are now widely, and in my view rightly, regarded by many philosophers as epistemologically irreducible to physical (that is, neurological) ones. In the mind–brain–body case, the idea that mental properties can be

[12] The physical scales of these levels (according to Patricia S. Churchland and T. J. Sejnowski, 'Perspectives in Cognitive Neuroscience,' *Science* 242 (1988): 741–745) are as follows: molecules, $10^{-10}$m.; synapses, $10^{-6}$m.; neurons, $10^{-4}$m.; networks, $10^{-3}$m.; maps, $10^{-2}$m.; systems, $10^{-1}$m.; central nervous system (CNS), 1m., in human beings.

[13] *Q.v.*, for example, the collection of papers in *Mental Causation*, eds. J. Heil and A. Mele (Oxford: Clarendon Press, 1993).

'physically realized' has also been much deployed in association with the non-reductive physicalist view of the mind–brain issue.

This view is usually represented as in Figure 12.2. (What M and P might refer to is discussed further below).

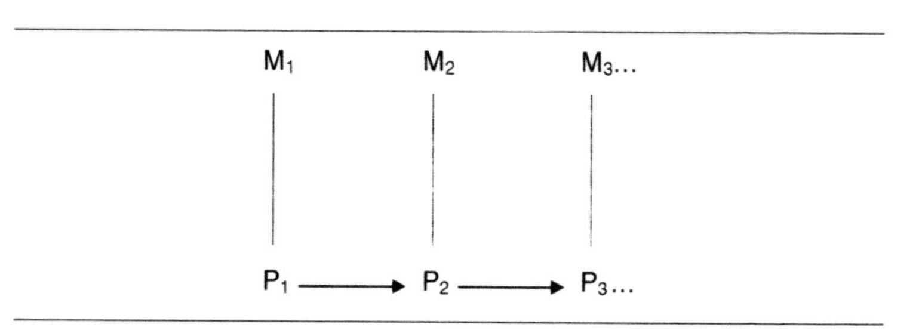


**Figure 12.2**

Kim has argued that this concept is paradoxical for, if it is taken to mean that a microstructure physically realizes a mental property by being a *sufficient* cause for that property, then there is complete causal closure at the physical level alone. Thus, mental properties cannot, in fact, have real causal powers irreducible to physical ones.[14] However, if for mental properties to be real is for them to have new, irreducible causal powers, then the *non-reductive* physicalist is thereby committed to downward causation from the mental to the physical.

What light can be thrown on this impasse for non-reductive physicalism by the above treatment of relations between higher- and lower-level states in complex systems? I suggest that it is Kim's assumption that when a physical microstructure 'physically realizes' a higher-level property (in this case, putatively, a mental one) then a *sufficient* description of the causal relations can be given in terms of microphysical events in the realizing level and entirely (if only eventually) explicated by the laws and theories of physics. However, as I have argued above, in the wider range of physical, biological, and other systems, the determining effects of the higher levels on the lower were real but different in kind from the effects the parts had on each other operating at the lower-level. The patterns of the higher levels make a real difference to the way the constituents behave. Hence, what happens in these systems at the lower level is the result of the *joint* operation of both higher- and lower-level influences. The higher and lower levels

[14] Kim, 'Non-Reductivism,' 202–205.

could be said to be *jointly* sufficient determinators of the lower-level events, a proposition which has also been developed philosophically in terms of higher- and lower-level properties by Carl Gillett.[15] This proposal can illuminate the paradox in non-reductive physicalism, as normally propounded, which Kim has accurately identified.

How can we apply this recognition to the relationship between the levels that operate in the mind–brain–body complex? Three graded possibilities suggest themselves working 'upwards,' as it were, from the purely physical. In Figure 12.1:

I. Levels H are states of the brain; levels L are individual neuronal events.
II. Levels H are mental-with-brain states; levels L are individual neuronal events.
III. Levels H are mental states; levels L are brain states.

Here by a 'state of the brain' is meant the 'temporarily coherent functional units distributed through different maps and nuclei'[16]—that is, the spatial and temporal *patterns* of activity at the brain level, as can be observed externally by empirical techniques. I have discussed these possibilities more fully elsewhere.[17] Suffice it to say here that (I) is an example of the same kind of purely physical systems involving whole–part influence that we have discussed above. In (II), mental states are regarded as brain states under another description (a dual-aspect or even an identitist understanding) and would therefore involve mental causation. (III) involves a recognition that mental states are an emergent property of the physical system of the brain and involves an extrapolation from understood physical systems to the non-understood, the mental. In all three cases a *joint* influence of the higher and lower levels on the succession of coincident higher- and lower-level states is regarded as operative, so that there is higher- to lower-level causation (H→L) as well as lower-level causation (L→L). Most 'non-reductive physicalists,' particularly in their talk of the

[15] Carl Gillett, 'Strong Emergence as a Defense of Non-Reductive Physicalism: a Physicalist Metaphysics for "Downward" Causation,' *Principia* 6 (2003): 83–114.

[16] Churchland and Sejnowski, 'Perspectives in Cognitive Neuroscience,' see note 12.

[17] In 'Emergence, Mind and Divine Action,' in *The Re-emergence of Emergence*, eds. P. Clayton and P. Davies (Oxford: Oxford University Press, 2006).

'physical realization' of the mental in the physical, when not supplemented by any further discussion of whole–part influences, hold a much less realistic view of these higher-level mental properties than I wish to affirm here in this emergentist 'monist' perspective—and also do not attribute determinative (causal) powers to that to which higher-level concepts refer. Just as the complex brain states can be said to emerge from the states of the individual neurons, so similarly mental states can coherently be regarded as emergent from brain states, as having causal efficacy and so as also being real. The content of 'consciousness' then putatively becomes how we describe to ourselves the holistic higher-level state of the component neurons, and so forth, of our brains. Perhaps the capacity of mentalistic language to self-refer to its own activity in the experience of consciousness may one day be understood but, meanwhile, it is legitimate to postulate, with respect to human persons, a whole–part determinative influence (top-down causation) of holistic mental states via lower brain states on the 'micro-physical' neuronal level, and so on the body. For, *that* mental events, such as intentions—whatever they are ontologically—have causal efficacy in the physical world can scarcely be doubted in view of the ability of human agents to act in that world (e.g. the intending, then effecting, of the picking up of an object); *how* this might be so, consistently with well-understood relations in complex systems, is the issue to which the preceding discussion is addressed.

### 12.3.2. *Persons*

Up to this point, I have been taking the term 'mental' to refer to that activity which is an emergent reality distinctive especially of human beings. But in many wider contexts, not least that of philosophical theology, a more appropriate term for this emergent reality would be 'person,' and its cognate 'personal,' to represent the total psychosomatic, holistic experience of the human being in all its modalities, conscious and unconscious, rational and emotional, active and passive, individual and social.

There is a strong case for designating the highest level, the whole, in that unique system which is the human-brain-in-the-human-body-in-social-relations as that of the 'person.' Hence to speak only of mental states as having downward causal efficacy on lower-level brain and neuronal events does not do justice to the rich complexity of the actual higher level of the *person*, namely, of the human-brain-in-the-human-body-in-social-relations. Persons as such experience themselves as causal agents with respect to their own bodies and to the surrounding world (including other persons) so that the exercise of personal *agency* by individuals transpires to be a paradigm case and supreme exemplar of whole–part influence. They can, moreover, report with varying degrees of accuracy to themselves (and also to others) on aspects of their internal mental states and so implicitly on the brain states concomitant with their actions. In other words, 'folk psychology' is unavoidable and the real reference of the language of 'personhood' is justified.

## 12.4. Theological Applications

The cluster of notions developed here regarding the real existence and causal efficacy of emergent entities can be relevant to theology. I suggest that the language and concepts referring to these realities in the natural world can serve to clarify the nature of theological language and concepts in their reference to theological realities.[18] For theological language refers to what one might call the apex in human experience of the hierarchy of complexity—namely, the threefold interaction of nature, persons, and God. When human beings are exercising themselves in religious (or should one better say, today, 'spiritual'?) activities they are in fact operating at a level in the hierarchy of complexity that is more intricate and cross-related than in any of the sub-systems of the natural and social sciences. In those religious activities whole human persons (representing multiple levels of the natural world—'body, mind, and soul,' traditionally) utilize every facet of their being in interacting with each other, with the natural world; as they discover the meaning in these interactions, they can further encounter and experience the presence of the transcendent yet immanent Creator who is the source of all-that-is. It seems to me that no higher level of integration in the hierarchy of natural systems can be envisaged, and theology is about the conceptual schemes that

[18] As I pointed out in my *Creation and the World of Science* (Oxford: Clarendon Press, 1979, 2nd edn. 2004), Appendix C.

articulate the content of this integrated activity. Theology, therefore, refers to this most holistic level in the hierarchy of natural and human relationships. So it should not be surprising if the theories and concepts which it develops to explicate the nature of this activity, both as experienced and as intellectually articulated, are uniquely specific to and characteristic of this level. It should not be surprising, too, if special methods, techniques, and languages had to be developed to describe this supremely integrating and highly complex activity. Moreover, it could also be the case that the pattern of relationships between higher and lower levels which we have been elaborating might serve to model relationships which are the focus of interest in theology. Some areas of theological concern that might be illuminated by these considerations are as follows.

### 12.4.1. *God's Interaction with the World*

A major, critical question in contemporary theology is: in a world that is a closed causal nexus, increasingly explicated by the sciences, how might God be conceived of as influencing particular events, or patterns of events, in the world without interrupting the regularities observed at the levels the sciences study?

I have proposed a model that is based on the recognition that the omniscient God uniquely knows, over all frameworks of reference of time and space, everything that it is possible to know about the state(s) of all-that-is, including the interconnectedness and interdependence of the world's entities, structures, and processes.[19] By analogy with the operation of whole–part influence in natural systems, the suggestion is that, because the 'ontological gap(s)' between the world and God is/are located simply *everywhere* in space and time, God could affect holistically the state of the world (the whole in this context). This is a pan*en*theistic[20] perspective for it conceives of the

[19] See *Theology for a Scientific Age*, 160–166; and *Creation and the World of Science*, Appendix C for an elaboration of this move. For the development of this proposal see my 'God's Interaction with the World: The Implications of Deterministic Chaos and Interconnected, Interdependent Complexity,' in *CC*, 215; and 'The Sound of Sheer Silence,' in *NP*, 215–247.

[20] For a contemporary discussion of panentheism see *In Whom We Live and Move and Have Our Being: Panentheistic Reflections on God's Presence in a Scientific World*, eds. Philip Clayton and Arthur Peacocke (Grand Rapids: Eerdmans, 2004).

world as, in some sense, being 'in' God who is 'more' than the world, so that the world can be subject to divine determinative influences not involving matter or energy (or forces). Thence, mediated by such whole–part influences[21] on the world-as-a-whole (as a *System-of-systems*), and so on its constituents, God could cause particular events and patterns of events to occur which express God's intentions. These would then be the result of 'special divine action,' as distinct from the divine holding in existence of all-that-is, and would not otherwise have happened had God not so intended.

This unitive, holistic effect of God on the world could occur without abrogating[22] any of the laws (regularities) which apply to the levels of the world's constituents, by analogy with the exercise of whole–part influence in the natural systems already discussed. Moreover, this action of God on the world may be distinguished from God's universal creative action in that particular intentions of God for particular patterns of events to occur are effected thereby and the patterns could be intended by God in response *inter alia* to human actions or prayers.

### 12.4.2. *Incarnation*

We have seen that new levels of reality can emerge in natural systems and that, when they do so, new concepts and languages have to be used to explicate these new realities. Although there is a natural continuity with the systems from which the new has emerged, there

[21] What passes across this 'interface,' I have also suggested (*Theology for a Scientific Age*, 161, 164), may perhaps be conceived of as something like a flow of information—a pattern-forming influence—but one has to admit that, because of the 'ontological gap(s)' between God and the world which must always exist in any theistic model, this is only an attempt at making intelligible that which we can postulate as being the initial effect of God seen, as it were, from our side of the boundary. Whether or not this use of the notion of information flow proves helpful in this context, we do need some way of indicating that the effect of God at this, and so at all, levels is that of pattern-shaping in its most general sense. I am encouraged in this kind of exploration by the recognition that the Johannine concept of the *Logos*, the Word, of God may be taken to emphasize God's creative patterning of the world and so as God's self-expression *in* the world.

[22] The same may be said of *human* agency in the world. Note also that this proposal recognizes more explicitly than is usually expressed that the 'laws' and regularities which constitute the sciences usually apply only to certain perceived, if ill-defined, levels within the complex hierarchies of nature.

is a discontinuity in the conceptual resources utilized in explaining the new situation. This has seemed to me to provide a pointer to making intelligible traditional Christian affirmations about the relation of the human to the divine in Jesus, that is, the doctrine of the Incarnation.[23]

The basic starting point for the Christian theist should, I suggest, be that creation is in God and that God is the agent of its events and processes—that God is *semper Creator*, giving existence to all-that-is in all time. In Jesus, his followers encountered, especially in light of his resurrection, a dimension of that transcendence which, as good monotheists, they attributed to God alone. But they also encountered him as a full human person, and in his personhood they experienced an intensity of God's immanence in the world. The fusion of these two aspects of awareness of the presence of God in the person of Jesus led to the conviction that something new had appeared in the world and they ransacked received concepts to try to give expression to this *dis*continuity—to that new non-reducible distinctive mode of existence, eventually designating it *inter alia* as 'incarnation.' We can now properly, I am suggesting, regard Jesus as a new kind of reality, an 'emergent,' for that manifestation of God in a human life must both have been a manifestation emanating from within creation and be a fulfillment of its inherent God-given potentialities. Because of the continuity of the creative activity of God throughout time acting through the inherent creativity of the world, we can come to see Jesus as a unique manifestation of a possibility always inherently there for human beings by virtue of their potential nature being created by God. The 'incarnation' which has occurred in Jesus can be regarded as an example of that emergence-from-continuity that characterizes the processes of the created order. There is continuity with all that preceded him yet with him a new mode of human existence emerges through a new openness and obedience to God—for this new relationship of God-humanity creates a new emergent reality. Such openness to God could arise only from his human response to an experienced initiative from God, so that he was, as it were, a God-informed human being.[24] Hence, God was, in

[23] *Creation and the World of Science*, 241ff.

[24] In the terminology of information theory, q.v., Arthur Peacocke, 'The Incarnation of the Self-expressive Word of God,' in *Religion and Science: History, Method,*

this sense, *acting* in the incarnation through the immanent processes of Jesus' human consciousness and will, and the fusion of the two engendered a new kind of reality.[25] Yet, for Christians, Jesus also manifests a unique *dis*continuity with what preceded him in that titles employing new, non-reducible language—of Christ (=Messiah, Anointed), *Logos* (=Word), Lord, Son of God—had to be applied to his person to represent the new kind of reality experienced in him.

### 12.4.3. *Worship*

The nature of relationships of persons to God may also be illuminated by understanding the emergence of new realities in complex systems. In many situations where God is experienced by humans we have by intention and according to well-winnowed experience and tradition complexes of interacting personal entities, material things and historical circumstances which are epistemologically not reducible to concepts applicable to these individual components. Could not new realities and so new experiences of God for humanity be seen to *emerge* in such complexes and even to be causally effective?

I am thinking, for example, of the Church's Eucharist (Holy Communion, the Mass, The Lord's Supper), in which there is present a distinctive complex of interrelations among its constituents. These latter could be identified *inter alia* (for it is many-layered in the richness of its meanings and symbols) as follows.

1. Individual Christians are motivated by a sense of *obedience* to the ancient, historically well-authenticated, command of Jesus, the Founder of their faith, at the actual Last Supper to 'Do *this*...'—that is to eat the bread and to drink the wine in the same way he did on that occasion and so to identify themselves with his project in the world.

2. Christians of all denominations have been concerned that their communal act is properly *authorized* as being in continuity with that original act of Jesus and its repetition, recorded in the New Testament,

*Dialogue*, eds. W. M. Richardson and W. J. Wildman (New York: Routledge, 1996), 321–339.

[25] Such an emphasis on continuity (immanence) as well as on emergence (incarnation) is vital, in my view, to any account of Jesus that is going to make what he *was* relevant to what we *might be*.

in the first community of Christians. Churches have differed about the character of this authorization but not about its importance.

3. The physical elements, as they are often called, of bread and wine are, of course, part of the matter of the world and so representative of the created order. So, Christians perceive in these actions, in this context and with the words of Jesus in mind, that a *new significance and valuation of the very stuff of the world* is being expressed in this action.

4. Because it is bread, and not wheat, wine, and not grapes, which are consecrated, this act has come to be experienced also as a new evaluation of the work of *humanity in co-creating with God in ordinary work.*

5. The broken bread and poured-out wine was explicitly linked by Jesus with his anticipated self-sacrificial offering on the cross in which his body was broken and blood shed to draw all towards unity of human life with God. Christians consciously acknowledge and identify themselves with Jesus' *self-sacrifice*, thereby offering to reproduce the same self-emptying love for others in their own lives and so to further his purposes of bringing in the Reign of God in the world.

6. They are also aware of the promise of Jesus to be present again in their re-calling and re-making of the historical events of his death and resurrection. This 'making present' (*anamnesis*) of the Jesus who is regarded as now fully in the presence of, and is, in some sense, identified with God is a unique and spiritually powerful feature of this communal act.

7. The action is to be undertaken in the *community* of the Church and both forms and strengthens it—a determinative influence.

Here do we not have an exemplification of the emergence of a new kind of reality since this complex situation is epistemologically not reducible? For what (if one dare so put it) *emerges* in the eucharistic event *in toto* can only be described in special non-reducible terms such as 'Real Presence' and even 'Sacrifice.' A new kind of reality is attributable to the eucharistic event, for in it there is an effect both on the individual and on the community that creates specifically Christian personhood and society. That the Eucharist has a causal efficacy in enhancing the spiritual life of the participants has been the concern of sacramental theology, which has been developed to interpret this special reality and the human experience of it. Since God is present 'in, with, and under' this holistic eucharistic event, God may properly be regarded as acting distinctively through it on the individual and community—an exemplification of God's non-intervening, but specific, 'whole–part' influence on the world.

I have taken this as one example, but I propose that the principle involved in trying to make clear what is special about this particular spiritual situation is broadly applicable to many other experiences of worship, of prayer, and of 'grace' (with its implications of determinative, 'causal', efficacy). The concept of emergence can contribute not only to a release for theology from the oppression of excessively reductionist interpretations by the sciences, such as psychology, but also to theological language and concepts accessible to the general exchanges of the intellectual life of our times. As I have asked elsewhere: Would it be too much to suggest that these new, emergentist 'monist' insights into the inbuilt creativity of our world through its complexifiying and self-organizing capacities open up a vista of continuity between the physical, the mental, and the spiritual which could, in this new century, break down the parallel barricades mounted in the last century between not only the two cultures—of the sciences and the humanities—but also between the developed experiences of nature and of God, that is, between those of the sciences and of religion?[26]

[26] Arthur Peacocke, 'Complexity, Emergence and Divine Creativity,' in *From Complexity to Life: On the Emergence of Life and Meaning*, ed. Niels Henrik Gregersen (Oxford: Oxford University Press, 2003), 187–205.

# 13

# Reduction and Emergence in Artificial Life: A Theological Appropriation

*Niels Henrik Gregersen*

## 13.1. Introduction

In discussions of emergence and reductionism, research on artificial life (A-Life) is an ambiguous case. On the one hand, A-Life is particularly concerned about the informational aspects of the material world—aspects that have often been neglected in traditional physics with its focus on the particle–energy duality of matter. On the other hand, A-Life research suggests that even complex informational patterns can be generated by simple algorithms. Proponents of A-Life are thus often sceptical about reducing information to physical entities, but happy to embrace a reduction of informational patterns to computational rules. This is particularly the case for proponents of strong A-life. Proponents of weak A-life, however, tend to see the study of the informational rules as intimations to a much more subtle and complex world.

The case of A-Life may thus show how reduction and emergence may be treated as complementary strategies for approaching the levels of complexity inherent in biological life-forms. But, by virtue of its focus on nature's capacities for pattern-formation, A-Life also touches upon metaphysical issues that are central to philosophical as well as religious interest: What is, after all, the relation between particles, energy, and information? There is, I argue, something to be learned from A-Life research for theological reflection on transformations in our current scientific worldview.

What has the idea of artificial life to do with religion, and how may religions perceive the research program of artificial life (A-Life or AL)? At first glance not very much. A-Life models are by definition human artifacts. But for religion it is the biological world in all its richness and serendipity that triggers both religious awe and a sense of resonance. Artificial systems do not immediately possess the same attraction. So what are the intersections between theology and A-Life? Indeed quite a few.

My first point will be quite simple and general. The artificiality of A-Life should not count against its relevance for religion; art and mathematics, likewise products of human imagination, have prompted religious reflection. Moreover, any religion with a zest for novelty will be interested in scientific inquiry, both as a value in itself and as something with potential bearing for our general worldview. In what follows, I argue that the particular relevance of A-Life is a new understanding of the world of nature as undergirded by informational patterns. A worldview is dawning of horizontally interacting, dynamical networks, a worldview that *grosso modo* replaces a pre-modern concept of nature as a hierarchy of natural beings and at least remolds the early modern view of nature as a clockwork of particles mechanically governed by general laws. In light of the new information sciences, 'matter' appears as an irreducible triad of mass, energy, and information. I further argue that if information or pattern-formation does turn out to be an inherent property of material systems (not separable from, but distinct from both particles and energy), then the world of nature is at least open for a religious interpretation in terms of the radical in-dwelling in the world of matter of a ubiquitous creative Logos.[1]

My second claim is that the natural points of contact between A-Life and religion are mostly of an indirect nature, mediated through a common interest in the phenomena of life. What Christopher Langton has written about the interest of A-Life within the field of

[1] See Niels Henrik Gregersen, 'Fra urvaerket til netvaerket: Teologi mellem fysik og informatik' [trans. 'From the Clockwork to the Network: Theology Between Physics and the Science of Information'], *Dansk teologisk Tidsskrift* 65/4 (2002): 272–295; *idem*, 'Complexity: What Is at Stake for Religious Reflection?,' in Kees van Kooten Niekerk and Hans Buhl, eds., *The Significance of Complexity: Approaching a Complex World Through Science, Theology and the Humanities* (Aldershot: Ashgate, 2004), 59–95.

biology—'[t]he acceptance of Artificial Life techniques within the biological community will be directly proportional to the contributions it makes to our understanding of biological phenomena'[2]—also applies to the theological reception: the relevance of A-Life for theology will finally depend on its power to enrich our understanding of real-life biology. I fully grant that diverse AL models have an intrinsic value in exploring 'possible biologies.' But from the perspective of theology, AL research programs will be particularly interesting if they evidence the existence of a deep mathematical order that pervades the complex world of biology, an order that contains both universal mathematical principles and a special fecundity to produce new patterns under the appropriate circumstances. Investigating the interplay between universal and more local features of the biological world will impact our understanding of the dynamic world in which we all—irrespective of beliefs—live and breathe.

There are both strong and weak programs in A-life. While some AL researchers pursue a strong A-Life program that intends to create a wholly new type of life, developed from scratch in digital media,[3] others more modestly aim to simulate real-world life by computer simulations. Note here that 'strong' and 'weak' refer to claims, and not to the justification of claims. It may well turn out that weak AL programs fare better than strong AL programs (as it has been the case in the parallel case of artificial intelligence, AI). What makes A-Life theologically exciting is the prospect of investigating the creative zones between what-is and what-is-not-yet, between actuality and possibility. AL computer models have the potential of revealing deep-seated structures in the real-world biology, while *also* exploring the 'adjacent possibilities' (Stuart Kauffman) of evolutionary options. Thereby, A-Life suggests that the biological world is embedded in a world of possibilities, not unlike the distinction in the Judeo-Christian tradition between the 'earth' (symbolizing the actual world of creation) and the 'heavens' (pointing

[2] Christopher Langton, 'Artificial Life,' in *The Philosophy of Artificial Life*, ed. Margaret A. Boden (Oxford: Oxford University Press, 1989), 92; page references are to the 1996 edition.

[3] Thomas S. Ray, 'An Evolutionary Approach to Synthetic Biology: Zen and the Art of Creating Life,' in Christopher G. Langton, ed., *Artificial Life: An Overview* (Cambridge, MA: MIT Press, 1995), 179–210; 179f.

to the yet-unseen possibilities of new creativity within the realm of creation).

Third, by programmatically contesting the commonly held distinctions between the natural and the artificial, the biological and the cultural, A-Life may help religious reflection to move beyond the understanding of divine creation as something which is confined to the 'naturally given.' Rather, I shall argue that a theology of creation has also to do with self-productive or autopoietic systems, the distinctive nature of which are only determinable as a result of temporal processes that involve the creative interactions between systems and their environments.[4]

In what follows, I begin with a cultural analysis of the AL movement (section 13.2), followed by a discussion of some topics of AL research that are internal to science and particularly pertain to the problem of reduction and emergence (sections 13.3–5). I then lay out some theological reflections on A-Life research on the heuristic role of computational metaphors for theology (section 13.6). Finally, I propose a Trinitarian interpretation of the concepts of information, energy, and matter (section 13.7).

## 13.2. Worldview Assumptions of A-Life: A Cultural Analysis

In *Silicon Second Nature*, anthropologist Stefan Helmreich has analyzed the cultural milieu surrounding the *Santa Fe Institute on Complexity Studies* (instituted in 1984).[5] Looking at the scientists from an informed outsider's perspective, Helmreich treats the AL community just as any other tribe. He notices that even though the AL researchers are highly ambivalent, if not hostile towards religion, a spiritual quest has surrounded AL research from the beginning. Here I shall thus focus on A-Life's public image, as it has been articulated by its practitioners and communicated by its popularizers, and draw on Helmreich's anthropological analysis. It should be noted that under the

[4] Niels Henrik Gregersen, 'The Idea of Creation and the Theory of Autopoietic Processes,' *Zygon: Journal of Religion and Science* 33/3 (1998): 333–367; and *idem*, 'Einheit und Vielfalt der schöpferischen Werke Gottes: Pannenbergs Beitrag zu einer trinitarischen Schöpfungslehre,' *Kerygma und Dogma* 2 (1999): 102–129.

[5] Stefan Helmreich, *Silicon Second Nature: Culturing Artificial Life in a Digital World* (Berkeley: University of California Press, 1998).

anthropologist's external perspective, science is taken as one cultural phenomenon among others, without epistemic privileges.

Helmreich's overall thesis is that AL researchers are 'powerfully inflected by their cultural conceptions and lived understandings of gender, kinship, sexuality, race, economy, and cosmology and by the social and political contexts in which these understandings take shape.'[6] This could well be the conclusion of any anthropological work studying any scientific movement. Nonetheless, I find Helmreich's material observations from his field study of AL researchers quite illuminating.

Helmreich points out that sensitivity toward the cultural context is part of the self-understanding of the AL community. 'The practice of science involves more than science,' acknowledges Langton, one of the prime movers behind the AL movement.[7] More precisely Helmreich detects in the AL community a sort of substitute religion that is thought to 'demolish religion,' though it also forms a new secular religion.[8] Using Clifford Geertz's definition of religion as a way of tuning 'human actions to an envisaged cosmic order' and of projecting 'images of cosmic order onto the plane of human experience,'[9] Helmreich shows the persistence of religious themes among A-Life scientists, themes that even 'resurrect very Christian themes.'[10] First, there is strong awareness of founding disclosure-experiences or 'revelations' among the pioneers of AL research. At least this is how it is presented in popularizing works such Mitchell Waldrop's best-seller *Complexity*,[11] or in Steven Levy's *Artificial Life.*[12] In an interview with Waldrop, Langton reported his first discovery

[6] Ibid., 11.

[7] Christopher Langton, 'Introduction,' in Christopher G. Langdon, Charles Taylor, J. Doyne Farmer, and Steen Rasmussen, eds., *Artificial Life II: Proceedings of the Workshop on Artificial Life held February, 1990, in Santa Fe, New Mexico* (Proceedings Volume X: Santa Fe Institute Studies in the Sciences of Complexity, 1992), 3–24. Cf. Helmreich, *Silicon*, 17.

[8] Helmreich, *Silicon*, 185.

[9] Clifford Geertz, *The Interpretation of Cultures: Selected Essays* (New York: Basic Books, 1973), 90; cf. Helmreich, *Silicon*, 182f.

[10] Helmreich, *Silicon*, 185.

[11] Mitchell M. Waldrop, *Complexity: The Emerging Science at the Edge of Order and Chaos* (New York: Simon & Schuster, 2002).

[12] Steven Levy, *Artificial Life: A Report from the Frontier Where Computers Meet Biology* (New York: Vintage Books, 1992).

of self-reproducing loops popping up on his screen in a night of computer experimentation at Massachusetts General Hospital: 'One time I glanced up,' he says. 'There is the Game of Life cranking away on the screen. Then I glanced back at my computer code—and at the same time, the hairs on the back of my neck stood up. I sensed the presence of someone else in my room.'[13] Levy interviewed Langton about the same primal moment, and interprets Langton's discovery in terms of a prophet turned into an evangelist: 'Langton had defined his mission; now he had to evangelize it.'[14] Langton, in fact, is regularly described as the 'guru' of AL research, thereby associating him with the New Age movement. However, the founding moment of revelation is not confined to charismatic leaders in religion. Helmreich reports a widespread feeling of being part of an almost Hegelian *Zeitgeist* that transcends the individual scholar and places him or her deep in the epic of evolution. 'There is a zeitgeist out there. No matter what we do, Artificial Life is going to happen,' one informant said.[15] There is also a prophetic sense of 'accelerating the process' that evolution is going to undertake anyway at some point of time, said another scientist.[16] In an interview with Levy, Langton was asked to comment on people who say that AL 'is almost an inevitable part of our evolution.' Langton responded: 'Well, I think so. All life that we know has evolved and passed on and changed.... My feeling is that [life] is out of our control; we are just little cogs in a much bigger evolutionary process.'[17]

Alongside this feeling of being embedded in a wider evolutionary network, there is also a feeling of creating something new from scratch, which naturally leads to the joy of being a divine-like creator. An informant said to Helmreich: 'If I had to declare a religion for myself, it would be basically the quest to become God.' However, aware that this statement could appear as a Western scientific hubris, he added that it was a kind of Zen-god he spoke about: a god that materializes in a person as soon as this person realizes that 'there's no distinction between you and the universe.'[18] In a Zen-like manner, one becomes divine by being the channel of evolutionary

[13] Waldrop, *Complexity*, 202. [14] Levy, *Artificial Life*, 102.
[15] Helmreich, *Silicon*, 191. [16] Ibid., 192.
[17] Quoted in Helmreich, *Silicon*, 191. [18] Ibid., 193.

enlightenment. The repeated reference to Zen fits with the feeling of instantiating an alternative worldview close to nature, a worldview that breaks down dualistic barriers between self and environment, and between biological life and artificial life.

But how does this reference to a non-dualistic Zen-awareness square with the Promethean consciousness of creating something wholly new (*de novo*), not out of the womb of nature (*de ovo*). 'Look. I've got a universe, and I have things which self-reproduce in it.' So Langton according to Levy.[19] Helmreich diagnoses this and similar statements as the articulation of a masculine creation story that is not in need of any physical stuff such as matter. In the Western tradition since Aristotle, 'matter' and 'soil' have usually had female connotations, while the 'form' and 'seed' have represented the masculine principles of making things move. This old mythology may not be a part of the conscious self-understanding of current AL researchers, but Helmreich gives evidence of the fact that themes of a purely masculine creation myth permeate the AL community, among both its practitioners and its popularizers.[20]

Here is one example from a research article. Thomas Ray (the creator of *Tierra* system) explicitly formulates his urge to move beyond the constraints of the physical and chemical world:

> Synthetic organisms evolving in other media, such as the digital computer, are not only not part of the same phylogeny, but they are not even of the same physics. Organic life is based on conventional material physics, whereas *digital life exists in a logical, not material, informational universe.* Digital intelligence will likely be vastly different from human intelligence; forget the Turing Test.[21]

Information is here understood as an a-material concept, independent not only of the carbon-based embodiment of physical chemistry, but apparently of all physical strictures. A-Life is understood to be dissimilar to real-world biology, and yet is claimed to be genuinely living, insofar as it is self-replicating and has the potential for evolution. Ray is thus a proponent of strong A-Life. He even demands a respect for the particularities of the digital medium.[22]

[19] Levy, *Artificial Life*, 103. [20] Helmreich, *Silicon*, 117.
[21] Ray, 'An Evolutionary Approach to Synthetic Biology,' 183; italics mine.
[22] Ibid., 204.

What Ray does not make explicit, however, is the fact that the computational concept of information cannot in reality be removed from 'ordinary' physics. After all, the computer hardware is a Newtonian machine, working only through the transmission of electricity through all its circuits. An A-Life model of a creation *ex nihilo*, however, would demand a full separation of information and physical stuff. As I shall argue below, this is an illusion.

The gravitation toward the concept of information may also provide an opening for a religious orientation in Christian terms, a possibility not reflected in Helmreich's analysis. Kevin Kelly, co-founder and editor-at-large of the computer magazine *Wired*, describes himself as a 'postmodern evangelistic Christian.' His general view on information seems similar to Ray's, though it is couched in Christian language. He again points to the almost disembodied character of information:

> The whole wired world we are building, this great big rush we are in right now to make technology, is founded on something called 'information.' In many ways, information is as weird and intangible as the Holy Spirit. When people talk about information, they could just as well be talking about spirits. They wave their hands the same way; they use the same kind of vocabulary. Presently, we don't have a good theory of information. Everything in the world is really information. John Wheeler calls it 'Its are Bits'. Go as far as you can into the cosmos, and you end up with information. Go down as deep as you want, and you find information. It's very much like spirit. Exploring the way nerds are getting information holds, I think, immense potential for theologians.[23]

However inviting this scenario might be for a theologian, it should always be remembered that even though the same informational pattern can sometimes be realized in various media, no information is ever disembodied. This is clearly acknowledged by Langton. He describes information as a predominant force, though not ever in reality freed from physical media: '[T]he most salient feature that distinguishes living organisms is that their behavior is clearly based on a complex dynamics of information. In living systems, information

[23] Kevin Kelly, 'Theology for Nerds,' in W. Mark Richardson, Robert John Russell, Philip Clayton, and Kirk Wegter-McNelly, eds., *Science and the Spiritual Quest: New Essays by Leading Scientists* (London: Routledge, 2002), 51.

processing has somehow gained the upper hand over the dynamics of energy that dominates the behavior of most non-living systems.'[24] What is here said about biological life also applies to digital life. Moreover, the informational pattern in biology and A-Life are far from always universal. Rather, A-Life compels us to look for a new understanding of the *local* rules pertaining to *local* physical domains.[25] Let us therefore proceed from an external to a more internal analysis of A-Life as a scientific research program.

## 13.3. Varieties and Ambitions of A-Life Programs

A-Life takes a variety of forms. Charles Taylor and David Jefferson have helpfully distinguished between three different media of AL.[26] The first medium is the real-world *wetware*; think of the many attempts to catalyze primitive life from chemical compounds into proto-RNAs since the early experiments of Stanley Miller and colleagues.[27] *Software* is another medium: by algorithmic computer programs, AL researchers aim to understand how multicellular assemblages can replicate themselves and even develop in unexpected ways. This is what I take to be the central field of A-Life. John von Neumann's early work on cellular automata from the end of the 1940s onwards provided the theoretical background for the field. The first simple, but impressively workable, A-Life model was created by John Conway in 1969.[28] Since the 1980s, Stephen Wolfram has undertaken a systematic investigation and categorization of cellular automata,[29] which has informed the organizing work of Langton and colleagues

[24] Langton, 'Life at the Edge of Chaos,' 41–92; 42f.

[25] Cf. Langton, 'Artificial Life,' in Boden, ed., *Philosophy of Artificial Life*, 64: 'The transition function for the automata constitutes a *local* physics for a simple, discrete space/time universe.'

[26] Charles Taylor and David Jefferson, 'Artificial Life as a Tool for Biological Inquiry,' in *Artificial Life: An Overview*, ed. Christopher G. Langton (Cambridge, MA: MIT Press, 1995), 1–14.

[27] Wolfgang Stegmüller, *Hauptströmungen der Gegenwartsphilosophie*, Vol. 3 (Stuttgart: Reclam, 1987), 160–164.

[28] See M. Gardner, 'Mathematical Games: The Fantastic Combinations of John Conway's New Solitaire Game "Life",' *Scientific American* 223/4 (1970): 120–123.

[29] Stephen Wolfram, 'Cellular Automata as Models of Complexity,' *Nature* 311 (1984): 419–24; summarized in Wolfram, *A New Kind of Science* (Champaign, IL: Wolfram Media, 2002).

since the end of the 1980s.[30] Third, we have examples of AL research that works on the level of the *hardware*, corresponding to the level of organisms such as robot ants or autonomous *mo*bile ro*bots* such as Genghis and other mobots developed at MIT since 1988.

One can also divide the field according to the claims of strong and weak A-Life programs. The former interprets software (and possibly also hardware) systems as genuine instantiations of biological life, whereas the latter understands AL computer models as tools for simulating real-world biology. However, there may also be interesting mixed cases that can be interpreted as medium strong A-Life. *Strong AL* proponents will contend that, for instance, computerized software models of 'vants' (*v*irtual *ants*) are as real and as living as real-life ants, the medium just being different. Ray, a proponent of strong A-Life, has even expressed an environmental concern for the engendered species of ALs. On 7 July, 1994, he gave a talk at MIT on A-Life with the title, 'A Proposal to Create a Network-Wide Biodiversity Reserve for Digital Organisms.' He is reported to have said: 'I think of these things as alive, and I'm just trying to figure out a place where they can live.'[31] *Medium-Strong AL* proponents claim that biological phenomena may possibly be recreated in the digitalized silicon world of computers. The *Santa Fe Institute for Complexity Studies* offered the following interesting definition of the aims of A-Life:

> Artificial Life ('AL' or 'Alife') studies 'natural' life by attempting to *recreate biological phenomena from first principles* within computers and other artificial media. Alife complements the analytical approach of traditional biology with a synthetic approach in which, rather than studying biological phenomena by taking apart living organisms to see how they work, researchers attempt to put together systems that *behave like living organisms.* Artificial life amounts to the practice of 'synthetic biology.'[32]

This understanding of A-Life preserves the distinction between real-world organisms and the artificial world ('behave *like* living organisms'). A-Life is only a re-creation, not an entirely new creation;

[30] See Christopher G. Langton, 'Artificial Life,' in Langton, ed., *Artificial Life*; *idem*, 'Life at the Edge of Chaos'; *idem*, 'Editor's Introduction,' in *Artificial Life: An Overview* (Cambridge, MA: MIT Press, 1995), ix–xi; and *idem*, 'Artificial Life,' in Boden, ed., *The Philosophy of Artificial Life*, 39–94.

[31] Helmreich, *Silicon*, 4.

[32] Santa Fe Institute, *1993 Annual Report on Scientific Programs* (Santa Fe, New Mexico: Santa Fe Institute, 1994), 38; my italics.

nonetheless, the program also expresses a strong demand of being able to recreate life 'from first principles'—from logical principles, as I take the quotation to mean. Here lies the difference to *Weak AL* proponents who understand AL as the attempt to model real-world life (RWL) by computer modeling in order to uncover, if possible, a deep underlying mathematical order in the realm of biology. Computer modeling is understood not as creating or recreating a new world of living beings, but as a tool to understand the much richer and more complex RWL. As expressed by Thiemo Krink, a research leader of the EVALife (*Ev*olution of *A*rtificial *Life*) project at Aarhus University, Denmark, 'the two main purposes [of AL computer modeling] are to improve the understanding of a system and to make predictions about the future states of a system.... The model must be able to mimic and explain all essential aspects of the [natural] system to elucidate a certain scientific question or purpose.'[33] There is still an important emphasis on the need for *representation* between the relatively simple computer models and the far more complex biological systems. For some purposes, a gross simplification is advantageous, especially when it comes to modeling large-scale universal structures.[34] For other purposes, a still more detailed simulation of the real-world objects is expedient, particularly when we are interested in modeling the behavior of specific species such as spiders, bees, or ants. Here a model seeks to imitate a biological reality as far as possible. I find this 'weak' program of A-Life the stronger and more promising one.

## 13.4. Reduction and Emergence: The Janus Face of A-Life

Whatever spin one takes on computational A-life, the aim is to uncover the *simple rules* that generate *complex patterns* of behavior

[33] Thiemo Krink, 'Complexity and the Computing Age: Can Computers Help Us to Understand Complex Phenomena in Nature?' in Kees van Kooten Niekerk and Hans Buhl eds., *The Significance of Complexity: Approaching a Complex World Through Science, Theology and the Humanities* (Aldershot: Ashgate, 2004), 47–74; 50.

[34] An example is Per Bak's theory of Self-Organized Criticality; see Per Bak, *How Nature Works: The Science of Self-Organized Criticality* (Oxford: Oxford University Press, 1999); cf. Niels Henrik Gregersen, 'Beyond the Balance: Theology in a Self-Organizing World,' in Gregersen and Ulf Görman, eds., *Design and Disorder: Perspectives from Science and Theology* (Edinburgh: T&T Clark, 2002), 53–92; esp. 65–72.

of natural phenomena. Simplicity often underlies complexity. With regard to *emergence*, AL therefore has something like a Janus face. Insofar as complex patterns are generated by simple and rigid mechanisms, AL must be seen as a reductionist research paradigm. This is acknowledged by Langton: 'The field of artificial life is unabashedly mechanistic and reductionist. However, this *new mechanism*—based as it is on the multiplicities of machines and on recent results in the fields of nonlinear dynamics, chaos theory, and the formal theory of computation—is very different from the mechanism of the last century.'[35]

The differences from microphysical reductionism are indeed worth noting. First, the general purpose of A-Life is to investigate the *social* interactions taking place in biological or digital space. As programmatically stated by Langton: 'Artificial Life is simply the synthetic approach to biology: rather than take living things apart, Artificial Life attempts to put living things together.'[36] Second, the reduction is to 'interaction rules,' and not to 'physical entities.' In this sense, it is again information that matters and not stuff. In some programs we may find examples of a 'downwards' feedback influence of higher-order rules on lower-level interaction-rules. An example in real-world biology is the nest building in social insects such as ants and termites; note, however, that these self-organizing systems presuppose stable rules, so that the holistic result (e.g. the completed nest) emerges out of the many rule-governed individual actions, all of which are fully determined from the outset. In other cases, for instance in John Holland's complex adaptive systems, one can make rules for how to change the rules when the internal operations plus the incoming external inputs reach a certain level. An interesting case of such self-productive or autopoietic systems is the development of language, where vocabularies are self-organizing, by being *generated* by individual agents, by being *propagated* in populations, and finally by *influencing the selection* of words in fluid communications by virtue of positive feedback.[37] Thus, AL is sometimes interpreted as an anti-reductionist research program.

[35] Langton, 'Artificial Life,' in Langton, ed., *Artificial Life*, 6.

[36] Langton, 'Artificial Life,' in Boden, ed., *The Philosophy of Artificial Life*, 40.

[37] Luc Steels, 'Self-Organizing Vocabularies,' in Christopher G. Langton and Katsunori Shimohara, eds., *Artificial Life V: Proceedings of the Fifth International*

In my view, this is correct insofar as new unexpected patterns appear in the course of long-run computer-experiments. Higher-order properties (such as the size of nest, or the career of vocabularies) may influence lower-level activities of individuals (such as ants or language users). But, as far as I can see there are no examples of a strong 'ontological' emergence in the sense of new independent causal powers.[38] As long as the set of the computer algorithms and input functions are pre-established, and as long as the algorithms run within a deterministic physical system (the computer) according to clear-cut sets of *if–then* mechanisms, then the future effects of the computer programs are determined by the decisions of the past (irrespective of the question whether the particular decisions are made at the highest organizational level or at lower levels). AL models should therefore be seen as *deterministic*, though many unforeseeable things happen due to the non-linear equations used by the computer. The indeterminacy, in other words, is epistemological rather than ontological. If an observer had time to follow the individual computational steps of a megacomputer, there would be nothing mysterious going on the A-Life screen; nothing genuinely new would happen. Emergence, in other words, is in the holistic results, but not in the causal fabric.

However, if one preserves a distinction between real-world complexity (RWC) and computational complexity (CC), A-Life and other software models may be taken as first intimations into a study of RWC-ontological emergence, that is, the view that higher-level properties can neither be adequately described nor sufficiently explained by lower-level properties.[39] Inside a computer, however, elements of ontological novelty could only be produced, say, by connecting a Geiger counter near a radioactive source to the computer, and then using the interval between the atomic clicks to determine the changes within the artificial world (as suggested in a thought experiment by Ray[40]). Thus, the distinction between CC and RWC remains. For the

*Workshop on the Synthesis and Simulation of Living Systems* (Cambridge, MA: MIT Press, 1997), 179–184.

[38] See the distinctions in Philip Clayton, 'Emergence,' in W. Van Huyssteen, ed., *Encyclopedia of Science and Religion*, Vol. 1 (New York: Macmillan, 2003), 156–159; and in chaps. 1–4, this vol.

[39] Peacocke, *Theology for a Scientific Age.*

[40] Ray, 'An Evolutionary Approach,' 138.

same reason one should not be led to argue that a CC reductionism in any sense implies a RWC reductionism: *Computer models are simply neutral to the ontological question, whether or not natural systems exhibit cases of strong ontological emergence.*

Furthermore, it should always be remembered that a computer model, even if it succeeds in creating a persuasive pictorial model of a natural system, *has not thereby uncovered the physical causes that have produced these real-world processes in the first place.* Consider the fractal patterns that appear on a computer screen when the processor is appropriately programmed: they may greatly resemble the intricate structures of plants and their leaves, but no such computer simulation can explain the intricate biochemical processes that underlie the morphogenesis of real-world plants. As rightly underlined by Claus Emmeche, 'the fact that a computer plant resembles its original forebear is far from a guarantee that the algorithm that has generated the image corresponds to the mechanism that has generated the plant.... Our obsession with computer-generated plants is just as much a matter of the observer as of what is observed.'[41]

Emmeche here points to the differences between the massive and variegated causal processes going on in the world of physics and biology, and the simple processes going on in A-Life scenarios. Indirectly, this is a plea for a weak A-Life interpretation. Computer models are not the world, nor is the world likely to be a computer. Computer experiments can be seen, as suggested by John Holland, as a halfway house between a theory and a real-world experiment.[42] A model is not the real thing; rather, computational models investigate the propagation of orders that actually *do* occur analogously in nature.

In this sense, one might hypothesize that if one opts for strong A-Life (and emphasizes the parity of wetware and software media), one will be inclined to assume an ontological reductionism. However, if one belongs to the camp of weak A-Lifers (and keeps up the fundamental distinction between CC and RWC), one may opt for an ontological emergence theory. However, the reasons for ontological

[41] Claus Emmeche, *The Garden in the Machine: The Emerging Science of Artificial Life* (Princeton: Princeton University Press, 1994), 77.

[42] John Holland, *Emergence: From Chaos to Order* (Oxford: Oxford University Press, 1998).

emergence should then be found in the real-world nature, and not in the artificial worlds of AL.

## 13.5. From the *Game of Life* to *Swarm Logic*: Reconnecting CC and RWC

Let me here go into some detail on the special case of cellular automata. A cellular automaton is an artificial world. Its 'space' is usually a grid consisting of uniform squares in a two- or three-dimensional lattice. The 'time' dimension of cellular automata is provided by the transition rules, which determine how the individual cells are to be changed, moved, removed, or reproduced at each computational step. Cellular automata thus use individual-based modeling: the 'organisms' are placed in square or cubic cells in the grid, and their 'actions' are specified, according to the rules, by the number and features of the cells in their immediate vicinity.

John Conway's cellular automaton, *Game of Life*, is especially well known.[43] It is a two-dimensional automaton with only two states. A cell's state transitions depend on the states of the other eight neighboring cells (including its diagonal neighbors). The general rule is a so-called totalistic rule, in that the rule is determined by the total number of neighboring colors, not by their particular positions relative to the cell. Each cell can be either full (black) or empty (white). The transition rules in Conway's system are then as follows:

1. If two of the neighboring cells are black, the cell remains unaltered (mimicking equilibrium).
2. If three are black, the cell becomes black (mimicking reproduction).
3. In all other cases, the cells become white (mimicking extinction).

Most would agree that this is simple, indeed very simple. It is either 'die' or 'multiply.' Nonetheless, the evolving features of this system can be highly complex. This becomes evident when one tries out different initial conditions and observes how the system proceeds. When the

[43] Gardner, 'Mathematical Games.'

program is run, it generates clusters of cells ('populations'), pulsating processes of near-extinction and sudden regeneration, populations meeting, coalescing, and reinforcing one another. All this is complex in itself. The most astonishing feature is the emergence of 'gliders,' that is, localized structures that develop in one general direction and create interesting self-organizing structures on their way—structures that are far from simple. Thus, the *Game of Life* may also model historical lines of descent, some of which continue to grow endlessly while eliciting ever-changing structures, new forms of order.

Conway's *Game* showed, as early as in 1969, that even a limited set of very simple if–then rules is able to produce complex behavior. But, Wolfram's systematic investigation of all possible cellular automata also shows that the phenomenon of creative gliders in Conway's model is a universal feature of a certain category of cellular automata, and does not depend on a specifically ingenious design. Moreover, cellular automata show that orders can emerge without being supported by 'higher levels.' This suggests that there may be *two sorts of organization in nature.* One is the hierarchical type of organization, in which lower levels are nested within higher levels (cells in tissues, tissues in organisms). The other form of organization is the associative pattern that arises due to local interactions without any need for overarching structures. A-Life focuses on the latter form because of its 'flat' design-structure.

An excellent example of this associative logic in real-world biology is found in the world of social insects, such as ants, termites, and some bee and wasp species.[44] Ancient writers were already aware of the amazing qualities of formic organization, and the book of Proverbs even points to the ant community as an exemplary model for human living: 'Go to the ant, you lazybones; consider its ways, and be wise. Without having any chief or officer or ruler, it prepares its food in summer, and gathers its sustenance in harvest' (Proverbs 6:6–8). While neo-Darwinian theory has been able to partly explain ant cooperation as a result of kinship selection (that is, preferential

[44] Around 6% of insects are eusocial in the sense of cooperatively caring for young, overlapping at least two generations, and reproductively dividing labor; see Henry Hermann, 'Social Organization in Insects,' in Sidney W. Fox, ed., *Selforganization: Proceedings of the Liberty Fund Conference on Selforganization, 1984, Key Biscayne, Florida* (Guilderland, NY: Adenine Press, 1986), 123–132; 124.

opting in favor of one's own kin), empirical studies increasingly show the extent to which there also exists mutualistic behavior unrelated to kinship.[45] In this field, complexity theory has been able to shed considerable light on many of the interactions that take place at the epigenetic level, that is, at the level of group interaction relatively independent of genetic constraints. The example of ant cooperation shows how the paradigm of computational complexity is empirically motivated, and offers powerful explanatory frameworks. If complexity theory needs empirical material to work on, biology can certainly supply it.

A case in point is the nest building of ants, in which several processes go hand in hand.[46] For one thing, ants use all sorts of heterogeneities in their natural environment (such as holes) as *templates* when determining where to begin building an anthill. They may, however, also use their own bodies as chemical templates for anthill construction. When the queen is physogastric (that is, filled with eggs), she emits a pheromone that attracts other ants. They accordingly begin to deposit the soil pellets around her and the anthill is constructed around her body, which serves as a pheromonal template. Another principle is *stigmergy*. That is, the soil pellets already placed together by some ant workers will stimulate other ant workers to do the same, so that the initial state $A_1$ stimulates response $R_1$, which reinforces $A_1$ into the next state $A_2$, which exerts an even stronger stimulus towards $R_2$, and so forth. Nest building through stigmergy is a self-amplifying process, often based on random initial conditions.

Furthermore, the size of the anthill's chambers is determined by the length of the ants, which puts a natural limit on the distance between the ground and the ceiling of each chamber. Jean-Louis Deneubourg and colleagues have demonstrated that the ant's nest-building behavior follows the general pattern of self-organization:

[45] As early as 1986, Henry R. Hermann ('Social Organization in Insects,' 132) commented: 'Based on comments by a few researchers in the literature and on my personal communication with some of them, it appears that it will be important for us in the future to avoid overstressing altruistic behavior as related to kinship theory and understressing the concept of mutualism.' In current biology, group selection theories have strongly re-entered the scene of theoretical biology.

[46] Guy Theraulaz, Eric Bonabeau, and Jean Louis Deneubourg, 'The Origin of Nest Complexity in Social Insects,' *Complexity* 3/6 (1998): 15–25.

1. *amplifications* of pre-given random fluctuations of the template,
2. *positive feedback* in terms of amplification of the cooperative processes of stigmergy, and
3. *negative feedback* in terms of limits to growth.[47]

In all cases, the rules that specify the interactions between the ants and their environments, and between the ants themselves, are based entirely on local information, without the ants needing to possess any awareness of a 'plan' for the ant colony as a whole. However, something the general structure of self-organization cannot explain is the causal trigger of the whole process, in this case primarily the pheromone. While complexity theory yields insights into the general formative patterns, the particular causes triggering the process can only be investigated through empirical study (in this case biochemistry). The relation between computational models and empirical studies is thus one of *complementarity*: without complexity, one loses sight of the bigger picture of the nest-building process as a whole, but without empirical studies one cannot identify the causal agents.

Local rules also apply to the activities and division of labor among the ants. Ant activities are highly regulated. In her study of the harvester ant *Pogonomyrmex barbatus* in Arizona, Deborah Gordon identifies four distinct tasks for adult ants: foraging, patrolling, nest maintenance, and midden work (i.e. trash removal).[48] By means of experiments, she has been able to demonstrate the rules governing task allocation and the switching of functions. Ants do not have castes, as earlier assumed, so there is no genetic bias for task allocation. All decisions are epigenetic, based on the very local information available. The pheromone once again plays an important role, functioning as a 'semiochemical' that instructs the ants to perform this or that task. The various tasks in an ant colony are interdependent, even though the individual ants seem to behave in a stimulus–response manner to the pheromone. There are specific rules for task switching: when the hill is attacked from the outside, nest maintenance workers switch into patrollers; when ample food is available, nest workers

[47] Théraulaz *et al.*, 'The Origin of Nest Complexity,' 18–21.

[48] Deborah Gordon, 'The Development of Organization in an Ant Colony,' *American Scientist* 83 (January–February 1995): 50–57.

switch into foragers. Hence, the performance of nest building tasks and midden work can only be ensured through the steady reproduction of young ants within the anthill community. Thus, *the ant colony exhibits a pattern formation that transcends the local informational inputs, although the pattern formation is causally produced merely by virtue of the local information processes.* As eloquently put by Steven Johnson:

> Call it swarm logic: ten thousand ants—each limited to a meager vocabulary of pheromones and minimal cognitive skills—collectively engage in nuanced and improvisational problem-solving. A harvester ant colony in the field will not only ascertain the shortest distance to a food source, it will also prioritize food resources, based on their distance and ease of access. In response to changing external conditions, worker ants switch from nest-building to foraging.... Their knack for engineering and social coordination can be downright spooky—particularly because none of the individuals is actually in 'charge' of the overall operation.[49]

So far we have been dealing with a real-world complex system that has been investigated both empirically (RWC) and through a computer modeling of the soft type (CC). Insofar as building up structures takes place from the bottom up and without any master ant overseeing the whole process, we have an organizational pattern of the associative type. The local rules of interaction between the ants and their environments can be modeled in a computer program, which is likewise governed by definite if–then rules. These informational strategies produce the life cycles of the surprisingly large ant colonies that can contain up to 12,000 members. In this case, there is a fruitful division of labor in empirical studies of ants and computer studies of the complexity of virtual 'aintz.' In addition, complexity researchers have been able to apply the rules of ant communication to the organization of human affairs, such as task allocation in companies, and the distribution of cargo in airports. The technological applications are many and varied.[50]

[49] Steven Johnson, *Emergence: The Connected Lives of Ants, Brains, Cities and Software* (London: Allen Lane/Penguin, 2001), 74.

[50] Eric Bonabeau and Guy Théraulaz, 'Swarm Smarts,' *Scientific American* 257/3 (2000): 74–79.

What is interesting from a more theoretical perspective is that we here find a different model for emergent processes than we do in the older concepts of emergence in the organicist tradition, which still presupposed the classic notion of a *scala naturae*, a hierarchy of beings.[51] Orders grow from the bottom up (from semiochemicals to ants to anthills). But once the ant society is established, it begins to exercise a causal feedback effect on its constituent parts—without providing these parts with further information. The distinction between higher level and lower level is replaced by localized processes that result in overall concerted behavior, beyond the knowledge of individual agents and without anyone designing the process.

In this bottom-up fashion the paradigm of computational complexity theory sets itself the aim of cracking the codes of naturally evolved phenomena. 'Cracking the code' means tracing the minimal set of computational steps necessary for regenerating the logic behind the phenomenon. These rules are highly localized and only evolve under certain highly contingent evolutionary conditions. But the road to analyzing these constraints goes through massive computer experiments that have different rules and different outcomes.

Of course, the local rules of any computer model will always be constrained by the phase space of the computer program. This also applies to real-world biology: during evolution, very specific informational pathways have been carved out. As a result, biological individuals and societies are highly specific—think of the very specific rules we found among the ants. These rules differ from species to species. All are consistent with physics and chemistry. But, none of these informational instructions can be explained fully in the vocabulary of basic physical processes. *The search for universal mathematical equations simply becomes uninteresting when we come to biology, and even more so when we arrive at the level of human societies. For what matters here is not matter, but embodied information.*

At least this is an epistemological irreducibility: local rules cannot be explained from the universal laws of physics. Yet, what cannot be subsumed under a general law of nature may nonetheless be

[51] Scott F. Gilbert and Sahotra Sarkar, 'Embracing Complexity: Organicism for the 21st Century,' *Developmental Dynamics* 219 (2000): 1–9.

compressible into a relatively limited number of interaction rules. This move from general equations to computer programs is perhaps the revolutionary essence in the science of complexity. It leads to rediscovery of the concrete particulars of the world of creation. Gravity applies everywhere; but the rules for task allocation and role switching are unique to the Arizona harvester ant.

So how may we interpret in religious terms this interplay between universal order and the continuous formation of new local orders? What is the relevance of informational structures and pattern formation for theology?

## 13.6. The Heuristic Value for Theology of Computational Thinking: The Network Imagery

Whereas the relevance of biological complexity for religion is immediate, insofar as the fecundity and variety of living beings continues to trigger a religious sense of wonder, I have so far argued that the relevance of A-Life for religious reflection is of a more indirect character, that is, mediated by a common interest in the informational structures that pervade biological life. But, computational thinking also informs the metaphors of lived religion, and thus influences the thought models of theology as a second-order reflection on religious life. The computer sciences influence the cultural atmosphere of our age, and the information sciences may supply religious language with some of the most persuasive metaphors at hand.

For example, we are used to thinking about 'codes'—not only encoded messages that have to be deciphered, but also genetic codes, dress codes, and codes of behavior. We think in terms of 'digits,' which can be assembled, taken apart, and recombined in multiple ways. Likewise, 'information' is no longer about knowledge alone. The idea of information has been creeping into biology and physics as well: DNA is seen as an informational device, and physical matter and energy can have different configurations and informational states. We are indeed living in *The Moment of Complexity*, as the theologian and cultural critic Mark C. Taylor has called it. He points to the extent to which many old binary oppositions (such as form and content, nature and culture) begin to dissipate. In particular, the dichotomy between information and matter seems obsolete. '[I]nformation is, in important ways, material, and matter is informational. From this expanded point of view, neither information nor materiality is what it seems to be when it is interpreted in simple oppositional terms.'[52] With Taylor, I believe that the sciences of complexity, including A-life, will continue to influence religious self-reflection via ideas such as bottom-up emergence, self-organization, and fluid networks. Against this background, and considering the prevalence of evolutionary theory, it is hardly likely that human beings will continue to be understood in opposition to nature, as was fashionable in twentieth-century personalism and existentialism, which influenced theology immensely (in my view to the detriment of Christian theology). It is also probable that half-forgotten ideas such as 'the communion of believers' will come to play a more prominent role in theological reflection over the coming decades, thereby questioning the modernized idea of believers as solitary, private believers.

The computer sciences will doubtless play an important heuristic role as a resource for finding new theological metaphors and rethinking traditional ones. While the scientific revolution of the seventeenth century was accompanied by the metaphor of God as the great *Designer* or *Watchmaker*, the sciences of complexity may be accompanied by metaphors of God as the great *Networker*.

At this interface, the half-forgotten Christian doctrine of God as Trinity has again begun to fuel theological reflection over the last decades. We find philosophical theologies espousing the idea of the triune God as a divine *Matrix* that generates and sustains all that exists.[53] In the current revival of Trinitarian thinking, God is seen as a divine community of love that gives room to, nourishes, and includes all created communities.[54] 'Salvation' is interpreted, in line with patristic tradition, as a 'participation' in a divine network that embraces the otherness of created existence.

On this interpretation, the Christian concept of God may be more 'Zen-like' than usually assumed. God is *not* perceived as an

[52] Mark C. Taylor, *The Moment of Complexity: Emerging Network Culture* (Chicago: University of Chicago Press, 2002), 106.

[53] Joseph Bracken, *The Divine Matrix: Creativity as Link Between East and West* (New York: Orbis, 1995).

[54] Gregersen, 'Fra urværket til netværket.'

individual substance that resides here or there, nor as an abstract infinite substance beyond the world. Neither is God an individual 'person.' Rather, God is a *living community* that only exists in the 'mutual indwelling' or *perichoresis* of the divine 'persons' (Father, Son, and Spirit). These divine 'persons' (*persona* originally means the 'mask' that an actor speaks through) are not understood as individual self-owned persons, but as self-conscious poles within the divine life of self-giving love. The divine 'persons' always live in communion within the divine life. The Zen-like aspect lies in the constant transcending of the characteristic features of Father, Son, and Spirit; the divine persons are semi-permeable.

What is not Zen-like about the Christian doctrine of Trinity is the persistent feature of divine love. The identity or essence of the divine community is certainly in flux, but only within the movement of a divine love that expires and inhales, gives and receives, within the eternal life of God. Also, what is not Zen-like is the assumption of the three Abrahamic traditions that while all created things are in a state of impermanence, God is eternal in that God is His/Her own source of being. The Father (or Mother) is giving birth to the Son, who in turn gives light to the world by expressing God's inner heart. The 'Son' or Word of God thus externalizes the inner 'Reason in God' (*logos*) or divine Wisdom (*sapientia*). In this function of God's expressive love, the Logos is oriented outward towards the world of creation. Accordingly, the Spirit is the divine life-giver who constantly energizes the mutual love between the Father and the Son, the Source and the Offspring. In the Spirit's outward orientation, the energizing Spirit passes on the divine Energy to the material world, thus giving life and inspiring the ever-new circulation of life.

I emphasize that my purpose is here not to develop a natural theology that starts from the world and then infers the existence of a triune God. Rather, it is to develop a theology of nature, a way of re-describing in religious terms the same world that is already described and to some extent also explained scientifically. My aim is to show that the synthetic orientation of A-Life—with its insistence on the creative interactions between autonomous agents and their environments—is congenial with the Trinitarian model of divine persons, whose properties can only be identified in relation to one another; mutuality is the key to the divine life.

## 13.7. The Triune God and the Triad of Matter

So far, I have only pointed to important linguistic intersections between the information sciences and theology. But beyond models, metaphors, and thought models, are there also ontological meeting points? From theological premises indeed there must be. For according to the Christian faith (and to Judaism, Islam, and Hinduism), God is creator of the very same world that the sciences are investigating. In what follows, I shall thus propose a theological ontology of the material world, which involves the Triune God being intimately present in the core of physical matter as described by the sciences, without conflating God and creation. I shall argue that the idea of a Triune God—Father, Son, and Holy Spirit—offers unique resources for developing a theological ontology which is congenial to the concept of physical matter as a field of mass, energy, and information. The following proposal thus stands or falls with the assumption that the concept of information should be accorded an ontological role in our current scientific concept of matter.

In the previous sections, I have argued that while the best A-Life programs presuppose a robust naturalism, they also suggest that information is as seminal to nature as are the matter-and-energy aspects of physical matter. In fact, this development is not entirely new, but has precursors in twentieth-century physics. Already Einstein's special theory of relativity de-materialized the concept of matter insofar as matter was described as transmutable into energy, and vice versa. Later on, quantum theory has shown that physical reality does not have the solid-state character that old-style materialists assumed.[55] With the computer sciences, the idea of matter as an assembly of isolated bricks has been further challenged. As the physicists Paul Davies and John Gribbin have put it, the matter myth is broken, for 'matter as such has been demoted from its central role, to be replaced by concepts such as organization, complexity and information.'[56]

[55] Max Jammer, 'Materialism,' in *Encyclopedia of Science and Religion*, Vol. 2 (New York: Macmillan, 2003), 538–543.

[56] Paul Davies and John Gribbin, *The Matter Myth: Dramatic Discoveries that Challenge our Understanding of Physical Reality* (New York: Simon & Schuster, 1992), 15.

I prefer to be more cautious and say that our concept of matter has been enlarged so that the substance of matter (that is, its basis in material stuff such as particles), the energy of matter (the potential and actual changeability of physical matter), and the organization of matter (its capacity for pattern formation) constitute three relatively independent, though inseparable, aspects of matter, which cannot be fully reduced to one another.[57] Especially in biology, the notion of information is important, since information is used to determine how to use the energy budget available for each organism and for interactive group behavior. As Langton argued, '[i]n living systems, information processing has somehow gained the upper hand over the dynamics of energy that dominates the behavior of most non-living systems.' Langton further argues that the exploration of Artificial Life is motivated by the following fundamental question: 'under what conditions can we expect a dynamics of information to emerge spontaneously and come to dominate the behavior of a physical system?'[58]

The question then arises, what is meant by information? John Puddefoot has helpfully distinguished between three types of information.[59] The first is *counting information*, or the mathematical concept of information as defined by Claude Shannon. Here 'information' means the minimal information content, expressed in bits (*b*inary dig*its*), of any state or event: '1' or '0.' In order, for example, to know which of sixteen people have won a car in a lottery, one would expect to need sixteen bits of information, namely the fifteen losers (symbolized by a '0') and the one winner (symbolized by a '1'). However, if one began to ask intelligently ('is the winner in this or that group of eight, and so on), one would only need four computational steps ($\log_2 16 = 4$).[60] Here 'information' means

[57] Jiri Zemann, 'Energie,' in *Europäische Enzyklopädie zu Philosophie und Wissenschaften*, ed. Hans Jörg Sandkühler *et al.* (Hamburg: Felix Meiner, 1990), vol. 1, 694–696; 695: 'Der Begriff der Materie hat drei Hauptaspekte, die untereinander untrennbar, aber relativ selbständig sind: der stoffliche (mit Beziehung zum Substrat), der energetische (mit Beziehung zur Bewegung) und der informatorische (mit Beziehung zur Struktur und Organisiertheit).'

[58] Langton, 'Life at the Edge of Chaos,' 42; Langton's italics.

[59] John C. Puddefoot, 'Information Theory, Biology, and Christology,' in Mark Richardson and Wesley J. Wildman, eds., *Religion and Science: History, Method, Dialogue* (New York: Routledge, 1996), 301–320.

[60] Warren Weaver, 'Recent Contributions to the Mathematical Theory of Communication,' in Claude E. Shannon and Warren Weaver, eds., *Mathematical Theory of Communication* (Urbana: University of Illinois Press, 1949), 94–117; 100f.

mathematically compressible information. This counting concept lies at the heart of CC, including the idea of Artificial Life. Different from this is what Puddefoot calls *shaping information*, which is the form or pattern of existing things. Here the interest lies in morphology, the study of forms. Shaping information may derive either from internal sources (such as a zygote) or from external constraints that cause something to have a definite pattern in relation to its environment. The third type of information is what we refer to in daily parlance: coming to know something of importance. In *meaning information*, information means something *to someone* in a given context. A facial expression, for instance, can convey kindness or aggression, the awareness of which makes a vast (and causal!) difference in a social context.

Against this background, Langton's question is ambiguous. For what is understood by 'conditions,' and by 'dynamics of information?' To proponents of strong AL, the question seems to imply that shaping information can be reduced to counting information. Similarly, the proponents of strong AI (who contend that machines can, or could, have thoughts and feelings comparable to ours) will even argue that meaning information, too, can be reduced to counting information. In both cases, we face a stark reductionism.

I do not think that leaps from one type of information to another are possible. It seems, however, that the concept of *shaping information is basic* in relation to the two other forms of information. Counting information seeks to model or simplify shaping information, and meaning information seeks to interpret shaping information for a specific purpose. The new role of information, and the corresponding idea of a 'local' or domain-specific physics offer, as far as I can see, two new possibilities for a theological understanding of nature.

First, there is in the world of nature a richness and heterogeneity—a natural complexity—that discourages the old-style reductionist claim that we and our fellow creatures are 'nothing but molecules' (J. Monod), or 'nothing but extended phenotypes of self-replicating DNA' (R. Dawkins). The reductionist idea that nature is uniform and can be comprehended from the perspective of one single overarching scientific theory (molecular or genetic) presupposes that information can be left out of the picture of reality. Because organization is what ultimately matters in biology, the physicalist versions of reductionism are dead ends for the purpose of understanding biological life, though it may well explain the origin of life. Reductionism can and should be pursued as a *method* for the many types of investigation that benefit from the isolation of parts. But reductionism can no longer be professed as a scientifically based *metaphysics*. If one were to do so, one would have to appeal to physical laws and initial conditions that are not yet known (and perhaps not knowable by means of the physical sciences).

Seen from the perspective of Christian tradition, one cannot help but notice the similarities between the three irreducible aspects of matter, and the doctrine of the Trinity of Father (stuff), Son (form), and Holy Spirit (energy). Let me therefore conclude with a theological experiment concerning the ontological relationship between God and the material world. I hereby hope to show not only why God and physical reality are one and united, but also how God differs from the world of matter.

Let me begin by clarifying elements of the meaning of Trinitarian faith. For just as scientists dislike speaking of 'matter' in general terms, but always prefer to discuss matter under specific conditions, so theologians avoid speaking in general terms about 'divinity' or 'spirituality.' In the fourth century, the Cappadocian Fathers (Basil the Great, Gregory of Nyssa, and Gregory of Nazianzus) interpreted the Christian confession of the Father, Son, and Holy Spirit as follows:

1. The three divine persons make up the divine being (Greek *ousia*) as a living communion (*koinonia*), which is realized only in the eternal reciprocal interactions (later called *perichoresis*) between the three divine persons (Greek *hypostasis*). The Father, Son/Logos, and Spirit are thus inseparable from one another in the one community of God's being, and yet not reducible to one another as three distinct persons or poles.

2. More specifically God is conceived of as a living community, in which the Father eternally generates (makes room for) the otherness of the Son, while the 'life-giving Spirit' proceeds from the Father in accordance with the Son or Logos. At the same time, the Father could not be Father without giving birth to the Son, who makes the Father a father (or mother), just as the relation of love between the Father and the Son would not be fulfilled without the Spirit as the living realization of personal love between them. The divine 'Persons' are thus co-constituted by one another.

3. While the Father is the eternal source or 'fountainhead' of the Trinity, the eternal Son and the eternal Spirit are the Divine Persons who issue forth from the pure, ineffable divinity and in the act of creation establish the world of creation. The Son and the Spirit are the 'two hands of the Father,' as Irenaeus had already put it around AD 200. But, neither would the Father be the creator, and thus not 'the Father of all' (Ephesians 4:6), without the divine Logos who incessantly informs the world of creation as its creative Pattern in constant communication with the Spirit, who continues to energize and transform the world.

How are we to envisage the relation between God and matter if we assume, even hypothetically, that God is an eternal Trinitarian community, and if we assume that the three aspects of matter—substance, energy, and organization—are inseparable but irreducible aspects of our universe?

One possibility is to think of God and matter as two names for the same underlying reality. This is the position of *pantheism*. Pantheism remains a possibility in the wake of complexity theory. A version was formulated by Samuel Alexander (one of the main emergentists in the 1920s) and has currently been rearticulated by Harold Morowitz, for whom the laws of nature are identical with the immanent God. But eventually the rules of selection give rise to the emergence of *Homo sapiens*. At some point in time, consciousness entered into the picture of nature, and nature was thereby transcended for the first time. By means of consciousness, we can project different futures and become responsible for our own actions. 'We, Homo sapiens, are the transcendence of God.'[61] God, by contrast, is purely immanent.

The question is whether the identification of God with the laws of nature comes even remotely close to what is usually meant by 'God' in religious language. In Morowitz's portrayal, God is dumb, without will, without knowledge, without feeling. God would not even be the cause of the world in the full sense of the word, since the laws of nature presuppose some initial conditions to have existed prior to the laws themselves. Matter would exist independently of God as the subvenient base of the supervenient divine spirit, which then emerges late in the history of the cosmos. Moreover, any physical, biological,

[61] Morowitz, *The Emergence of Everything*, 196.

or historical event would be an immediate expression of God's underlying nature. The Holocaust would be a divine revelation in the same sense as faith, hope, and love. Although I admit the logical possibility of pantheism, and even though I also acknowledge the intellectual economy of the idea, I do not find it an attractive hypothesis.

I suggest another route. First, God and matter are so deeply interconnected that one could say, with Anselm of Canterbury: 'Where God is not, there is nothing' (*Monologion* 14). This is to say that wherever something exists, God is present in it. Furthermore, if one takes a position of minimal naturalism (saying that all-that-exists within the world of creation is materially constituted), then God, by implication, is present in matter itself. This is what is suggested in Jewish scripture (Jeremiah 23:24; Psalm 139) as well as in the New Testament (Acts 17:28). In the writings of Martin Luther, we find perhaps the most striking affirmation of the unity between God and material beings. In discussing how God can 'really' be present in the bread and wine of the Eucharist, Luther takes a strong view in favor of God's omnipresence in the world of matter: 'For it is God who, through his almighty power and right hand, creates, works in and contains all things.... Therefore, He must fully exist in each individual creature, in their intimacy as well as around them, through and through, below and above, in front and behind, since nothing can be more present or intimate in all creatures than God with His power.'[62] Luther here expresses a theological ontology, but he does not develop a natural theology: The claim that God is present in all material parts of the universe does not mean that God is revealed in all things. On the contrary, Luther held the view that the God who is everywhere in the cosmos is usually hidden to us, but made graspable in God's own Word: Jesus Christ. For God is in all things without being identical with any thing. God is omnipresent without being 'omnimanifest.'

With the new concept of matter provided by information theory, I believe that this theological position can be made more precise. *On the Trinitarian view, God is not identical with matter, nor is the Father identical with the stuff of matter (mass)*. For if this were the case, the

[62] Martin Luther, 'Das diese Wort Christi "Das is mein Leib" noch feststehen, wider die Schwärmgeister,' in *D. Martin Luthers Werke* (Weimar: Hermann Böhlaus, 1964), vol. 23, 133f; my translation; cf. *Luther's Works* (Philadelphia: Fortress Press, 1961), vol. 37, 57f.

identity of God as an eternal communion of love would not be safeguarded; all things that happen in our world would immediately be identical with God's being. Rather, the contention is that God makes room for the otherness of matter in creation, similar to the manner in which God, within the internal Trinitarian life, makes room for the otherness of the 'Son.' The Son is thus the principle of otherness within God's own life.

The term 'Son' is derived from the Jesus history. However, we should bear in mind that the term 'Son' is a familial metaphor for the manner in which the eternal Logos expresses the mind of God in an uncompromising manner, while remaining distinct from the Father (or the Source of Being) by having his own will and self-awareness. In this sense, the eternal Father makes room for the eternal Logos, and this Logos ('in God') is the principle of otherness, who freely distinguishes itself from the Father in order to be God's 'hand' in creation. This is why it can be said in the New Testament that all things are 'from' the Father (who is the 'whence' of physical stuff), but that all things have come into being 'through' the Son (1 Corinthians 8:6). For the Logos is the Creative Pattern for all that is 'logical' in the world of creation. *The Son/Logos is the free, outward expression of the inner mind of God the Father, and as such the Creative Pattern that releases pattern formation in the midst of creation.*

It is helpful to bear in mind the full range of the Logos-Christology. Indeed, from the perspective of Trinitarian faith, the Logos has to do with the informational aspects of all that is. Christ does not enter history with the man Jesus, in whom the Word became incarnate, but the Logos or the eternal Son was ever-present in God. For Logos is God's readiness to let the world be and let Godself become flesh in this world. Logos is the creative Pattern of the universe, the informational Matrix from which all patterns arose, arise, and will continue to emerge throughout history. *As such, Logos is not identical with matter's informational capacities, but the wellspring of order in the universe.*

What, then, is the corresponding role of the Holy Spirit? It is precisely to energize the world of creation. The first thing said about the Spirit in the Nicene Creed is that it is the vivifying spirit, 'the giver of life,' who proceeds from the Father (and the Son). Only then is it said that the Spirit also spoke in historical time to the prophets, and thus has been guiding humankind. Just as Christ has a cosmic function as co-creator before and after the time of Jesus, so the Spirit has a function in nature before and apart from the more spiritual aspects of human existence. Therefore, we again find that the Spirit is the Giver of life, but not itself a biological being with DNA, and so on. And yet, *as the transcendent source of life, the giver is present in the gifts of biological existence, though without being identical with specific life-forms, for if this were the case the divine Spirit would die with the end of biological existence.* The Spirit is the eternal energizer of divine life, who is present in all vital functions of all living beings, and who continues to release the ambience of mutuality in the world.

Now the triadic structure of divine life also shows its inner logic. If God the Father had only one 'hand,' be it the Son or the Spirit, the complex world could not come into being. Imagine if only the Logos had been sent into the world, but not the Spirit: there would be structure everywhere, but no life and no evolution. And imagine if only the Spirit had come into the world, but not the Logos: there would be chaotic processes everywhere, but no structure anywhere from which the world of organized complexity could grow forth.

It is exactly in the interplay between Logos and Energy that creativity unfolds itself, and emergence appears—between pure order and pure randomness.

# 14

# Toward a Constructive Christian Theology of Emergence[1]

*Philip Clayton*

Can one construct a viable Christian theology that uses the theory of emergence as its systematic principle? The question is vital: if emergence is in fact a fundamental feature of the natural world, as authors in this volume are arguing, then one wants to know how Christian theology might be reformulated to be responsive to these new insights. What would a theology of emergence look like if fleshed out in detail? Where is it consistent with traditional theology and where do the two diverge? Does this theology do a better or worse job of doing justice to the Christian scriptures, the Christian faith experience, and the traditional emphases of theologians through the ages? As a systematic theology, is it able to reduce the tensions between Christian belief and the world as we have come to know it through scientific study?

The goal here cannot be to prove the rational superiority of the theology of emergence to all other theological options—how could one

[1] The text that appears here was first prepared for inclusion in my *Mind and Emergence* (Oxford: Oxford University Press, 2004). At the time the book was entitled *The Emergence of Spirit*, with the idea that its argument would be extended to encompass a theology of emergence. It proved a daunting enough task, however, to convince scientists and philosophers that there were good reasons for them to take the emergence concept seriously; hence the theological project was deferred to the present volume. Note that accepting an emergentist theory of the natural world does not *require* one to accept a theology of emergence. Scholars may agree on the relevance of emergence for science while disagreeing on its metaphysical implications. I dedicate this essay to Arthur Peacocke, who first inspired its vision.

do that without a full discussion of the competing positions? Indeed, given the necessary brevity, it's not even possible to trace all the different ways in which this theology diverges from the other major options in the Christian tradition. Our goal must be a much more modest one: to show how a theology that is responsive to emergence insights might address the standard *loci* of systematic theology, and to ask whether these responses are consistent, coherent, and convincing.

## 14.1. Christian Theology and Theories of Agency[2]

The question of agency is of central importance for theologians. Whenever one appeals to the notion of divine action, one implicitly evokes the concept of agency, for God must be the agent of God's own actions. By contrast, one does not speak of the actions of *karma*. Karma is a law-like regularity that has a moral dimension. It entails that actions of a certain sort will have causal effects, for example, that your hostile action toward another, which is a sort of negative force, will inevitably affect you either in this life or in a later one. An agent, by contrast, is one who initiates and carries out her own actions.[3] Moreover, God is universally understood by believers as an agent who is worthy of worship. But the practices of praying to and worshiping the ultimate reality are valid only if this reality has attributes that render it *worthy* of this behavior, which include not only moral attributes but also cognitive attributes such as autonomy, rationality, intentionality, and so forth.[4] The practice of worship implies the belief that God is the sort of agent who possesses at least all the positive qualities we have as person-agents, and (presumably) others as well, and possesses them to the highest possible degree.

*Which* position one takes on consciousness, mind, or spirit in humans will have a major effect on what one will be able to say about

[2] I am grateful to Stephen Kosslyn, Steve Kosik, Sarah Coakley, and Wesley Wildman for their criticisms of an earlier draft of this portion of the paper, presented at Boston University in April 2002.

[3] See the four conditions of agency in Christoph Schwöbel, *God: Action and Revelation* (Kampen: Kok Pharos, 1992), 36.

[4] Steven Knapp, correspondence.

the nature of God.[5] If we are unable even to give a coherent account of what it is to be, think, and act as a human agent, our odds of making sense of a divine agent are pretty slim. Conversely, if we can develop an adequate account of human agency, then the door is at least open for theologians to specify the ways in which God as divine agent is like human agents and the ways that God is distinct from them.

Consider five conclusions one might reach about human mind or spirit and, in each case, what will follow for the question of God:

(1) If one makes *physicalism* fundamental, then (as far as I can see) mental states will ultimately have to be epiphenomenal.[6] Such a metaphysic would clearly not be very congenial to the existence and activity of a God who is pure Spirit and whose existence precedes the material world altogether.

Likewise, (2) *dual aspect theories* are no more helpful than theories of double agency for making sense of divine actions. As is well known, dual aspect theories set the mental and the physical side by side, as two 'attributes' of one underlying substance or 'stuff,' without expressing how or why the two attributes are related. This philosophical move certainly seems to eliminate *ab initio* any scientific research into the evolution of mind or the relationship between mind and brain. Theologically, such theories link most naturally with monist or pantheist positions, such as the anti-dualist positions of Ramanuja and Spinoza.

(3) *Emergentist monism* denies that reality is primarily to be characterized as *either* 'physical' *or* 'mental' or, for that matter,

[5] Indeed, correlations between theories of self and theories of ultimate reality are ubiquitous. Vedantic philosophies of the self in the Indic traditions correlate with the philosophies of *brahman* or ultimate reality; Buddhist and Zen Buddhist theories of the *no-self* correlate with teachings on *sunyata* or Emptiness; and physicalist or reductionist moves in the philosophy of mind correlate with agnostic or atheist metaphysics. Theists will see in mind or spirit or soul an anticipation of the nature of God—and must so anticipate, since according to all the theistic traditions, humankind is made 'in the image of God' (*imago dei*). Epicurus once noted the parallels between the values of a culture and the qualities of its god or gods; here we seem to have encountered a principle no less fundamental.

[6] Jaegwon Kim is usually credited as the most significant critic of 'non-reductive physicalism'; see e.g. *Supervenience and Mind*; *Mind in a Physical World*; and Kim, ed., *Supervenience*. Although Kim's criticisms have been influential, they have also come in for their own share of criticism; see e.g. Nancey Murphy, 'The Problem of Mental Causation: How Does Reason Get its Grip on the Brain?' *Science and Christian Belief* 14/2 (October 2002): 97–192.

simultaneously both. Reality manifests itself at various levels, and at each level it is (more or less) as the sciences particular to that level describe it to be. Moreover, these various levels—and thus the objects and properties particular to each—have arisen one by one over the course of cosmic evolution. At one point reality was physical; later it was both physical and biological; and on at least one planet there are now entities that are simultaneously physical, biological, and mental. 'Higher' (i.e. later) entities, causes, and properties are dependent on the 'lower' levels out of which they arose.[7] Thoughts and other emergent mental properties are thus dependent on neurophysiological states, though not reducible to them.[8]

(4) *Substance dualism* is probably the standard theological position in Western history. For dualists, God is the absolute substance or perfect being (*ens perfectissimum*). Yet, on their view, God also created a world of independently existing substances, which have their own existence even while they, being contingent, continue to depend on God as their necessary ground. There is fairly good evidence that biblical anthropology does not treat a mental substance as the essence of humanity, if only because of its stress on the resurrection of the body. Hence in the biblical traditions, at least, there is strong reason to question an overly sharp separation of mind and body.[9]

Finally, (5) one would have to draw a different set of theological inferences from the more *idealist* positions on the relationship of consciousness and brain, either in their Eastern forms (especially in the Vedantic philosophies) or in their Western forms (Berkeley,

[7] In science-based discussions one will tend to speak of mental properties, since there are no theories in the natural sciences today that offer a place for mental entities. Metaphysical and theological discussions, not working under the same strictures, tend to use the substantive 'mind.' What is crucial is whether distinctively and ultimately irreducible mental causes exist, or whether micro-physical objects and forces are, in the end, the real causal agents. It seems to me that the physicalist has to wager on the latter outcome and the emergentist on the former.

[8] Critics of emergentist monism often argue that emergence theory comes closest to *property dualism*. Actually, as suggested above, it would be better to say that it is a form of *property pluralism*: many different and intriguing properties emerge in the course of natural history, and conscious experience is only one of them. The substrate out of which the various properties emergence is however neither physical nor mental.

[9] This represents a core premise of the essays in Warren S. Brown, Nancey Murphy, and H. Newton Malony, eds., *Whatever Happened to the Soul? Scientific and Theological Portraits of Human Nature* (Minneapolis: Fortress Press, 1998).

German Idealism, panpsychism, panexperientialism). In metaphysical discussions in both East and West, defenses of the primacy of mind are much more attractive and harder to refute than they are in science-dominated discussions of the sort represented in the present volume. Idealist positions, one might say, turn the 'hard problem' of consciousness upside-down; their problem is not to explain how mental experiences could arise out of a physical world, but to explain how physical reality (or, at minimum, the belief that there *is* a physical world) might arise from a reality that is fundamentally mental or spiritual—whether that foundation is universal spirit, *brahman*, or the God who is pure Spirit.

Fortunately, it's not our task at present to resolve the debate among these five views of agency, but rather to gain a clearer sense of what kind of theology fits most naturally with option (3). Still, the broader list serves as a good reminder: one's theory of human agency will have a strong impact on how one understands divine agency, and *vice versa*. In fact, Christian theology can be approached, and has been approached, from the standpoint of all five of these positions. It's certainly true that the vast majority of Christian theologians through the centuries have been substance dualists (option 4). As a result, it's much easier to sound orthodox, and perhaps to *be* orthodox, as a dualist compared with any of the other positions—despite the fact that this dualism is all but impossible to integrate with contemporary science. But there are also Christian theologians who work with option (2), the dual aspect theory of mind (e.g. Ian Barbour), and many Christian theologians and philosophers whose position on mind is fundamentally idealistic (option 5), whether in the British Idealist tradition (e.g. Bishop Berkeley, Bradley) or the German Idealist tradition.[10]

Incidentally, one might well suppose at first that physicalism (option 1) and Christian theology would mix like oil and water, hence

[10] Dieter Henrich's publications offer a thorough exploration of the resources that German Idealism offers to theology, and those of Werner Beierwaltes trace the history of idealism through the Greek, Roman, and medieval periods. See also Falk Wagner, *Was ist die Theologie?: Studien zu ihrem Begriff und Thema in der Neuzeit* (Gütersloh: G. Mohn, 1989); and Friedrich Wilhelm Graf and Falk Wagner, eds., *Die Flucht in den Begriff: Materialien zu Hegels Religionsphilosophie* (Stuttgart: Klett-Cotta, 1982). A summary of these debates is offered in Clayton, *The Problem of God in Modern Thought* (Grand Rapids: Eerdmans, 2000).

that a physicalist could not be a Christian theologian. But there are a number of Christian theologians today who work with physicalist assumptions; at least one of them, Nancey Murphy, has made 'non-reductive physicalism' the linchpin of her theological program.[11] This rather unlikely-seeming conjunction may perhaps be made consistent. But, I suggest, the consistency comes at the cost of an uncomfortable dualism between the nature of God (who is pure Spirit) and the nature of human beings (who on this view are, ontologically speaking, purely physical).[12]

## 14.2. Radically Emergentist Theism and Its Less Radical Competitors

What then of option (3), emergentist monism? The emergentist view of human agency—the view that the qualities of human agency arise from complex physiological structures but subsequently exercise a distinctive 'agential' causal agency of their own—is well presented in other chapters. Our task is to formulate a systematic theology of emergence that is both consistent with this view of agency and adequate to the Christian tradition.

Immediately a difficulty arises, however. The view of God that maximizes similarities with the emergentist view of persons—let's call it *radically emergentist theism*—conflicts at a number of points with basic tenets of the theological tradition. For example, Christian

[11] See Nancey Murphy's contributions to Brown *et al.*, eds., *Whatever Happened to the Soul?*, as well as Murphy's essay, 'Supervenience and the Downward Efficacy of the Mental: A Nonreductive Physicalist Account of Human Action,' in *NP*. See further her 'Physicalism Without Reductionism: Toward a Scientifically, Philosophically, and Theologically Sound Portrait of Human Nature,' *Zygon* 34/4 (1999): 551–572; and 'Emergence and Mental Causation,' in Clayton and Davies, eds., *The Re-emergence of Emergence*.

[12] Murphy comments, citing Don Gelpi, that the key question is whether God is to be understood as the chief exemplar of, or as the chief exception to, the metaphysical categories that one employs to describe the world (personal communication). The choice is not dichotomous, however: the history of philosophical theology evidences a spectrum of positions with these two options as endpoints. A God whose every quality is an exception to our categorial scheme would be an utterly incomprehensible God; a theology that posited such a God could not be rationally discussed. Consequently, I here argue for at least some common metaphysical categories (such as agency) that apply to both God and finite creatures.

theology has traditionally taught that God pre-exists the world, created it *ex nihilo*, guides its development in some way, and will be involved in bringing about its final culmination. The best way to deal with these tensions is first to present them in their strongest possible form and then to expand the range of what might count as a theology of emergence.

The obvious starting point for an emergentist is to draw the strongest possible analogy between the divine agent and human agents. If the mental side of human existence is emergent from a complex physical system (the brain and central nervous system), yet remains dependent upon it, and if divine subjectivity is understood on analogy with this result, wouldn't one also have to construe the divine as an emergent property of the physical world? Radically emergentist theism holds that at one time there was no God, only the physical world, and that God is gradually emerging in the process of natural history. On this view, 'the divine' is another emergent property of the universe, alongside life and mind, gradually appearing as the universe reaches certain stages of complexity. This was the position taken by Samuel Alexander, for example. According to Alexander's form of radically emergent theism,

> As actual, God does not possess the quality of deity but *is the universe as tending to that quality*. ... Only in this sense of straining towards deity can there be an infinite actual God. ... Thus there is no actual finite being with the quality of deity; but there is an actual infinite, the whole universe, with a nisus to deity; and this is the God of the religious consciousness ... God as an actual existent is always becoming deity but never attains it. He is the ideal God in embryo.[13]

Certainly radically emergentist theism offers the closest possible parallel between divine and human agents. Yet note that *accepting emergence theory does not compel one to endorse this particular conclusion.* One can agree that the most plausible contemporary account of human agency involves the emergence of mental phenomena from brains interacting with their environments, while still insisting that this account does not exhaust the nature of the divine agent. For example, an emergentist can hold (as I do) that God was present from

[13] Samuel Alexander, *Space, Time, and Deity*, the Gifford Lectures for 1916–18, 2 vols. (London: Macmillan, 1920), 2:361–362, 364.

the very beginning as the Ground of all things, and that the essential divine nature remains unchanged throughout cosmic history, even if many aspects of the agency of God—God's actual responses to actually existing beings—are only gradually manifested as the universe proceeds to develop life, consciousness, and spiritual experience.

An emergentist theology of this type implicitly posits two 'poles' within the divine: an antecedent nature, pre-existing the cosmos, which is responsible for its creation, and a consequent nature, which arises in relation and response to the creative activity of humans and other living things within the world. On this 'moderately emergentist' view the use of emergence vocabulary in describing human subjectivity thus draws theology closer to one or another form of process thought, with its well-known stress on the emerging, responsive nature of the divine experience.

## 14.3. Six Theories regarding the Significance of Emergence for Theology

Now it's nothing new in the history of theology to revise one's theological stance in light of a systematic principle derived from science or 'natural philosophy.' Almost every great systematic theologian brought a particular organizing principle to the biblical documents and to the tradition. Thomas Aquinas, for example, drew on many of the central metaphysical principles of Aristotle's philosophy in writing his *Summa* of theology; Augustine was in many ways a disciple of Plato; Karl Barth was deeply influenced by Hegel. Nor is it unusual for theologies to revise the tradition that preceded them. Virtually every systematic (re)formulation of Christian belief involves changes of accent and emphasis, if not the outright dismissal of certain beliefs and doctrines held in the past. Although people often ask 'How much revision is okay, and when do the revisions go too far?' no simple formula exists for answering the question. Different persons and traditions have different standards for 'how much is too much.' Beliefs that seem wild or 'over the top' in the present have been defended with passion and logic by theologians of the past, and views that seem absolutely clear and unobjectionable to one group are attacked just as resoundingly by others.

Thus, the response to radically emergentist theists cannot involve the complaint that they have reformulated theology in light of a new systematic principle. Rather, the big question concerns the *manner and extent* of one's use of this principle. Does one follow the lead of the emergentist theory of human mentality *all the way*, deriving one's doctrine of God directly, even exclusively, from its principles? Or can one endorse an emergentist anthropology while supplementing it with other principles on the way to a complete doctrine of God?

What hasn't yet been recognized in contemporary discussions of theology and emergence is that there is a wide variety of ways in which a theology can be emergentist, and that not all of these imply the same conclusions as the radically emergentist theism of Samuel Alexander. I have identified six distinct ways to relate emergence theory to theology, running from a merely nominal use of emergence to the most radical use:

1. One can take a traditional Christian theology and simply *call* it 'emergentist,' perhaps because there was a time when the world did not exist and then a later time when it did, or because one accepts the truth of evolution and wants to make God responsible for what emerges through evolution. Since there is no interesting sense in which such views represent an emergentist doctrine of God, I shall consider them no further.
2. One could accept the facts of emergence in the natural order and then try to show that traditional theism helps to explain these emergent phenomena, without otherwise revising one's doctrine of God vis-à-vis the tradition. So, for example, one might emphasize that God's experience is enriched by interacting with humans, or that God intends the moral or spiritual evolution of the human race, and even of all living things. I do believe that there is a place for features of emergence within traditional theism, and that it is fruitful to explore these features. But they do not exhaust the significance of emergence for theology.
3. One can modify traditional theism to bring it in line with emergence insights. Call it 'emergentist theism.' This is the position that I explore most extensively in the remaining pages.

4. One could endorse the radically emergentist theism already described.
5. One could argue that only a post-theistic metaphysic is adequate to the demands of contemporary science, and in particular to the emergence insight. Thus some argue, for instance, that theism entails a God who has a purpose in creating the universe and who acts to bring about that purpose in history. But, they continue, theism so defined is incompatible with the process of random emergence that the biological sciences reveal to us.[14] Even a radically emergentist theology does not go far enough; only some form of post-theism, such as a religiously tinged naturalism or 'sacred atheism,' will do justice to what we have now learned about the evolution of the biosphere.[15]

   On this view, one would presumably attempt to build one's metaphysic out of the data and theories of emergence alone. Of course, one wouldn't know in advance that the result would have any theological dimension at all. It might turn out to be a theory that includes minds and mental properties, but one in which they simply cease to exist when bodies die; it might stop short of minds, yielding only basic biological processes and agents; or it might be that an ontology of physical forces and particles is all that the science supports. On this view one is committed to supporting only those metaphysical position(s) that the sciences of emergence support, and nothing more.
6. Finally, one could eschew the metaphysical road altogether. Besides summarizing the scientific theories and interpreting them, one might insist, there's really nothing more to be said. The idea of a metaphysics of emergence, of *whatever* ilk, is misguided.

Except perhaps for the first, all of these are genuinely 'emergentist' positions. They range from a more-or-less unmodified traditional theology to viewpoints so radically divergent from traditional theism

[14] See e.g. Willem B. Drees, *Religion, Science and Naturalism* (New York: Cambridge University Press, 1996).

[15] See for example the recent papers by Wildman: 'The Divine Action Project,' 31–75, and his response to his critics, 'A Reply to My Respondents on "The Divine Action Project."'

that one is hard-pressed to call them 'theologies' any longer. The term 'theology of emergence' thus turns out to be a much broader concept, encompassing a wider variety of options, than one would have suspected.

## 14.4. The Two Sides of God

Once again, in the interest of formulating a constructive theology I move away from the comparativist task, focusing in on the third option. Once a particular theology of emergence is on the table, it will be easier to assess its strengths and weaknesses relative to its competitors. Concreteness has an additional merit. Seeing a plausible theology of emergence developed, complete with christology, ecclesiology, and the rest, may help to allay the suspicions of those who are worried that emergence will inevitably compromise Christian theism. Conversely, those who are skeptical that *any* plausible form of theism can still be found—say, because of the tensions between traditional theism and contemporary science—may be encouraged by encountering a more progressive view of theism that at least tries to respond to the standpoint of emergence.

If I am to get to a concrete theology of emergence, I must assume for now the viability of some form of theism. And if it is to be a Christian theology of emergence, I must pay particular attention to the biblical texts, first as they describe the nature of God, and subsequently as they convey the narrative and teachings of Jesus of Nazareth, who was called the Christ.

But what form of theism will it be? One encounters what looks like a painful incongruity in the notion of an unchanging, self-sufficient God who is supposed to express the divine nature in a ubiquitously interrelated, constantly emerging world. I thus find Aristotle's notion of *nous noetikos* ('thought thinking itself') a bad fit with the world of pervasive emergence as contemporary science has described it. Unfortunately, Aristotle's timeless, non-relational God has deeply impacted the history of Western theology. As a result, much of the Western theological tradition, with its emphasis on 'the God of infinite perfection' and on the '*omnis*' (omnipotence, omniscience, omnipresence, etc.), now stands in tension with the world that cosmology, biology, and the cultural sciences have now revealed to us. There are equally grave tensions between the biblical picture of God and the notion of an unchanging God, for whom relations with contingent creatures remain external and non-essential.

Abandoning the non-responsive God leaves a number of other options, of which I here consider only two. God could be full and complete in God's own nature, and fully relational within God's self, prior to the creation of any world at all (classical trinitarian theology); or God could have an eternal nature and a *potential* for relation, but only actualize that relation fully in the context of God's relationship with a created world.

I admit that, in choosing between these two options, I am strongly influenced by the biblical picture of the relational and responsive God. I cannot do the exegesis here, other than to note that the responsiveness of God seems to be an important theme both in the Hebrew Bible and in the New Testament.[16] I thus urge caution about any positions in systematic theology that tend to undercut the two-sidedness of the relationship with God that occurs in actual historical moments of encounter. A theology of genuinely two-sided relations entails real effects on God: the divine experience is different after the encounter than before. This stance inclines me toward the second of the two views.

Twentieth-century theology by and large made important advances in conceiving God's genuine relationship with and responsiveness to creatures in the world. What biblical and systematic theologians did not do as well in the same period is to think through some of the other, more metaphysical roles that God also plays. On the view I am defending, and in much of traditional theism, God serves a variety of functions that are not direct products of God's personal nature: the world as contingent is ontologically dependent on God as its non-contingent ground; the world as finite in time

[16] I cannot here attempt to resolve the heated debate regarding the precise nature of the authority of scripture for theologians. It seems obvious that the Hebrew Bible and New Testament will remain points of reference for Christian theologians, who seek to bring those texts into dialogue with our contemporary context. But specifying exactly what balance should be found between literal and metaphorical or allegorical uses of scripture is no easy task. See Kevin J. Vanhoozer, *Is There a Meaning in this Text?: The Bible, the Reader, and the Morality of Literary Knowledge* (Grand Rapids, MI: Zondervan, 1998).

is dependent on a being with no beginning in time; the world as a mixture of good and evil (of positive and negative moral attributes) depends on a ground which is goodness itself; and the finite world must be included within the all-encompassing infinite (as Hegel showed in his Logic).

It is difficult but not impossible to encompass both of these types of function within a single doctrine of God. Minimally, one has to conceive God's nature as having two sides or 'poles': an antecedent (pre-existing) pole, which represents God's eternal and unchanging nature, and a consequent pole, which emerges in the course of God's interaction with the world. Process theologians such as Charles Hartshorne, John Cobb, and David Griffin have done an effective job of constructing a sophisticated philosophical theology that elaborates 'dipolar theism,' and I have sought to supplement their accounts by drawing more heavily on the tradition of German Idealism, especially Schelling.[17] A dipolar theology allows one to conceive God both as the ongoing Ground of the process of emergence *and* as involved in and responsive to the entities that emerge within that process. The adequacy of this philosophical theology is presupposed in what follows.

## 14.5. 'God with us' and the Transition to Christology

It is time now to consider the features of a distinctively Christian theology. There is only one way to make this transition: to establish conceptual links between the philosophical theology sketched to this point and christology, the doctrine of Christ.[18] Again, the goal cannot be to survey the entire field of christological options, but only to provide an example of the sort of christology that would be consistent with a theology of emergence.

[17] I have developed the argument in *The Problem of God in Modern Thought* and in 'Pluralism, Idealism, Romanticism: Untapped Resources for a Trinity in Process,' in Joseph Bracken, SJ, and Marjorie Hewitt Suchocki, eds., *Trinity in Process: A Relational Theology of God* (New York: Continuum, 1997).

[18] The following argument has been developed in collaboration with Steven Knapp at Johns Hopkins University. Parts of this portion of the argument appeared as 'Can Liberals Still Believe that God (Literally) Does Anything,' *CTNS Bulletin* 20/3 (Summer 2000): 3–10.

From a Christian perspective, the personal or 'consequent' nature of God represents 'God with us.' To conceive God in this way is to attempt to understand God in God's identification with the world. It is to think of God in terms of a love that encompasses all creation, that is willing to embrace humanity and other living things, to identify with us, to suffer with us. (Think of the description of God's Spirit as *Paraclete* in the Gospel of John: the one who comes alongside.) For an infinite God to enter into relation with finite creatures in this way is for God to have emptied Godself of certain divine attributes. An all-powerful, always-acting God (*actus purus*) appears to leave no place for separate agents to freely initiate independent creaturely actions. The Greek word for this self-emptying is *kenosis*. Hence, the christology compatible with this position will be above all a kenotic christology.

The question is, How far is this kenosis, this 'coming alongside' and suffering-with, to be extended? When does belief end and metaphor take over? Let me describe the possibility in terms of a series of 'what ifs.' What if divine action involves more than a passive responsiveness to the world and human experience? What if God took some initiative in the process? Indeed, what if God responded to the incredible concreteness of human experience—our need to get to universal truths (if such there be) via individual experiences—by making Godself known in particular persons, or even, possibly, in an especially clear way in one individual?

Perhaps you can't quite develop the belief that this has happened. But perhaps you conceive it as an attractive possibility to explore and even endorse. In earlier publications Steven Knapp and I have described this approach as *possibilistic theology*: the theology of a compelling possibility that one can accept (and live by) even while lacking the requisite historical proofs that it has actually occurred.[19] If this is correct, what is indispensable for Christian theology is not historical proof of the uniqueness of Jesus, but rather some way to think of the Jesus event as involving an act of God. We seek an understanding of 'God in Christ' that involves a two-way relationship

[19] See Clayton and Steven Knapp, 'Belief and the Logic of Religious Commitment,' in Godehard Bruntrup and Ronald K. Tacelli, eds., *The Rationality of Religious Belief* (Dordrecht: Kluwer Academic Press, 1999), 61–83.

between God and an individual person, expressed through acts on the part of both.[20] There are actions that Jesus performed and attitudes that he manifested toward God that helped to constitute his uniqueness before God, and there are actions that God performed that helped to make the Jesus event revelatory of the divine nature.

Who was Jesus, then, and what set him apart from others in his apprehension of God? Some want to say, more minimally, that Jesus accessed a divine power or knowledge that is available to all humans, perhaps as someone with good antennae can receive the radio waves that are continually streaming along the surface of the earth. Or others, using a different metaphor, claim that Jesus tapped into a spiritual reservoir in the way that someone might dig a well and access a rich water source in an underground aquifer.[21] Call this the *religious genius* view. These views imagine God as a resource that is continually available to humanity (or, for that matter, to all living things); for example, the unchanging solace and strength of the divine presence is available to any persons who are able to access it. Jesus knew how to reach these depths and in his teaching shared what he had learned with others. In any case, on the religious genius view *God* does not respond to an individual or situation; rather, these religious truths are always and at all times available for those who know how to plumb the depths.

The difficulty with accepting religious genius views of the act of God in Christ is that they do not really allow one to speak of the result of Jesus' tapping into this reservoir of spiritual truth as *God's* act. Suppose that Jesus should happen to be the greatest religious genius of all time, describing the nature of God more accurately than anyone before or since. This would be a commendable act on his part, but it would not thereby count as an act of divine revelation. Nor could God be held responsible for what Jesus (or anyone else) did or did not ascertain of the divine nature. In fact, it appears, no direct

[20] In contrast to much of the tradition, I do not define the incarnation first in terms of the ontological status of Jesus Christ. Most of the two-natures doctrines of the incarnation are based on the categories of a substance-based metaphysic that is foreign to how most people today think. These doctrines also require a pre-existent logos christology that remains in tension with the fundamental humanity of Jesus, 'tempted in every way, just as we are—yet . . . without sin' (Hebrews 4:15).

[21] See e.g. Matthew Fox, *One River, Many Wells: Wisdom Springing from Global Faiths* (New York: Jeremy P. Tarcher/Putnam, 2000).

role for God in the Christ event would remain—only the indirect role of creating humans with certain generic capacities to apprehend spiritual truths.

So what *would* be necessary to constitute the life of Jesus as also involving an act of God? God would have to invite, or in some way respond to, those who seek to connect with the divine. If God makes some knowledge or experience available to those who seek it, a divine act is involved. And if a human agent such as Jesus is able to teach or act in certain ways in part as a result of this divine act, then the actions that he carries out can properly be spoken of as both human and divine. I do not here present proofs that such special divine acts actually occurred in the case of Jesus; the goal for now is merely to understand what it would *mean* for such acts to have occurred.

## 14.6. Toward a Kenotic Christology

Let's now explore what appears to be the most attractive solution. St. Paul writes, 'Have this mind among yourselves, which is yours in Christ Jesus, who, though he was in the form of God, did not count equality with God a thing to be grasped, but emptied himself...' (Philippians 2:5–7 RSV). The disciples experienced Jesus as the most powerful individual they had ever met—so powerful that, they concluded, the source of his power could only be divine. And yet he did not 'lord this power' over others 'like the Gentiles.' Instead, he '[took] on the form of a servant, being born in the likeness of man' (Philippians 2:7). The sentences about Jesus' self-knowledge that John inserts just before Jesus begins to wash the disciples' feet (John 13:1–4)[22] continue to resonate as a uniquely potent expression of this strange juxtaposition of power and self-emptying. What was it like to be in the presence of Jesus? It was to experience a unique form of (personal, moral, intellectual, religious) power and, at the same

[22] 'Jesus knew that the time had come for him to leave this world and go to the Father. Having loved his own who were in the world, he now showed them the full extent of his love.... Jesus knew that the Father had put all things under his power, and that he had come from God and was returning to God; so he got up from the meal, took off his outer clothing, and wrapped a towel around his waist. After that, he poured water into a basin and began to wash his disciples' feet' (John 13:1–5).

time, to hear repeatedly 'not me, but my Father who is in me. . . .'[23] How could the disciples not draw the conclusion—which Jesus may well have intended—that Jesus' incredible charisma and power were a direct result of his submitting his will to the divine will? Remember that it would have been beyond question to the disciples that God was directly responding to Jesus' prayers.

But then Jesus died. Although he somehow seemed powerful to the end, and although he continued to call out to his Father, that Father remained silent.

Scholars hold different beliefs about which events were the first catalysts for the resurrection belief.[24] The process might have involved visions or hallucinations, or inferences drawn from celebrating the Eucharist (Luke 24), or an inner divine leading that the disciples interpreted as the direct voice of Jesus. What happened on the human side, certainly, was that the disciples began to live the life they had seen modeled in Jesus, more intensely and more confidently than they had ever done while he was with them in the flesh. In living in the way of the Christ, they seemed to experience the same divine power that Jesus had apparently drawn on. How could they *not* believe that God's Spirit, and even Jesus himself in resurrected form, was present to them and acting through them? But the theologians of emergence (as defined above) are not satisfied with the human dimension alone; we want to know how the resurrection and the disciples' response might be understood as involving a divine act as well.

In the more liberal versions of the theology of emergence one finds affirmations of the resurrection of 'the Spirit of Christ' but not a physical resurrection of the individual man Jesus. The act of God that produces the resurrection raises the Spirit of Christ, the Spirit or Counselor 'whom the Father will send in my name, [who] will teach you all things and will remind you of everything

[23] 'My teaching is not my own. It comes from him who sent me' (John 7:16). 'The words I say to you are not just my own' (John 14:10). 'These words you hear are not my own; they belong to the Father who sent me' (John 14:24). 'Father, if you are willing, take this cup from me; yet not my will, but yours be done' (Luke 22:42). 'Your kingdom come, your will be done, on earth as it is in heaven' (Matthew 6:10). Or, most simply, 'may your will be done' (Matthew 26:42).

[24] For those who are not faint-hearted I recommend John Dominic Crossan, *The Birth of Christianity: Discovering What Happened in the Years Immediately after the Execution of Jesus* (San Francisco: HarperSanFrancisco, 1998).

I have said to you' (John 14:26). Because of this act of God, after Jesus' death the 'mind of Christ'—Jesus' surrender of his will to the Father's will, for which God 'highly exalted him' (Philippians 2:9)—remained available to his disciples and later to their followers in the church.

Some readers will not be able to form a belief in the continuing personal existence of Jesus, whereas others will insist upon the physical resurrection of Jesus and hence his continuing existence as a person. I will not be able to resolve that debate here. Note, however, that in either case one may still believe that the divine power and presence is manifested whenever a person adopts, internalizes, and thereby shares in the 'mind of Christ'—the mind of the one who prayed 'not my will, but thine be done' (Luke 22:42). Allowing the will of God to work in one's own actions and 'mind' can produce a mode of being in which not the individual human ego but the divinely intended outcome is decisive (Philippians 2:5). To be 'in Christ'—Paul's favorite description of the mode of Christian existence[25]—means to subordinate one's own will to the will of the divine, echoing Jesus' basic prayer, 'may your will be done' (Matthew 26:42; cf. Matthew 6:10).

What's attractive about this approach to christology is that it already incorporates the divine act into Jesus' God-consciousness. It's not as if there is a description of Jesus' will and actions on the one hand and, on the other, a separate metaphysical superstructure of divine intervention added on top of it. Instead, to describe the historical Jesus as I have done *just is* to give an account of divine involvement. In virtue of continually subsuming his will to the divine will, Jesus caused his actions to become *part of* the divine act. There are not two actions, but one: Jesus manifests the divine power by subsuming his will to God's; at the same time (or for the same reason, or in virtue of the same act) God acted through Jesus to manifest God's will and bring about God's intentions. This fusion of human and divine is what was right about traditional 'two natures' christologies and the traditional doctrine of 'incarnation;' it's just that emergentists now locate the fusion in *shared action and attitude* rather than in some a priori ontological story. Just as process

[25] 'If anyone is in Christ, he is a new creation; the old has gone; the new has come!' (2 Corinthians 5:17). The phrase 'in Christ' is used some ninety times in the New Testament epistles.

theologians describe divine action as 'the lure of God' (Lewis Ford), and just as traditional theology refuses to separate the act of God and the revelation of God, we too might understand God to be genuinely revealed through human agents who seek to submit themselves to God's will.[26]

We're back to the 'what ifs.' What if God, by an intentional divine self-limitation, chooses to bring about God's purposes in the world only through the actions of worldly creatures? And what if a particularly clear model, or *the* uniquely clear model, for this kenotic revelation comes in the self-submission of Jesus to what he called 'the will of the Father?'[27] Finally, what if Jesus' acts of intentional self-submission are matched not by an impersonal and undifferentiated transmission of divine energy (as in the radio metaphor above), but rather by an equally personal and specific act of acceptance and guidance on God's part?

Naturally there are more liberal positions than the one I have defended here, positions in which the entire Jesus narrative is understood as a series of purely human events, with no divine act involved. I have not here shown that the biblical and historical evidence *requires* a stronger sense of divine action than on the more liberal readings, though perhaps this could be shown. Instead, the goal of this brief argument has been to show that another view is also possible: that the Jesus event—and by extrapolation, other events in the natural world—can also credibly be understood as representing one or more divine acts. And out of this combination of human and divine emerges a combined act or series of actions which Christians call the revelation of God in Jesus Christ.

## 14.7. A Systematic Theology of Emergence in Outline

Three major factors have influenced the formulation of Christian theologies: the Bible, conceptual frameworks or philosophical systems,

[26] As a woman described it to me recently at a Quaker Meeting, 'Perhaps God's only act is to make manifest the divine love. Is this not enough?'

[27] I have told the story in Christian terms, as behooves a Christian theologian. But I presume that similar accounts could be developed within other traditions. In fact, resources already exist. Examples include references to the Buddha mind, the emulation of the Buddha by boddhisatvas, and the state of perfect receptiveness shown by God's prophet as he received the dictation of the Koran (as reflected, e.g., in the ideals of Sufi mysticism).

and the traditional categories or topics to which theologians must respond. Systematic theologians of the patristic, medieval, and Reformation eras combined core biblical teachings with the dominant philosophical systems of their day. The questions of Peter Lombard, which dominated the Scholastic era, became by the time of Melanchthon the major systematic *loci*, the topics that each theologian was expected to address: christology, pneumatology, ecclesiology, eschatology, and so forth. As the modern era progressed, however, and as radically different philosophical systems gained in credibility, many theologians came to believe that the philosophical frameworks that had once guided systematic theologians were no longer adequate.[28] The story of modern theology is the story of the whirlwind of confusion that has resulted.

One response to the crisis lies in the approach I have taken: to attempt to address the traditional questions or *loci* using a different conceptual framework—in this case, the framework of emergence. What would a systematic theology look like if one used this framework as its organizing principle? Consider these responses to eight of the traditional *loci*:

1. *The doctrine of creation*. The doctrine of the created world is the place where the whole range of scientific data concerning emergence can be incorporated. It is not the place of a theological doctrine of creation to take the place of science but to acknowledge its findings.[29] Thus, all the scientific detail on emergent systems in evolution found elsewhere in this book and in other publications[30] becomes relevant to a revised doctrine of creation.

But emergence represents a sort of upwardly ascending arrow, since it is hard to avoid the meta-scientific questions that it raises.

[28] In *The Problem of God in Modern Thought* I tried to show why philosophical systems do matter to systematic theology and why modern thinkers were right to search for new ones.

[29] Cf. Wolfhart Pannenberg, *Systematic Theology*, 3 vols., trans. Geoffrey W. Bromiley (Grand Rapids: Eerdmans, 1991–98), vol. 2, chap. 7.

[30] For a summary of the sciences of emergence, see Clayton and Davies, eds., *The Re-emergence of Emergence*. See also Morowitz, *The Emergence of Everything*; Niels Gregersen, ed., *From Complexity to Life: On the Emergence of Life and Meaning* (New York: Oxford University Press, 2003); Holland, *Emergence*; Roger Lewin, *Complexity: Life at the Edge of Chaos* (Chicago: University of Chicago Press, 1992); Albert-Laszlo Barabási, *Linked: The New Science of Networks* (Cambridge, MA: Perseus, 2002); and Johnson, *Emergence*. I summarize the theories in *Mind and Emergence*, e.g. chap. 3.

Is there a further stage of emergence beyond the level of mind, such as spirituality or spirit? Is mind utterly without precedent, or does the emergence of mind reflect something of the nature of whatever Cause preceded the Big Bang and helped to produce it? If mind is not a strange anomaly in the universe but somehow reflects the nature of its ultimate Source, could it be that the Source also includes other personal qualities—or, at least, that it is not less than personal? If so, might that Source, *qua* personal cause, not also have intentions regarding the evolution of the universe, and might it sometimes act in such a way as to further those purposes? The fusion of emergence and belief in a Creator God, I have argued, represents the strongest available answer to these questions.

2. *The doctrine of God*. As I have argued elsewhere, the doctrine of God that best allows belief in divine action to be synthesized with modern science in general, and with the sciences of emergence in particular, is panentheism.[31] Panentheism involves the belief that God includes the world within Godself, although God is also more than the world. The title of this essay might therefore more appropriately read 'Toward a Constructive Christian Theology of Emergentist Panentheism.'

Note that panentheism is consistent with most of the traditional attributes of God.[32] No major modifications need to be made to traditional understandings of divine wisdom, goodness, love, grace, mercy, long-suffering, righteousness, or veracity. The doctrines of divine omnipotence and sovereignty do however need to be modified, insofar as panentheism implies that God has freely chosen to limit God's a priori omnipotence in order to allow other free centers of conscious agency to come into being. Features of the created world that might have been otherwise, such as certain of the laws of nature, represent free divine decisions. But if God is to remain consistent with God's own nature, God is now constrained by those decisions, which means

[31] See *God and Contemporary Science* (Edinburgh: Edinburgh University Press, 1997); and *The Problem of God*. For further references, definitions, and arguments for the viability of panentheism, see Clayton and Peacocke, eds. *In Whom We Live and Move*.

[32] See my four-article series in *Dialog*: 'The Case for Christian Panentheism,' *Dialog* 37 (1998): 201–208; 'A Response to My Critics,' *Dialog* 38 (1999: 289–93); 'Panentheist Internalism: Living within the Presence of the Trinitarian God,' *Dialog* 40 (2001): 208–215; and 'Kenotic Trinitarian Panentheism,' *Dialog* 44/3 (Fall 2005): 243–248.

that God's present power is further limited. Even more clearly, the doctrine of God's immutability can no longer be asserted of God unequivocally. The primordial nature of God is indeed unchanging, but the personal responsive side of God is genuinely emergent, in the sense that it comes to encompass new experiences as it enters into new relations with creatures.

In comparison to many traditional theologies, emergentist panentheism gives a much greater role to knowledge of the world for understanding the nature of God. What one takes to be the primary sources for theological knowledge will depend greatly on the philosophical framework that one employs. For example, the more Platonic one's framework, the more one will tend to employ rationalist and a priori arguments to derive knowledge of God. By contrast, advocates of emergentist panentheism put much more weight on the contributions of the sciences, the historical disciplines, and the humanities. Scientific results are not by themselves sufficient to reach theological conclusions; as we have seen, there is a point at which the ladder of emergence itself begins to raise meta-scientific (metaphysical and theological) questions. Still, on this view theological reflection comes at the end of the knowledge process, rather than controlling it from the outset.

The emergentist doctrine of God thus does not share Karl Barth's reticence about knowledge of God through analogies. If panentheism represents a viable way to conceive the God–world relation, as I have argued, then epistemic inferences can be drawn both from God to world (e.g. *imago Dei* arguments) and from features of the world to the nature of God. For example, I have elsewhere defended the Panentheistic Analogy as a means for comprehending God's relationship to the world.[33] The Panentheistic Analogy suggests that the relationship of God and the world is in some ways analogous to the relationship of our minds and our bodies. 'Mind' is not merely a part of the body, and 'mental' means more than the brain taken as a whole; yet mind is also not totally separate from the body. Similarly, panentheists do not conceive God as outside or separate from the world; rather we understand divine immanence in the strongest possible sense, such that relatedness to the world becomes (and for some

[33] Clayton, *God and Contemporary Science*, e.g. 233–242, 257–265.

panentheists, always was) an essential part of the divine-being-in-relation.

Invariably, analogies such as the Panentheistic Analogy must be adapted, of course, to insure that they are appropriate to the relationship between the world and its infinite Creator. Still, if one is to avoid complete skepticism about the nature of God, such analogies will perforce play a rather significant role in any statements about the nature of God, it being impossible for us to have knowledge of God that bypasses all human epistemic and cognitive abilities and limitations. It goes without saying that at least some aspects of the nature of an infinite God will remain eternally unknowable, beyond the reach of the best analogies theologians can ever conceive.

3. *Christology*. We have already explored the christology of emergentist panentheism in some detail. As we saw, on this view Jesus becomes the exemplar for humanity in a no less profound sense than in classical theology. According to Christians, Jesus actualized the possibility that each human enjoys as one who is made in the image of God, living a life of perfect devotion to God and acknowledging the true relationship of creature to Creator in every thought and action. Panentheists believe that all are located within the divine presence in no less a sense than Jesus was. But gaps remain between what God wills and what we will; this is the core insight of the doctrine of sin. By contrast, according to Christian belief no such gaps existed in the life of Jesus; he alone perfectly lived a life of perfect union with God.

4. *Pneumatology*. The understanding of Spirit is central to emergentist panentheism. Insofar as it accepts the emergence of mind and spirit within evolutionary history, this view diverges from panpsychism and panexperientialism, which make Spirit an inherent element of the world from the beginning. It likewise eschews all dichotomies between Spirit and matter or between Spirit and body, following the lead of emergentist theories of human personhood. Even if the divine Spirit precedes all creation, every manifestation of Spirit in the world depends essentially on the evolutionary process.

Nor can the divine Spirit be a timeless entity standing immutably outside the flow of cosmic history. That aspect of the divine being—call it Spirit—that correlates with the spirit of which we have knowledge in ourselves must also be temporal, the emergent result of a long-term process of intimate relationship with beings in the world.

On this view, then, Spirit is not a fundamental ontological category but an emergent form of complexity that living things within the world begin to manifest at a certain stage in their development. Two theological correctives must be made to the 'straight emergence' view, however: the Spirit that emerges corresponds to the Spirit who was present from the beginning; and this Spirit's actions—both its initial creation and its continual lure—help bring about the world and its inhabitants as we know them. Because the theology in question is emergentist *panentheism*, it holds that the physical world was already permeated by and contained within the Spirit of God, long before cosmic evolution gives rise to life and mind.

Gone, on this view, are Spirit–body dualities and those claims for immutability that stem from the world of Greek metaphysics, which once served as the philosophical authority for theology's fundamental categories. No place remains for an initial creation of humanity with the dual substances of *res cogitans* and *res extensa*. Aside from this change, many of the attributes that theologians commonly associate with Spirit can still be affirmed. It's just that these qualities now become features of ever-changing entities in the world, rather than necessary features of an unchanging substrate from which the world's qualities are drawn.

5. *Anthropology*. In light of the foregoing points it's clear that the doctrine of anthropology will play a larger role than in many traditional theologies. As we saw above in connection with the knowledge of God, the theology of emergence begins with an understanding of human beings as bio-cultural agents in the context of evolutionary history. Our bio-cultural existence gives us some initial understanding of the nature of God as agent, though it must then be 'corrected' to be appropriate to the role of God as the ground and destiny of all things. In other words, emergentist panentheism as a form of dipolar theism includes two stages: the stage of deriving what can be known of the divine agent through our own experience of agency (the consequent nature), and the stage of modifying that understanding in order to (seek to) make it appropriate to the nature of a creator who preceded the universe as a whole (the antecedent nature).

This modified view of God in turn contributes an essential element to anthropology. Emergentist theologians will never deny or overrule what is revealed of human nature through the study of biological and cultural evolution; they will however supplement that understanding by including the perspective of the *imago Dei*, the image and nature of God. By taking this step, theology inevitably adds a normative dimension that natural science by itself could never supply. Natural science tells us who we now are and how we have come to be this way. But theology, by viewing humanity from the standpoint of a perfect creator, holds out a standard for what we *should* be, as creatures made in the image of God. For Christians, this model is inherently christological; it involves emulation of the one who was fully human insofar as he was fully related to God.

6. *Ecclesiology*. The church is defined, as formerly, in terms of those who follow Christ and live 'in Christ.' But the church, like reality in general, is an emergent entity. Thus the original horizon of understanding cannot control all present understandings of the church, its founder, and its role.[34] Because Jesus remains the exemplar of *kenosis*, the reports on and early interpretations of his life and teachings remain crucial. But such an exemplar role will not quite manage to support the doctrine of the 'plenary inspiration' of scripture, as if the context of the origination of the church were by itself sufficient to determine its actions today. One is confronted continually with new situations and ideas never before encountered by humanity or the church, and one must respond to them with new forms of the self-emptying that Paul thinks of as the mind of Christ: 'Have this mind among yourselves, which is yours in Christ Jesus' (Philippians 2:5).

If the church is an emergent, and continually emerging, community, it shares a situation and fate with all of humanity. Panentheists refuse to separate God from the world in the interest of preserving the divine purity; hence for similar reasons they decline to separate the church from the world. The church must be distinguished by behaviors that mirror the nature and being of God, as well as by a corporate identity that seeks to emulate the personal identity of the one who said, 'Not my will, but Thine be done.' Again following the lead of panentheism, the church will seek not to be more-than-human, but to be human in light of that toward which humanity is being lured; not to be anti-scientific, but to find through science some

[34] One will note overtones of the 'two horizons' hermeneutics of Hans-Georg Gadamer's *Truth and Method* (New York: Continuum, 1995).

intimations of the nature of God; not to be anti-intellectual, but to be more insightfully intellectual; not to eschew human moral reflection and striving, but to incorporate that striving within a theistic framework; not to be world-transcending, but to embrace a world that is in turn embraced by God.

7. *Soteriology*. Like Tillich's theology, emergentist panentheism begins not with a specific narrative of fall and redemption, but with the structural difference between God and creation, between infinite and finite, between perfect and imperfect. As we saw in considering anthropology, theology embraces the descriptive account of the human being but supplements it with a normative account of how humanity would look if it were true to its nature as *imago Dei*. This structural difference, and not the primordial myth of an original offense and punishment, provides the basis for introducing the concept of sin. The narrative of the fall remains indispensable for its symbolic functions but not doctrinally foundational. As Tillich also saw, the structural difference between infinite and finite is mirrored in an intense existential experience of sin. Not only with our minds do we conceive the structural difference between God and ourselves; every human being knows *akrasia* (weakness of will), self-centeredness, and the inability to do as one wishes to do (as potently described by Paul in Romans 7).

An emergentist theology cannot endorse an absolute dichotomy between the 'old man' and the 'new man' of the sort sometimes presupposed in the Pauline doctrine of salvation. Emergence theories are anti-dualistic, seeking always for the continuities that underlie even the appearance of radical novelty in the world. We no longer need to divide the world into two sharply opposed camps, the reprobate and the redeemed. Rather, the doctrine of salvation now takes on a structure closer to the traditional doctrine of *sanctification*: the new self, the mind of Christ, is continually emerging as individuals align their wills with the will of God. Such an alignment is a matter of degree. The process of living more and more of one's life in greater and greater conformity to the will of God is a gradual and never-ending process.

Clearly, this approach to soteriology represents some revision of traditional Christian teachings. There are greater or lesser costs associated with any revision of the tradition; this is one in particular that we should be more than willing to pay. In its day the 'in-or-out' dichotomy contributed to a view of the church that separated the lives of Christians much too sharply from the lives of other persons, and, sadly, it continues to do so today. Ascribing to an individual the eternal state called 'being saved' may have fit well within the mostly atemporal framework of pre-modern metaphysics and anthropology, but it accords poorly with the insights into continual process that we owe to emergence theory.[35] The understanding of the church as a collection of redeemed individuals, ontologically distinct from all others, also produced a view of the relationship between Christianity and the other religions, particularly Judaism, that strikes us today as deeply morally disturbing.

To comprehend the cosmic process of the emergence of Spirit in its full richness and complexity, one must draw on every available source, including all the world's religious and wisdom traditions. The thought that God would reveal the divine nature only to one people or group, who subsequently enjoy a state of blessing and closeness to God not matched by any other individuals or group, represents a triumphalism inappropriate to the Christian self-understanding. In the history of the church this triumphalism also served to justify the most atrocious of actions. The entire picture ill accords with the kenotic self-emptying of Jesus, whose 'mind' is to be the example for his church.

I find no place within a theology of emergence for substitutionary atonement, ransom metaphors, or the focus on the need for a sacrifice to propitiate the wrath of an angry God.

8. *Eschatology*. According to Wolfhart Pannenberg, the end of history already took place in the death and resurrection of Jesus Christ ('*prolepsis*'), such that the outcome is already certain. Eschatology is not less important for theologians of emergence, but on our view predictions of the final outcome are rather less certain. As theists we believe that God set the process of emergence in motion, intending that the creation would take on higher and higher levels of complexity, and hence more subtle and intimate relationships with the

[35] I already provided arguments against the 'inside–outside dichotomy' in *Explanation from Physics to Theology: An Essay in Rationality and Religion* (New Haven: Yale University Press, 1989), 134–143.

divine, thereby allowing more and more of the nature of God to be experienced by the creation. Still, the present understanding of the process of emergence does not prove that there will be an ultimate culmination of this history, one in which God brings all things to Godself.

Nonetheless, were history merely to end with the 'heat death' of the universe, it would be rather difficult to conceive what God might have intended by creating this process at the outset, only to allow it to be condemned in the end to ultimate futility. Why would God engage in relationship with this world—a relationship that on our view affects the very being of God—if the whole process will one day simply evaporate into nothingness? It is the impossibility of resting content with such a response that intensifies the hope of Christians (and others) for a future of the universe in which God becomes 'all in all.'

Panentheism fits naturally in such an eschatological perspective. John Polkinghorne has recently argued that only at the end of time can we conceive a state where all things are within God and God is all things.[36] I disagree with his contention that such a panentheistic closeness must be confined to the end of history, but I do concur that panentheism presupposes a telos in which the world conforms more and more fully to the divine character.

How fully will our eschatological hope be fulfilled? How fully *should* it be fulfilled? We hope at present not only that God will preserve what is most valuable in the human race (and the world as a whole), conforming it more fully to God's character; we also hope that those whom we love will be preserved in something like their present state (including, on Polkinghorne's view, perhaps even the family pet![37]) and that we will be able to continue in relationship with them. We cannot imagine existence without something like our present body, so we hope that it too will be preserved or in some sense replicated. At what point do these hopes become the projection of the only way of life we know, and to what extent are they theologically supported? How much of what you currently know as your identity

[36] See John Polkinghorne, *The God of Hope and the End of the World* (New Haven and London: Yale University Press, 2002). Paul Tillich also famously held this position (*Systematic Theology*, vol. 3), as have many mystical theologians through the ages.

[37] See Polkinghorne, *The God of Hope*, 123.

could disappear before that future state becomes so dissociated from you that it's no longer reasonable for you to long for it? According to classic versions of process theology stemming from Whitehead's *Process and Reality*, the only thing one can reasonably hope for is 'objective immortality,' that is, that all of one's thoughts, feelings and individual reactions will be preserved eternally in the unchanging and unending experience of God. Marjorie Suchocki has sought to extend Whitehead's thought so that it can include also subjective immortality, the continuing existence of the individual subjective principle.[38] Whether or not Suchocki is successful in integrating the hope for subjective immortality with process eschatology, she has correctly seen that the eschaton must be conceived in such a fashion that it remains relevant to individual agents in the present.

The key point to underscore about eschatology is the epistemic point: we do not know the universe's final fate. We find ourselves immersed in a process of continual emergence that evidences tendencies but does not produce final certainties. One need only consider the staggering emergence in technology over the last (say) twenty years, or the dismal failure of the alleged science of futurology, to realize that the novelty and uncertainty of cultural evolution far exceeds the capacity of human reason, or even imagination! Epistemic humility is a virtue in all branches of theology; in the field of eschatology, this virtue becomes a necessity. 'Dear friends, now we are children of God, and what we will be has not yet been made known. But we know that when he appears, we shall be like him, for we shall see him as he is' (1 John 3:2).

## 14.8. Conclusion

In these pages we have seen how particular positions on human subjectivity point toward particular theories about the nature of God. Thus construing the human person as an emergent result of the evolutionary process will have deep effects on one's theology. I have

[38] See Marjorie Hewitt Suchocki, *The End of Evil: Process Eschatology in Historical Context* (Albany: State University of New York Press, 1988); and Joseph Bracken, ed., *World Without End: Essays in Honor of Marjorie Suchocki* (Grand Rapids: Eerdmans, 2005).

sought to do justice to those effects by outlining a consistent theology of emergence and showing how it might respond differently to eight of the major *loci* of traditional theology. At the same time, the science–theology debate is a two-way street, involving what Robert Russell calls a 'creative mutual interaction.' One's theology also affects one's view of human nature, one's interpretation of the scientific results, and one's beliefs about the long-term fate of the universe.

Humans live in a world of pervasive process, a world historical to the core; each of our distinctive attributes is the result of millions of years of evolution. In us the evolutionary process has produced self-conscious, reasoning, questioning creatures. We are the animal that has begun to pose to itself metaphysical questions—questions about the universe's ultimate origin and final fate, and questions about the meaningfulness of our own existence. The strangeness of our plight stems in part from the realization that we may have emerged from an origin radically different from ourselves. This is true if we as conscious beings are the products of unconscious physical laws and blind forces, but it's no less true if we as finite beings owe our existence to infinite conscious Spirit.

The Christian tradition is based on the daring wager that the latter option is true. But if we are the products of a conscious creative choice and not the results of a random process alone, then our Creator cannot *also* have emerged through evolution in the same way. This means that, in the midst of a world of pervasive emergence and decay, we set our hopes on the existence of One whose purposes span the Before and After. Here, in this grey realm between the permutations of matter and the becoming of Spirit, and in a dialectic of knowledge and hope, lies Christian faith and Christian proclamation.[39]

[39] I am grateful to Stephen Knapp for extensive conversations, formulations, and criticisms, which have contributed significantly to the development of this theology of emergence.

# Postscript

*William R. Stoeger, SJ*

The theological contributions in this volume have clearly demonstrated the relevance of reductionism and emergence to theology and to the interaction between the natural sciences and theology. If a thoroughgoing causal reductionism really holds, genuine emergence of new causal powers is not possible, and neither theology nor relating theology to the natural sciences as separate disciplines oriented towards understanding complementary spheres of experience has any objective basis. For then theology's objects and points of reference are bereft of any epistemological or ontological status. If, on the other hand, there is genuine emergence of properties, or at least of causal powers, along with ontological reducibility (not to be confused with atomistic reductionism; see Murphy, this volume) then, though reality is physically based, avoiding both vitalism and dualism, a nuanced and mutually enlightening interaction between science and theology becomes both possible and desirable. This enables the intricate connections and structures revealed by the sciences to fill out and modulate theological understandings, as exemplified in Peacocke's, Gregersen's, and Clayton's chapters. At the same time, theological intuitions and insights endow nature and its manifestations with meaning, depth, and direction. This interaction, in turn, further opens up and discloses more natural, more credible, and more profound philosophical and theological insights and meanings.

Underlying the theological reflections on reductionism and emergence are interlocking scientific and philosophical investigations of these issues, focusing upon the relationships between lower-level components in various systems and complex entities and their higher-level realizations or manifestations. A variety of paradigmatic instances have been considered here, from the quantum mechanical description of single particles to that of systems of particles, where entanglement is strongly implicated, to the constraining and biasing action of macroscopic conditions on micro-relationships in the formation of snowflakes, the formation of systems of pre-biotic proteins and nucleic acids, to the emergence of life, the probable symbiogenesis of eukaryotic cells, and most surprisingly the emergence of different levels of mind and self-consciousness. Even though there has yet to be any general agreement in the philosophical analysis of these paradigmatic situations, the scientific and philosophical contributors to this volume exhibit significant convergence in their approaches and conclusions. The more purely scientific explorations nuance and exemplify the philosophical treatments, and the philosophical treatments illuminate, deepen, and formulate more precisely the scientific approaches.

This convergence moves toward an ontological reductionism consistent with versions of nonreductive physicalism or emergent monism—maintaining that the world is basically physical and material in the broadest sense of those terms (that there are no other entities in the world than those which are composed of particles and systems of particles). This is accompanied by the genuine emergence of causal powers as systems complexify in cybernetic sensitivity towards their environment. Higher-level entities and systems are thus 'real,' exercising causal influences which are not reducible to those of their components. Though lower-level causes certainly provide necessary conditions for complex entities and systems, they are incapable as such of supplying sufficient conditions. Essential contributions to the capabilities, behavioral characteristics, and causal powers of complex systems derive from highly differentiated synchronic and diachronic relationships with their environment and with the complex systems' antecedents, which are mediated by nested hierarchies of feedback, or action, loops. In these, causal factors are dynamically entangled in many different ways, inducing genuine top-down causal influences to complement the basic bottom-up causal substrate. These phenomena force the abandonment of causal reductionism and of determinism, and show the need to go beyond fundamental physics to model causal factors in nature. Both internal and external constitutive relationships render higher-order systems real and effective entities, sources of causal action in their own right. Virtually all the contributors support this philosophical position, and thus would be either nonreductive physicalists (depending what is comprehended by 'the physical') or emergent monists. Though rejecting causal reducibility, they would repudiate vitalism and any dualism other than that distinguishing God from 'not-God.'

Although there are these substantial agreements among the commitments and perspectives represented here, much philosophical work remains to be done. There is, as already emphasized, broad agreement concerning causal unity and effectiveness as the criteria for the 'reality' of a system or an entity and the rejection of causal reducibility as it pertains to higher levels of organization, including those involving life and consciousness. Thus, all the contributors embrace genuine emergence of new properties and/or causal powers in complex systems and entities. In particular, there *are* top-down causal influences, which are not reducible or completely explainable by bottom-up causes. This is strongly supported scientifically by Deacon in his description and analysis of three general categories of emergent phenomena. Brown successfully applies these insights to mental causation and to levels of consciousness in animals and humans. Scott's, Ellis's, and Hewlett's articles, each in a different way, significantly reinforce this conclusion from mathematical, physical, and biological points of view. Scott emphasizes nonlinearities and especially the crucial role of 'medium downward causality,' in which higher-level dynamics modify the local features of lower-level phase spaces. Philosophically, this corresponds to the 'selection' that Van Gulick and Murphy suggest as central to top-down causation. Hewlett shows that, in the development of the endomembrane system of a cell from more primitive component cells, selection is precisely for an emergent property of the system—the relationship in the network among the primitive cells themselves. In such situations, as Scott points out, top-down action is always dynamic and therefore developing, enforcing constraints and selection—or as Murphy describes them, 'context-sensitive constraints'—leading to the emergence of new properties and capacities. Furthermore, as Deacon, Scott, Ellis, and Brown all emphasize, such top-down action always involves hierarchies of entangled causal sequences—such as feedback and action loops, in which causes and effects are continually reversing roles within their dynamical systems. Thus, even rather simple systems resist straightforward linear causal analysis. This convincingly undermines any attempt at causal reducibility.

These informed, carefully elaborated, and scientifically well-motivated approaches to emergence, however, demand further philosophical refinement, development, and critique, as well as additional scientific support and confirmation. In particular, careful scientifically informed analysis of the networks of intertwined causal or determining factors in complex systems is needed, building on what Deacon, Brown, Scott, Ellis, and others such as Alicia Juarrero have already presented. For example, we do not yet have an adequate philosophical account of the way or ways in which macroscopic selection of the microscopic features to be amplified in an emergent situation is accomplished, or how the context-sensitive constraints arise, persist, and undergo modification. Stressing the highly differentiated external constitutive relationships which effect the generation, preservation, reproduction, and implementation of information within systems (such as in biological organisms), along with the constitutive relationships internal to systems, holds great promise in more adequately elaborating these scenarios. Deacon, Ellis, and Brown move in this direction scientifically, while Murphy, Van Gulick, Howard, Peacocke, and Clayton have begun to marshal these insights into philosophically compelling arguments. Hopefully, these encouraging beginnings will stimulate robust resolution of these and related issues.

Finally, implicit in all these philosophical approaches is an underlying presupposition that the sciences and philosophical reflections upon them do give us limited but reliable access to the entities and relationships which make up our world. Thus, we have some warrant for maintaining that ontological reducibility holds, but causal reducibility does not, and that the perceived emergence of properties and causal influences in systems is real and rooted in the internal and external constitutive relationships, including the deeply nested feedback and action loops, which characterize them. Examples of such emergence are ontological, and not merely epistemological. Much more work is needed in confirming and justifying these underlying assumptions.

A great deal has been accomplished in what has been presented here, not simply by the contributors, but by all those whose insights, clues, work, and encouragement provided the stimulus and the foundation for what has issued from this project. We look forward with expectation and hope to the exciting developments which will follow.

# *Bibliography*


Ackoff, R. *Ackoff's Best: His Classic Writings on Management*. New York: John Wiley, 1999.

Albert, Réka, Hawoong Jeong, and Albert-Lászl6 Barabási. 'Diameter of the World Wide Web.' *Nature* 401 (1999): 130–131.

Alexander, Samuel. *Space, Time, and Deity*. The Gifford Lectures 1916–18, 2 vols. London: Macmillan, 1920.

Andersen, P. B., *et al.*, eds., *Downward Causation: Minds, Bodies and Matter*. Aarhus, Denmark: Aarhus University Press, 2000.

Anderson, John R. *Cognitive Psychology and Its Implications*. New York: Worth Publishers, 2000.

Anselm of Canterbury. *Proslogion: St. Anselm's 'Proslogion' with 'A Reply on Behalf of the Fool' by Gaunilo and 'The Author's Reply to Gaunilo'*. Trans. M. J. Charlesworth. Oxford: Clarendon Press, 1965.

Aristotle. *The Physics*. Trans. P. H. Wicksteed and F. M. Cornford. Cambridge, MA: Harvard University Press, 1953.

Ashby, R. *An Introduction to Cybernetics*. London: Chapman and Hall, 1958.

Atkins, Peter. *Creation Revisited*. New York: W. H. Freeman, 1992.

____ *The 2nd Law: Energy, Chaos, and Form*. New York: Scientific American Books, 1994.

Ayala, Francisco J., and Theodosius Dobzhansky, eds. *Studies in the Philosophy of Biology: Reduction and Related Problems*. Berkeley and Los Angeles: University of California Press, 1974.

Baas, N. A. 'Emergence, Hierarchies, and Hyperstructures.' In *Artificial Life III*. Ed. C. G. Langton. Reading: Addison-Wesley, 1994.

Bak, Per. *How Nature Works: The Science of Self-Organized Criticality*. Oxford: Oxford University Press, 1996.

Barabási, Albert-Lászl6. *Linked: The New Science of Networks*. Cambridge, MA: Perseus, 2002.

Barbour, Ian G. *Religion and Science: Historical and Contemporary Issues*. San Francisco: HarperSanFrancisco, 1997.

Bargh, J. A. and T. L. Chartrand. 'The Unbearable Automaticity of Being.' *American Psychologist* 54 (1999): 462–479.

Baron-Cohen, S., A. Leslie, and Uta Frith. 'Does the Autistic Child Have a "Theory of Mind"?' *Cognition* 21 (1985): 37–46.

Barrow, J. D., and F. J. Tipler. *The Anthropic Cosmological Principle*. Oxford: Oxford University Press, 1986.

Batterman, Robert W. *The Devil in the Details: Asymptotic Reasoning in Explanation,* Reduction and Emergence. Oxford: Oxford University Press, 2002.

Bechtel, W., and A. Abrahamson. *Connectionism and the Mind*. Oxford: Blackwell, 1991.

—— and Robert C. Richardson. 'Emergent Phenomena and Complex Systems.' In *Emergence or Reduction? Essays on the Prospects of Nonreductive Physicalism*. Eds. Ansgar Beckermann, Hans Flohr, and Jaegwon Kim. New York: Walter de Gruyter, 1992.

Beckermann, Ansgar, Hans Flohr, and Jaegwon Kim, eds. *Emergence or Reduction? Essays on the Prospects of Nonreductive Physicalism*. New York: Walter de Gruyter, 1992.

Beer, S. *Brain of the Firm*. New York: John Wiley, 1964.

—— *Decision and Control*. New York: John Wiley, 1966.

Benedict, R. *Patterns of Culture*. Boston: Houghton Mifflin, 1934, 1989.

Bennett, John. *Linguistic Behavior*. 2nd edn. Indianapolis, IN: Hackett, 1990.

Bickhard, M. H., and D. T. Campbell. 'Emergence.' In *Downward Causation: Minds, Bodies and Matter*. Eds. P. B. Andersen *et al.* Aarhus, Denmark: Aarhus University Press, 2000.

Block, N., and R. Stalnaker. 'Conceptual Analysis, Dualism, and the Explanatory Gap.' *Philosophical Review* 108 (1999): 1–46.

Bohr, Niels. 'Can Quantum-Mechanical Description of Physical Reality Be Considered Complete? *Physical Review* 48 (1935): 696–702.

Bonabeau, Eric, and Guy Théraulaz. 'Swarm Smarts.' *Scientific American* 257/3 (2000): 74–79.

Boyd, R. 'Materialism with Reductionism: What Physicalism Does Not Entail.' In *Readings in Philosophy of Psychology*. Vol. 1. Ed. N. Block. Cambridge, MA: Harvard University Press, 1980.

Bracken, Joseph A. *Society and Spirit: A Trinitarian Cosmology*. Selingsgrove: Susquehanna University Press, 1991.

—— *The Divine Matrix: Creativity as Link Between East and West*. New York: Orbis Books, 1995.

Brillouin, L. *Science and Information Theory*. New York: Academic Press, 1956.

Bromberger, S. 'Why-Questions.' In *Mind and Cosmos: Essays in Contemporary Science and Philosophy*. Ed. R. Colodny. Pittsburgh, PA: University of Pittsburgh Press, 1966.

Brown, Harvey R., and Rom Harré, eds. *Philosophical Foundations of Quantum Field Theory*. Oxford: Clarendon Press, 1988.

Brown, Warren S., Nancey Murphy, and H. Newton Malony, eds. *Whatever Happened to the Soul? Scientific and Theological Portraits of Human Nature.* Minneapolis: Fortress Press, 1998.

Bunge, Mario. *Causality and Modern Science.* 3rd edn. New York: Dover, 1979.

—— *Ontology II: A World of Systems.* Dordrecht: D. Reidel, 1979.

Cairns, John, Gunther Stent, and James Watson. *Phage and the Origins of Molecular Biology.* Plainview, NY: Cold Spring Harbor Laboratory Press, 1992.

Callender, Craig, and Nick Huggett. *Physics Meets Philosophy at the Planck Scale: Contemporary Theories in Quantum Gravity.* Cambridge: Cambridge University Press, 2001.

Campbell, Donald T. '"Downward Causation" in Hierarchically Organised Biological Systems.' In *Studies in the Philosophy of Biology: Reduction and Related Problems.* Eds. Francisco J. Ayala and Theodosius Dobzhansky. Berkeley and Los Angeles: University of California Press, 1974.

Campbell, N. A. *Biology.* Redwood City, CA: Benjamin Cummings, 1996.

Cartwright, Nancy. *The Dappled World: A Study of the Boundaries of Science.* Cambridge: Cambridge University Press, 1999.

Chaitin, Gregory J. 'Randomness and Mathematical Proof.' In *From Complexity to Life: On the Emergence of Life and Meaning.* Ed. Niels Gregersen. New York: Oxford University Press, 2003.

Chalmers, D. *The Conscious Mind.* Oxford: Oxford University Press, 1996.

Chech, Thomas. 'A Model for the RNA-Catalyzed Replication of RNA.' *Proceedings of the National Academy of Sciences USA* 83 (1986): 4360–4363.

Churchland, Patricia S., and T. J. Sejnowski. 'Perspectives in Cognitive Neuroscience.' *Science* 242 (1988): 741–745.

Churchland, Paul M. 'Eliminative Materialism and the Propositional Attitudes.' *Journal of Philosophy* 78 (1981): 67–90.

—— 'Reduction, Qualia, and the Direct Introspection of Brain States.' *Journal of Philosophy* 82 (1985): 8–28.

Churchman, C. W. *The Systems Approach.* New York: Delacorte Press, 1968.

Clark, Andy. *Being There: Putting Brain, Body, and World Together Again.* Cambridge, MA: Bradford Books, 1997.

Clayton, Philip. *Explanation from Physics to Theology: An Essay in Rationality and Religion.* New Haven: Yale University Press, 1989.

—— 'Pluralism, Idealism, Romanticism: Untapped Resources for a Trinity in Process.' In *Trinity in Process: A Relational Theology of God.* Eds. Joseph Bracken, SJ, and Marjorie Hewitt Suchocki. New York: Continuum, 1997.

—— *God and Contemporary Science.* Edinburgh: Edinburgh University Press, 1997.

____ 'The Case for Christian Panentheism.' *Dialog* 37 (1998): 201–208.
____ 'A Response to My Critics.' *Dialog* 38 (1999): 289–93.
____ *The Problem of God in Modern Thought*. Grand Rapids, MI: Eerdmans, 2000.
____ 'Panentheist Internalism: Living within the Presence of the Trinitarian God.' *Dialog* 40 (2001): 208–215.
____ 'Emergence.' In *Encyclopedia of Science and Religion*. Vol. 1. Ed. W. Van Huyssteen. New York: Macmillan, 2003.
____ 'Kenotic Trinitarian Panentheism.' *Dialog* 44/3 (2005): 243–248.
____ and Paul Davies, eds. *The Re-Emergence of Emergence: The Emergentist Hypothesis from Science to Religion*. Oxford: Oxford University Press, 2006.
____ and Steven Knapp. 'Belief and the Logic of Religious Commitment.' In *The Rationality of Religious Belief*. Eds. Godehard Bruntrup and Ronald K. Tacelli. Dordrecht: Kluwer Academic Press, 1999.
____ ____ 'Can Liberals Still Believe that God (Literally) Does Anything.' *CTNS Bulletin* 20/3 (Summer 2000): 3–10.
____ and Arthur Peacocke, eds. *In Whom We Live and Move and Have Our Being: Panentheistic Reflections on God's Presence in a Scientific World*. Grand Rapids, MI: Eerdmans, 2004.
Crandall, R. E. 'The Challenge of Large Numbers.' *Scientific American* (February 1997): 72–78.
Crossan, John Dominic. *The Birth of Christianity: Discovering What Happened in the Years Immediately after the Execution of Jesus*. San Francisco: HarperSanFrancisco, 1998.
Cushing, James T., and Ernan McMullin. *Philosophical Consequences of Quantum Theory: Reflections on Bell's Theorem*. Notre Dame, IN: University of Notre Dame Press, 1989.
Davidson, Donald. 'Actions, Reasons, and Causes.' *Journal of Philosophy* 60 (1963): 685–700.
____ 'Mental Events.' In *Experience and Theory*. Eds. L. Foster and J. Swanson. Amherst, MA: University of Massachusetts Press, 1970.
Davies, Paul, and John Gribbin. *The Matter Myth: Dramatic Discoveries that Challenge our Understanding of Physical Reality*. New York, NY: Simon & Schuster, 1992.
Dawkins, Richard. *The Blind Watchmaker*. New York: Norton, 1986.
____ *Climbing Mount Improbable*. New York: Norton, 1996.
Deacon, T. *The Symbolic Species: The Co-Evolution of Language and the Human Brain*. London: Penguin, 1997.
____ 'The Hierarchic Logic of Emergence: Untangling the Interdependence of Evolution and Self-Organization.' In *Evolution and Learning: The*

*Baldwin Effect Reconsidered*. Eds. B. Weber and D. Depew. Cambridge, MA: MIT Press, 2003.

Dennett, Daniel. *The Intentional Stance*. Cambridge, MA: MIT Press, 1987.

—— 'Quining Qualia.' In *Consciousness in Contemporary Science*. Eds. A. Marcel and E. Bisiach. Oxford: Clarendon Press, 1988.

Descartes, René. *Meditations on First Philosophy*. Paris, 1642.

d'Espagnat, Bernard. *Conceptual Foundations of Quantum Mechanics*. 2nd edn. Reading, MA: W. A. Benjamin, 1976.

Donald, Merlin. *A Mind So Rare: The Evolution of Human Consciousness*. New York: Norton, 2001.

Drees, Willem B. *Religion, Science and Naturalism*. New York: Cambridge University Press, 1996.

Dretske, Fred. *Knowledge and the Flow of Information*. Cambridge, MA: MIT Press, 1981.

—— *Explaining Behavior: Reasons in a World of Causes*. Cambridge, MA: MIT Press, 1988, 1991.

Dulles, Avery, SJ. 'The Meaning of Faith Considered in Relationship to Justice.' In *The Faith That Does Justice*. Ed. John C. Haughey. New York: Paulist Press, 1977.

Eddington, A. S. *The Nature of the Physical World*. Cambridge: Cambridge University Press, 1928.

Edelmann, Gerald M. *Neural Darwinism*. Oxford: Oxford University Press, 1990.

Edelman, G., and Giulio Tononi. *A Universe of Consciousness: How Matter Becomes Imagination*. New York: Basic Books, 2000.

Eigen, Manfred. *Steps Towards Life: A Perspective on Evolution*. New York: Oxford University Press, 1992.

—— and P. Schuster. *The Hypercycle: A Principle of Natural Self-Organization*. Berlin: Springer-Verlag, 1979.

Einstein, Albert. 'Über einen die Erzeugung und Verwandlung des Lichtes betreffenden heuristischen Gesichtspunkt.' *Annalen der Physik* 17 (1905): 132–148.

—— Boris Podolsky, and Nathan Rosen. 'Can Quantum-Mechanical Description of Physical Reality Be Considered Complete?' *Physical Review* 47 (1935): 777–780.

Ellis, G. F. R. 'True Complexity and its Associated Ontology.' In *Science and Ultimate Reality: Quantum Theory, Cosmology, and Complexity*. Eds. J. D. Barrow, P. C. W. Davies, and C. L. Harper. Cambridge: Cambridge University Press, 2004.

—— and J. Toronchuk. 'Affective Neural Darwinism.' In *Consciousness and Emotion: Agency, Conscious Choice, and Selective Perception*. Eds.

Ralph D. Ellis and Natika Newton. Amsterdam: John Benjamins, 2005.

____ U. Kirchner, and W. Stoeger. 'Multiverses and Physical Cosmology: Philosophical Issues.' *Monthly Notices of the Royal Astronomical Society* 347 (2004): 921–936.

Elsasser, W. M. *Reflections on a Theory of Organisms: Holism in Biology*. 1987. Baltimore, MD: Johns Hopkins University Press, 1998.

Emmeche, Claus. *The Garden in the Machine: The Emerging Science of Artificial Life*. Princeton: Princeton University Press, 1994.

____ S. Køppe, and F. Stjernfelt. 'Levels, Emergence, and Three Versions of Downward Causation.' In *Downward Causation: Minds, Bodies and Matter*. Eds. P. B. Andersen *et al.* Aarhus, Denmark: Aarhus University Press, 2000.

Farrer, Austin. *The Freedom of the Will*. The Gifford Lectures, 1957. London: Adam and Charles Black, 1958.

Feigl, H. 'The Mental and the Physical.' In *Minnesota Studies in the Philosophy of Science Volume II*. Eds. H. Feigl, M. Scriven, and G. Maxwell. Minneapolis: University of Minnesota Press, 1958.

Fewell, Jennifer. 'Social Insect Networks.' *Science* 301 (2003): 1867–1870.

Feyerabend, P. *Against Method*. London: New Left Books, 1975.

Fodor, Jerry. 'Special Sciences, or the Disunity of Science as a Working Hypothesis.' *Synthese* 28 (1974): 77–115.

____ *Representations*. Cambridge, MA: MIT Press, 1981.

____ 'Making Mind Matter More.' *Philosophical Topics* 17 (1989): 59–80.

Fontana, W., and L. W. Buss. 'The Arrival of the Fittest: Toward a Theory of Biological Organization.' *Bulletin of Mathematical Biology* 56 (1994): 1–64.

Fox, Matthew. *One River, Many Wells: Wisdom Springing from Global Faiths*. New York: Jeremy P. Tarcher/Putnam, 2000.

Fransén, E., and A. Lansner. 'Low Spiking Rates in a Population of Mutually Exciting Pyramidal Cells.' *Network* 6 (1995): 271–288.

Fraser, J. T. *The Genesis and Evolution of Time*. Brighton, UK: Harvester Press, 1982.

____ *Of Time, Passion, and Knowledge: Reflections on the Strategy of Existence*. 2nd edn. Princeton: Princeton University Press, 1990.

Freeman, W. J. *How Brains Make Up Their Minds*. New York: Columbia University Press, 2000.

Frege, G. 'On Sense and Reference.' In *Translations from the Philosophical Writings of Gottlob Frege*. Eds. P. Geach and M. Black. Trans. Max Black. Oxford: Oxford University Press, 1952.

Fuster, Joaquin M. 'Prefrontal Neurons in Networks of Executive Memory.' *Brain Research Bulletin* 42 (2000): 331–336.

Gadamer, Hans-Georg. *Truth and Method*. New York: Continuum, 1960, 1995.

Gardner, M. 'Mathematical Games: The Fantastic Combinations of John Conway's New Solitaire Game "Life".' *Scientific American* 223/4 (1970): 120–123.

Geertz, Clifford. *The Interpretation of Cultures: Selected Essays*. New York: Basic Books, 1973.

Ghirardi, Gian Carlo, and Alberto Rimini. 'Old and New Ideas in the Theory of Quantum Measurement.' In *Sixty-two Years of Uncertainty: Historical, Philosophical, and Physical Inquiries into the Foundations of Quantum Mechanics*. Ed. Arthur I. Miller. New York and London: Plenum, 1990.

Gilbert, Scott F. *Developmental Biology*. Sunderland, MA: Sinauer, 1991.

—— and Sahotra Sarkar. 'Embracing Complexity: Organicism for the 21st Century.' *Developmental Dynamics* 219 (2000): 1–9.

Gillett, Carl. 'Strong Emergence as a Defense of Non-Reductive Physicalism: A Physicalist Metaphysics for "Downward" Causation.' *Principia* 6 (2003): 83–114.

Giot, L., *et al.* 'A Protein Interaction Map of Drosophila Melanogaster.' *Science* 302 (2003): 1727–1736.

Goodwin, B. *How the Leopard Changed its Spots: The Evolution of Complexity*. New York: Princeton University Press, 1994, 2001.

Gordon, Deborah. 'The Development of Organization in an Ant Colony.' *American Scientist* 83 (January–February 1995): 50–57.

Gould, Stephen Jay. *Wonderful Life: The Burgess Shale and the Nature of History*. New York: Norton, 1989.

—— *The Structure of Evolutionary Theory*. Cambridge, MA: Belknap, Harvard University Press, 2002.

—— See: http://www.stephenjaygould.org/library/gould_nonmoral.html.

—— and Niles Eldredge. 'Punctuated Equilibria: An Alternative to Phyletic Gradualism.' In *Models in Paleobiology*. Ed. T. J. M. Schopf. San Francisco: Freeman, Cooper, and Co., 1972.

—— and Richard Lewontin. 'The Spandrels of San Marco and the Panglossian Paradigm: A Critique of the Adaptationist Programme.' *Proceedings of the Royal Society of London*, Series B, 205 (1979): 581–598.

Graf, Friedrich Wilhelm, and Falk Wagner, eds. *Die Flucht in den Begriff: Materialien zu Hegels Religionsphilosophie*. Stuttgart: Klett-Cotta, 1982.

Greene, B. *The Elegant Universe: Superstrings, Hidden Dimensions, and the Quest for the Ultimate Theory*. New York: Norton, 2003.

Gregersen, Niels Henrik. 'The Idea of Creation and the Theory of Autopoietic Processes.' *Zygon: Journal of Religion and Science* 33/3 (1998): 333–367.

____ 'Einheit und Vielfalt der schöpferischen Werke Gottes: Pannenbergs Beitrag zu einer trinitarischen Schöpfungslehre.' *Kerygma und Dogma* 2 (1999): 102–129.

____ 'Beyond the Balance: Theology in a Self-Organizing World.' In *Design and Disorder: Perspectives from Science and Theology*. Eds. Niels Gregersen and Ulf Görman. Edinburgh: T & T Clark, 2000.

____ 'Fra urværket til netværket. Teologi mellem fysik og informatik' [From the Clockwork to the Network: Theology between Physics and the Sciences of Information]. *Dansk teologisk Tidsskrift* 65/4 (2002): 272–295.

____ 'Complexity: What Is at Stake for Religious Reflection?' In *The Significance of Complexity: Approaching a Complex World Through Science, Theology and the Humanities*. Eds. Kees van Kooten Niekerk and Hans Buhl. Aldershot: Ashgate, 2004.

____ ed. *From Complexity to Life: On the Emergence of Life and Meaning*. New York: Oxford University Press, 2003.

____ and Ulf Görman, eds. *Design and Disorder: Perspectives from Science and Theology*. Edinburgh: T & T Clark, 2000.

Guénault, Tony. *Basic Superfluids*. London: Taylor & Francis, 2003.

Haken, H. *Principles of Brain Functioning: A Synergetic Approach to Brain Activity, Behavior and Cognition*. Berlin: Springer-Verlag, 1996.

Hardwick, Charley. *Events of Grace: Naturalism, Existentialism, and Theology*. New York: Cambridge University Press, 1996.

Hasker, W. *The Emergent Self*. Ithaca: Cornell University Press, 2001.

Haught, John. *God After Darwin: A Theology of Evolution*. Boulder, CO: Westview Press, 2000.

____ *Deeper than Darwin: The Prospect for Religion in the Age of Evolution*. Boulder, CO: Westview Press, 2003.

Healey, Richard. *The Philosophy of Quantum Mechanics: An Interactive Interpretation*. Cambridge: Cambridge University Press, 1989.

Hebb, D. O. *Organization of Behavior: A Neuropsychological Theory*. New York: John Wiley & Sons, 1949.

____ 'The Structure of Thought.' In *The Nature of Thought*. Eds. P. W. Jusczyk and R. M. Klein. Hillsdale, NJ: Lawrence Erlbaum Associates, 1980.

____ *Essay on Mind*. Hillsdale, NJ: Lawrence Erlbaum Associates, 1980.

Heisenberg, Martin. 'Voluntariness (Willkürfähigkeit) and the General Organization of Behavior.' In *Flexibility and Constraint in Behavioral Systems*. Eds. R. J. Greenspan and C. P. Kyriacou. New York: John Wiley & Sons, 1994.

Helmreich, Stefan. *Silicon Second Nature: Culturing Artificial Life in a Digital World*. Berkeley: University of California Press, 1998.

Hermann, Henry. 'Social Organization in Insects.' In *Selforganization: Proceedings of the Liberty Fund Conference on Selforganization, 1984, Key Biscayne, Florida*. Ed. Sidney W. Fox. Guilderland, NY: Adenine Press, 1986.

Hill, C., and B. McLaughlin. 'There is Less in Reality than Dreamt of in Chalmers' Philosophy.' *Philosophy and Phenomenological Research* 59 (1999): 445–454.

Hoddeson, Lillian, *et al.*, eds. *Out of the Crystal Maze: Chapters from the History of Solid-State Physics*. New York: Oxford University Press, 1992.

Hogan, C. 'Why the Universe is Just So.' *Review of Modern Physics* 72 (2000): 1149–1161.

Holland, John. *Emergence: From Chaos to Order*. Oxford: Oxford University Press, 1998.

Howard, Don. 'Einstein on Locality and Separability.' *Studies in History and Philosophy of Science* 16 (1985): 171–201.

—— ' "Nicht sein kann was nicht sein darf," or the Prehistory of EPR, 1909–1935: Einstein's Early Worries about the Quantum Mechanics of Composite Systems.' In *Sixty-two Years of Uncertainty: Historical, Philosophical, and Physical Inquiries into the Foundations of Quantum Mechanics*. Ed. Arthur I. Miller. New York and London: Plenum, 1990.

—— 'What Makes a Classical Concept Classical? Toward a Reconstruction of Niels Bohr's Philosophy of Physics.' In *Niels Bohr and Contemporary Philosophy*. Eds. Jan Faye and Henry Folse. Boston: Kluwer, 1994.

—— 'Who Invented the Copenhagen Interpretation? A Study in Mythology.' In *PSA 2002*, Part II: *Symposium Papers*. Proceedings of the 2002 Biennial Meeting of the Philosophy of Science Association, Milwaulkee, Wisconsin, 7–9 November 2002. A special issue of *Philosophy of Science* 71 (2004).

Hutchins, E., and B. Hazelhurst. 'Learning in the Cultural Process.' In *Artificial Life II*. Eds. C. Langton *et al.* Redwood City, CA: Addison-Wesley, 1991.

Huxley, Julian. *Evolution, The Modern Synthesis*. London: Allen and Unwin, 1942.

Isham, C. J. *Lectures on Quantum Theory: Mathematical and Structural Foundations*. London: Imperial College Press, 1997.

Jackson, F. 'Epiphenomenal Qualia.' *Philosophical Quarterly* 32 (1982): 127–136.

—— 'What Mary Didn't Know.' *Journal of Philosophy* 32 (1986): 291–295.

—— 'Postscript on Qualia.' In *Mind, Method and Conditionals*. Ed. F. Jackson. London: Routledge, 1998.

Jammer, Max. 'Materialism.' In *Encyclopedia of Science and Religion*. Vol. 2. Ed. Wentzel van Huyssteen. New York: Macmillan, 2003.

Johnson, Steven. *Emergence: The Connected Lives of Ants, Brains, Cities and Software*. London: Allen Lane/Penguin Press, 2001.

Johnston, Wendy, *et al.* 'RNA-Catalyzed RNA Polymerization: Accurate and General RNA-Templated Primer Extension.' *Science* 292 (2001): 1319–1325.

Juarrero, Alicia. *Dynamics in Action: Intentional Behavior as a Complex System*. Cambridge, MA: Bradford Books, MIT Press, 1999.

Kadanoff, L. *From Order to Chaos: Essays Critical, Chaotic and Otherwise*. Singapore: World Scientific, 1993.

Kauffman, Stuart. *The Origins of Order: Self Organization and Selection in Evolution*. New York: Oxford University Press, 1993.

Kelly, Kevin. 'Theology for Nerds.' In *Science and the Spiritual Quest: New Essays by Leading Scientists*. Eds. W. Mark Richardson *et al.* London: Routledge, 2002.

Kemeny, J., and P. Oppenheim. 'On Reduction.' *Philosophical Studies* 7 (1956): 6–17.

Kim, Jaegwon. 'Psychophysical Supervenience.' *Philosophical Studies* 42 (1982): 51–70.

____ 'Explanatory Exclusion, and the Problem of Mental Causation.' In *Information, Semantics and Epistemology*. Ed. E. Villanueva. Oxford: Basil Blackwell, 1990.

____ ' "Downward Causation" in Emergentism and Nonreductive Physicalism.' In *Emergence or Reduction?: Essays on the Prospects of Nonreductive Physicalism*. Eds. A. Beckermann, H. Flohr, and J. Kim. New York: Walter de Gruyter, 1992.

____ 'Multiple Realization and the Metaphysics of Reduction.' In *Supervenience and Mind*. Ed. J. Kim. Cambridge: Cambridge University Press, 1993.

____ *Supervenience and Mind*. Cambridge: Cambridge University Press, 1993.

____ 'Non-Reductivism and Mental Causation.' In *Mental Causation*. Eds. J. Heil and A. Mele. Oxford: Clarendon Press, 1993.

____ *Mind in a Physical World: An Essay on the Mind–Body Problem and Mental Causation*. Cambridge, MA: MIT Press, 2000.

____ 'Making Sense of Downward Causation.' In *Downward Causation*. Eds. P. B. Andersen *et al.* Aarhus, Denmark: Aarhus University Press, 2000.

____ 'Being Realistic about Emergence.' In *The Re-Emergence of Emergence: The Emergentist Hypothesis from Science to Religion*. Eds. Philip Clayton and Paul Davies. Oxford: Oxford University Press, 2006.

Kitcher, Philip. 'Explanatory Unification.' *Philosophy of Science* 48 (1981): 507–531.

Köhler, W. *The Mentality of Apes*. New York: Harcourt, Brace, 1927.
Kolb, B., and I. Q. Whishaw. *Fundamentals of Human Neuropsychology*. 5th edn. New York: Worth, 2003.
Krink, Thiemo. 'Complexity and the Computing Age: Can Computers Help Us to Understand Complex Phenomena in Nature?' In *The Significance of Complexity: Approaching a Complex World Through Science, Theology and the Humanities*. Eds. Kees van Kooten Niekerk and Hans Buhl. Aldershot: Ashgate, 2004.
Kripke, S. *Naming and Necessity*. Cambridge, MA: Harvard University Press, 1972.
Kuhn, T. *The Structure of Scientific Revolutions*. Chicago: University of Chicago Press, 1962, 1970.
Küppers, B.-O. *Information and the Origin of Life*. Cambridge, MA: MIT Press, 1990.
Langton, Christopher G. 'Life at the Edge of Chaos.' In *Artificial Life II: Proceedings of the Workshop on Artificial Life held February, 1990, in Santa Fe, New Mexico*. Eds. C. G. Langdon *et al.* (Proceedings Volume X: Santa Fe Institute Studies in the Sciences of Complexity), 1992.
____ *Artificial Life: An Overview*. Cambridge, MA: MIT Press, 1995.
____ 'Artificial Life.' In *The Philosophy of Artificial Life*. Ed. Margaret A. Boden. Oxford: Oxford University Press, 1996.
Laughlin, Robert B., and David Pines. 'The Theory of Everything.' *Proceedings of the National Academy of Sciences* 97 (2000): 28–31.
Levine, J. 'Materialism and Qualia: The Explanatory Gap.' *Pacific Philosophical Quarterly* 64 (1983): 354–361.
____ 'Could Love Be Like a Heat Wave?: Physicalism and the Subjective Character of Experience.' *Philosophical Studies* 49 (1986): 245–261.
____ 'On Leaving Out What It's Like.' In *Consciousness*. Eds. M. Davies and G. Humphreys. Oxford: Blackwell, 1993.
Levy, Steven. *Artificial Life: A Report from the Frontier Where Computers Meet Biology*. New York: Vintage Books, 1992.
Lewes, G. H. *Problems of Life and Mind*. Vol. 2. London: Kegan Paul, Trench, Tubner and Co., 1875.
Lewin, Roger. *Complexity: Life at the Edge of Chaos*. Chicago: University of Chicago Press, 1992.
Lewis, D. 'Postscript to Mad Pain and Martian Pain.' In *Philosophical Papers Volume I*. Ed. D. Lewis. Oxford: Oxford University Press, 1982.
Liljeros, Fredrik, *et al.* 'The Web of Human Sexual Contacts.' *Nature* 411 (2001): 907–908.
Loar, B. 'Phenomenal States.' In *Philosophical Perspectives*, Vol. 4: *Action Theory and the Philosophy of Mind*. Ed. J. Tomberlin. Atascadero, CA: Ridgeview, 1990.

Lonergan, Bernard, SJ. *Insight: A Study of Human Understanding*. New York: Philosophical Library, 1957.

____ 'Cognitional Structure.' In *Collection*. Ed. F. E. Crowe, SJ. New York: Herder and Herder, 1967.

Longino, Helen. *Science as Social Knowledge*. Princeton, NJ: Princeton University Press, 1990.

Lycan, W. 'What is the "Subjectivity" of the Mental?' In *Philosophical Perspectives*, Vol. 4: *Action Theory and the Philosophy of Mind*. Ed. J. Tomberlin. Atascadero, CA: Ridgeview, 1990.

MacKay, Donald M. *Behind the Eye*. Oxford: Basil Blackwell, 1991.

Mandelbrot, Benoit B. *The Fractal Geometry of Nature*. New York: Freeman, 1983.

Margulis, Lynn, and R. Fester. *Symbiosis as a Source of Evolutionary Innovation: Speciation and Morphogenesis*. Cambridge, MA: MIT Press, 1991.

Massimi, Michela. 'Exclusion Principle and the Identity of Indiscernibles: A Response to Margenau's Argument.' *British Journal for the Philosophy of Science* 52 (2001): 303–330.

McGinn, C. 'Can We Solve the Mind–Body Problem?' *Mind* 98 (1989): 349–366.

____ *The Problem of Consciousness*. Oxford: Blackwell, 1991.

McMullin, Ernan. *The Inference that Makes Science*. Milwaukee, WI: Marquette University Press, 1991.

____ 'Indifference Principle and Anthropic Principle in Cosmology.' *Studies in the History and the Philosophy of Science* 24 (1993): 359–389.

Metzinger, Thomas. *Being No One: The Self-Model Theory of Subjectivity*. Cambridge, MA: MIT Press, 2003.

Milgram, Stanley. 'The Small World Problem.' *Psychology Today* 2 (1967): 60–67.

Miller, Arthur I., ed. *Sixty-Two Years of Uncertainty: Historical, Philosophical, and Physical Inquiries into the Foundations of Quantum Mechanics*. New York and London: Plenum, 1990.

Miller, Stanley. 'A Production of Amino Acids Under Possible Primitive Earth Conditions.' *Science* 117 (1953): 528–529.

Milsum, J. H. *Biological Control Systems Analysis*. New York: McGraw Hill, 1966.

Moore, G. E. *Principia Ethica*. Cambridge: Cambridge University Press, 1902.

Morowitz, Harold J. *The Emergence of Everything: How the World Became Complex*. New York: Oxford University Press, 2002.

Moyers, B. *Healing and the Mind*. New York: Doubleday, 1993.

Murphy, Nancey. *Anglo-American Postmodernity: Philosophical Perspectives on Science, Religion, and Ethics*. Boulder, CO: Westview Press, 1997.

____ 'Supervenience and the Downward Efficacy of the Mental: A Nonreductive Physicalist Account of Human Action.' In *NP*.

____ 'Physicalism Without Reductionism: Toward a Scientifically, Philosophically, and Theologically Sound Portrait of Human Nature.' *Zygon* 34/4 (1999): 551–572.

____ 'Emergence and Mental Causation.' In *The Re-Emergence of Emergence: The Emergentist Hypothesis from Science to Religion*. Eds. Philip Clayton and Paul Davies. Oxford: Oxford University Press, 2006.

____ and Warren S. Brown. *Did My Neurons Make Me Do It?: Philosophical and Neurobiological Perspectives on Moral Responsibility and Free Will*. Oxford: Oxford University Press, 2007.

____ and George F. R. Ellis. *On the Moral Nature of the Universe: Theology, Cosmology, and Ethics*. Minneapolis, MN: Fortress Press, 1996.

Nagel, E. *The Structure of Science*. New York: Harcourt, Brace and World, 1961.

Nagel, T. 'What Is It Like to Be a Bat?' *Philosophical Review* 83/4 (1974) reprinted in *Mortal Questions*. Ed. T. Nagel. Cambridge: Cambridge University Press, 1979.

Nicolelis, M. A. L., E. E. Fanselow, and A. A. Ghazanfar. 'Hebb's Dream: The Resurgence of Cell Assemblies.' *Neuron* 19 (1997): 219–21.

Nicolis, J. S. *Dynamics of Hierarchical Systems: An Evolutionary Approach*. Berlin: Springer-Verlag, 1986.

Oppenheim, P. and H. Putnam. 'Unity of Science as a Working Hypothesis.' *Minnesota Studies in the Philosophy of Science Volume II*. Eds. H. Feigl, M. Scriven and G. Maxwell. Minneapolis: University of Minnesota Press, 1958.

Panksepp, J. *Affective Neuroscience: The Foundation of Human and Animal Emotions*. Oxford: Oxford University Press, 1998.

Pannenberg, Wolfhart. *Faith and Reality*. Trans. John Maxwell. Philadelphia: Westminster Press, 1977.

____ *Systematic Theology*. 3 vols. Trans. Geoffrey W. Bromiley. Grand Rapids: Eerdmans, 1991–98.

Pattee, Harold. 'The Physical Basis and Origin of Hierarchical Control.' In *Hierarchy Theory*. Ed. Pattee. New York: George Braziller, 1973.

Peacocke, Arthur R. *God and the New Biology*. Glouster, MA: Peter Smith, 1986.

____ *The Physical Chemistry of Biological Organization*. Oxford: Clarendon Press, 1983, 1989.

____ *Theology for a Scientific Age: Being and Becoming—Natural, Divine and Human*. London: SCM Press, 1990; 2nd enlarged edn. Minneapolis: Fortress Press, 1993.

____ 'The Incarnation of the Self-Expressive Word of God.' In *Religion and Science: History, Method, Dialogue*. Eds. Mark Richardson and Wesley J. Wildman. New York: Routledge, 1996.

____ 'God's Interaction with the World: The Implications of Deterministic Chaos and Interconnected, Interdependent Complexity.' In *CC*.

____ 'The Sound of Sheer Silence: How Does God Communicate with Humanity?' In *NP*.

____ 'Complexity, Emergence and Divine Creativity.' In *From Complexity to Life: On the Emergence of Life and Meaning*. Ed. Niels Gregersen. New York: Oxford University Press, 2003.

____ *Creation and the World of Science*. Oxford: Clarendon Press, 1979, 2004.

____ 'Emergence, Mind and Divine Action: The Hierarchy of the Sciences in Relation to the Human Mind–Body–Brain.' In *The Re-emergence of Emergence: The Emergentist Hypothesis from Science to Religion*. Eds. Philip Clayton and Paul Davies. Oxford: Oxford University Press, 2006.

Pearle, Philip. 'Toward a Relativistic Theory of Statevector Reduction.' In *Sixty-Two Years of Uncertainty: Historical, Philosophical, and Physical Inquiries into the Foundations of Quantum Mechanics*. Ed. Arthur I. Miller. New York and London: Plenum, 1990.

Penrose, R. *The Large, the Small and the Human Mind*. Cambridge: Cambridge University Press, 1997.

Peskin, M. E., and D. V. Schroeder. *An Introduction to Quantum Field Theory*. Reading, MA: Perseus Books, 1995.

Pickover, C. A. *Visualizing Biological Information*. Singapore: World Scientific, 1995.

Pitaevskii, Lev, and Sandro Stringari. *Bose-Einstein Condensation*. Oxford: Clarendon Press, 2003.

Place, U. T. 'Is Consciousness a Brain Process?' *British Journal of Psychology* 47 (1956): 44–50.

Polanyi, Michael. *Personal Knowledge*. New York: Harper Torchbooks, 1964.

____ *The Tacit Dimension*. Garden City, NY: Doubleday Anchor Books, 1967.

____ *Knowing and Being*. Ed. Marjorie Grene. Chicago: University of Chicago Press, 1969.

____ and Harry Prosch. *Meaning*. Chicago: University of Chicago Press, 1975.

Polkinghorne, John. *The God of Hope and the End of the World*. New Haven and London: Yale University Press, 2002.

Pols, Edward. *Mind Regained*. Ithaca: Cornell University Press, 1998.

Popper, K., and J. Eccles. *The Self and its Brain: An Argument for Interactionism*. Berlin: Springer-Verlag, 1977.

Prigogine, I., and I. Stengers. *Order Out of Chaos*. London: Heinemann, 1984.

Puddefoot, John C. 'Information Theory, Biology, and Christology.' In *Religion and Science: History, Method, Dialogue*. Eds. Mark Richardson and Wesley J. Wildman. New York: Routledge, 1996.

Putnam, Hillary. 'Philosophy and Our Mental Life.' In *Mind, Language and Reality, Philosophical Papers*. Vol. 2. Ed. Hillary Putnam. Cambridge: Cambridge University Press, 1975.

Ray, Thomas S. 'An Evolutionary Approach to Synthetic Biology: Zen and the Art of Creating Life.' In *Artificial Life: An Overview*. Ed. Christopher G. Langton. Cambridge, MA: MIT Press, 1995.

Rees, M. J. *Our Cosmic Habitat*. Princeton, NJ: Princeton University Press, 2001.

Richardson, Mark, and Wesley J. Wildman, eds. *Religion and Science: History, Method, Dialogue*. New York: Routledge, 1996.

Rorty, R. 'In Defense of Eliminative Materialism.' *Review of Metaphysics* 24 (1970): 112–121.

Russell, Robert John. 'Special Providence and Genetic Mutation: A New Defense of Theistic Evolution.' In *EMB*.

—— 'Divine Action and Quantum Mechanics: A Fresh Assessment.' In *QM*.

—— 'Ian Barbour's Methodological Breakthrough: Creating the "Bridge" between Science and Theology.' In *50 Years in Science and Religion: Ian G. Barbour and his Legacy*. Ed. R. J. Russell. Aldershot: Ashgate, 2004.

Santa Fe Institute. *Annual Report on Scientific Programs*. 1993. Santa Fe, New Mexico: Santa Fe Institute, 1994.

Schrödinger, Erwin. 'Die gegenwärtige Situation in der Quantenmechanik.' *Die Naturwissenschaften* 23 (1935): 807–812, 823–828, 844–849.

—— 'Discussion of Probability Relations Between Separated Systems.' *Proceedings of the Cambridge Philosophical Society* 31 (1935): 555–662.

—— 'Probability Relations Between Separated Systems.' *Proceedings of the Cambridge Philosophical Society* 32 (1936): 446–452.

—— *What is Life?: The Physical Aspects of the Living Cell*. Cambridge: Cambridge University Press, 1944.

Schweber, Silvan. 'Physics, Community and the Crisis in Physical Theory.' *Physics Today* 46 (1993): 34–40.

Schwöbel, Christoph. *God: Action and Revelation*. Kampen: Kok Pharos, 1992.

Scott, A. C. *Stairway to the Mind*. New York: Springer-Verlag, 1995.

—— *Neuroscience: A Mathematical Primer*. New York: Springer-Verlag, 2002.

——*Nonlinear Science: Emergence and Dynamics in Coherent Structures*, 2nd edn. Oxford: Oxford University Press, 2003.

——ed. *Encyclopedia of Nonlinear Science.* New York: Routledge, 2004.

Searle, John. *Minds, Brain and Science.* Cambridge, MA: Harvard University Press, 1984.

——*The Rediscovery of the Mind.* Cambridge, MA: MIT Press, 1992.

Seife, J. *Zero: The Biography of a Dangerous Idea.* London: Penguin, 2000.

Sellars, R. W. *The Philosophy of Physical Realism.* New York: Russell and Russell, 1932.

Shannon, C. E., and W. Weaver. *The Mathematical Theory of Communication.* Urbana, IL: University of Illinois Press, 1963.

Silberstein, M. 'Reduction, Emergence, and Explanation.' In *The Blackwell Guide to the Philosophy of Science.* Eds. Peter Machamer and Michael Silberstein. Oxford: Blackwell, 2002.

——and J. McGeever. 'The Search for Ontological Emergence.' *The Philosophical Quarterly* 49 (1999): 182–200.

Smart, J. J. C. 'Sensations and Brain Processes.' *Philosophical Review* 82 (1959): 141–156.

Smith, John Maynard. *Symbiosis as a Source of Evolutionary Innovation: Speciation and Morphogenesis.* Eds. L. Margulis and R. Fester. Cambridge, MA: MIT Press, 1991.

Sokal, A. see http://www.physics.nyu.edu/faculty/sokal/index.html.

Sperry, Roger W. *Science and Moral Priority: Merging Mind, Brain, and Human Values.* New York: Columbia University Press, 1983.

——'Psychology's Mentalist Paradigm and the Religion/Science Tension.' *American Psychologist* 43/8 (August 1988): 607–613.

——'In Defense of Materialism and Emergent Interaction.' *Journal of Mind and Behavior* 12 (1991): 221–245.

Steels, Luc. 'Self-Organizing Vocabularies.' In *Artificial Life V: Proceedings of the Fifth International Workshop on the Synthesis and Simulation of Living Systems.* Eds. Christopher G. Langton and Katsunori Shimohara. Cambridge, MA: MIT Press, 1997.

Stegmüller, Wolfgang. *Hauptströmungen der Gegenwartsphilosophie.* Vol. 3. Reclam: Stuttgart, 1987.

Stephan, Achim. 'Emergence: A Systematic View on Its Historical Facets.' In *Emergence or Reduction? Essays on the Prospects of Nonreductive Physicalism.* Eds. Ansgar Beckermann, Hans Flohr, and Jaegwon Kim. New York: Walter de Gruyter, 1992.

Sternberg, E. *The Balance Within: The Science Connecting Health and Emotions.* New York: W. H. Freeman, 2000.

Stoeger, William R., SJ. 'What Contemporary Cosmology and Theology Have to Say to One Another.' *CTNS Bulletin* 9/2 (Spring 1989): 1–15.

____ 'Contemporary Cosmology and Its Implications for the Science–Religion Dialogue.' In *Physics, Philosophy, and Theology: A Common Quest for Understanding*. Ed. Robert John Russell, William R. Stoeger, SJ, and George V. Coyne. Vatican City State: Vatican Observatory Press, 1988.

____ 'Describing God's Action in the World in Light of Scientific Knowledge of Reality.' In *CC*.

____ 'The Immanent Directionality of the Evolutionary Process, and its Relationship to Teleology.' In *EMB*.

____ 'The Mind–Brain Problem, the Laws of Nature, and Constitutive Relationships.' In *NP*.

____ 'Cultural Cosmology and the Impact of the Natural Sciences on Philosophy and Culture.' In *The End of the World and the Ends of God: Science and Theology on Eschatology*. Eds. John Polkinghorne and Michael Welker. Harrisburg, PA: Trinity Press International, 2000.

____ 'Science, the Laws of Nature, and Divine Action.' In *Interdisciplinary Perspectives on Cosmology and Biological Evolution*. Eds. Hilary D. Regan and Mark Worthing. Adelaide: Australian Theological Forum, 2002.

____ 'The Anthropic Principle Revisited.' In *Philosophy in Science*. Vol. 10. Eds. William R. Stoeger, SJ, Michael Heller, and Józef M. Życiński Tucson, AZ: Pachart, 2003.

____ 'Contemporary Physics and the Ontological Status of the Laws of Nature.' In *QCLN*.

Suchocki, Marjorie Hewitt. *The End of Evil: Process Eschatology in Historical Context*. Albany: State University of New York Press, 1988.

Tanenbaum, A. S. *Structured Computer Organization*. Englewood Cliffs, NJ: Prentice Hall, 1990.

Taylor, Charles, and David Jefferson. 'Artificial Life as a Tool for Biological Inquiry.' In *Artificial Life: An Overview*. Ed. Christopher G. Langton. Cambridge, MA: MIT Press, 1995.

Taylor, Mark. *The Moment of Complexity: Emerging Network Culture*. Chicago: University of Chicago Press, 2002.

Tegmark, M. 'Is the Theory of Everything Merely the Ultimate Ensemble Theory?' *Annals of Physics* 270 (1998): 1–51.

Teilhard de Chardin, Pierre. *The Human Phenomenon*. Trans. Sarah Appleton-Weber. Portland, OR: Sussex Academic Press, 1999.

Terhal, Barbara M., Michael M. Wolf, and Andrew C. Doherty. 'Quantum Entanglement: A Modern Perspective.' *Physics Today* 56/4 (April 2003): 46–52.

Theraulaz, Guy, Eric Bonabeau, and Jean Louis Deneubourg. 'The Origin of Nest Complexity in Social Insects.' *Complexity* 3/6 (1998): 15–25.

Tillich, Paul. *Systematic Theology*. 3 vols. Chicago: University of Chicago Press, 1963.

Tomberlin, J., ed. *Philosophical Perspectives*, Vol. 4: *Action Theory and the Philosophy of Mind*. Atascadero, CA: Ridgeview, 1990.

Tracy, Thomas F. 'Creation, Providence, and Quantum Chance.' In *QM*.

Turing, A. M. 'The Chemical Basis of Morphogenesis.' *Philosophical Transactions of the Royal Society of London* B237 (1952): 37–72.

van Fraassen, Bas C. *The Scientific Image*. Oxford: Clarendon Press, 1980.

Van Gulick, Robert. 'Physicalism and the Subjectivity of the Mental.' *Philosophical Topics* 13/3 (1985): 51–70.

____ 'Nonreductive Materialism and Intertheoretic Constraint.' In *Emergence or Reduction? Essays on the Prospects of Nonreductive Physicalism*. Eds. Ansgar Beckermann, Hans Flohr, and Jaegwon Kim. New York: Walter de Gruyter, 1992.

____ 'Conceiving Beyond Our Means: The Limits of Thought Experiments.' In *Toward a Science of Consciousness III*. Eds. S. Hameroff, A. Kazniak, and D. Chalmers. Cambridge, MA: MIT Press, 1999.

Voorhees, B. H. 'Axiomatic Theory of Hierarchical Systems.' *Behavioral Science* 28 (1983): 24–34.

Wagner, Falk. *Was ist die Theologie?: Studien zu ihrem Begriff und Thema in der Neuzeit*. Gütersloh: G. Mohn, 1989.

Waldrop, Mitchell M. *Complexity: The Emerging Science at the Edge of Order and Chaos*. New York: Simon & Schuster, 2002.

Wallace, Alan B. *The Taboo of Subjectivity: Toward a New Science of Consciousness*. New York: Oxford University Press, 2000.

Watts, Duncan. *Six Degrees: The Science of a Connected Age*. New York: Norton, 2003.

Weaver, Warren. 'Recent Contributions to the Mathematical Theory of Communication.' In *The Mathematical Theory of Communication*. Ed. Claude E. Shannon and Warren Weaver. Urbana: University of Illinois Press, 1949.

Weinberg, S. *Dreams of a Final Theory: The Search for the Fundamental Laws of Nature*. New York: Pantheon Books, 1992.

Whitehead, Alfred North. *Science and the Modern World*. New York: Free Press, 1925.

____ *Modes of Thought*. New York: Free Press, 1968.

____ *Process and Reality*. Corrected edn. Eds. David Ray Griffin and Donald W. Sherburne. New York: Free Press, 1968.

Wiener, N. *Cybernetics*. New York: John Wiley & Sons, 1961.

Wildman, Wesley J. 'The Divine Action Project, 1988–2003.' *Theology and Science* 2/1 (April 2004): 31–76.

——'A Reply to My Respondents on "The Divine Action Project".' *Theology and Science* 3/1 (March 2005): 71–83.

——and Robert J. Russell. 'Chaos: A Mathematical Introduction with Philosophical Reflections.' In *CC*.

Wilkes, K. '—, Yishi, Duh, Urn and Consciousness.' In *Concsciousness in Contemporary Science*. Ed. A. Marcel and E. Bisiach. Oxford: Clarendon Press, 1988.

——'Losing Consciousness.' In *Conscious Experience*. Ed. T. Metzinger. Thorverton: Imprint Academic, 1995.

Wilson, Edward O. *Consilience: The Unity of Knowledge*. New York: Vintage Books, 1998.

Wimsatt, W. C. 'Robustness, Reliability and Multiple-Determination in Science.' In *Knowing and Validating in the Social Sciences: A Tribute to Donald T. Campbell*. Eds. M. Brewer and B. Collins. San Francisco, CA: Jossey-Bass, 1981.

Winfree, A. T. *When Time Breaks Down: The Three-Dimensional Dynamics of Electrochemical Waves and Cardiac Arrhythmias*. Princeton: Princeton University Press, 1987.

——*The Geometry of Biological Time*. New York: Springer-Verlag, 2001.

Wolfram, Stephen. 'Cellular Automata as Models of Complexity.' *Nature* 311 (1984): 419–424.

——*A New Kind of Science*. Champaign, IL: Wolfram Research, 2002.

Wolpert, L. *Principles of Development*. Oxford: Oxford University Press, 1998.

Yablo, S. 'Concepts and Consciousness.' *Philosophy and Phenomenological Research* 59 (1999): 455–463.

Yi, Ang Wook. *How to Model Macroscopic Worlds: Towards the Philosophy of Condensed Matter Physics*. Ph.D. Dissertation. London School of Economics/University of London, 2000.

Zemann, Jiri. 'Energie.' In *Europäische Enzyklopädie zu Philosophie und Wissenschaften*. Hamburg: Felix Meiner, 1990.

# Index